Fake News and Information Disorder in the Era of Advance Information Technology

Current Topics in Library and Information Practice

Fake News and Information Disorder in the Era of Advance Information Technology

Edited by
Adeyinka Tella and Oluwole O Durodolu

DE GRUYTER
SAUR

ISBN 978-3-11-117448-8
e-ISBN (PDF) 978-3-11-117634-5
e-ISBN (EPUB) 978-3-11-117799-1
ISSN 2191-2742

Library of Congress Control Number: 2024945559

Bibliographic information published by the Deutsche Nationalbibliothek
The Deutsche Nationalbibliothek lists this publication in the Deutsche Nationalbibliografie; detailed bibliographic data are available on the internet at http://dnb.dnb.de.

Typesetting: Integra Software Services Pvt. Ltd.
Printing and binding: CPI books GmbH, Leck

www.degruyterbrill.com
Questions about General Product Safety Regulation:
productsafety@degruyterbrill.com

MIX
Papier | Fördert gute Waldnutzung
FSC® C083411

Contents

John Jackson Iwata

Fake News and Information Disorder in Tanzania's Social Media Platforms: Demystifying Truth in the Advanced Information Technology Era

Abstract: This chapter explores the proliferation of fake news and information disorder in Tanzania. It discusses the rise of social media platforms and their influence on communication and information sharing. The challenges posed by fake news and information disorder are examined while emphasizing the need for collective and affirmative actions and multi-stakeholder collaboration to address the challenges. The impact of fake news on society, including its threats to democracy and societal cohesion is also highlighted within this chapter. The chapter also delves into the realm of fake news and misinformation, discussing its various types, forms, and motivations. Furthermore, it explores the psychology behind fake news consumption, emphasizing cognitive biases and emotional triggers. The importance of media literacy, critical thinking, and digital resilience in combating fake news is also emphasized. The chapter concludes by highlighting the importance of strengthening regulation and legislation for the purpose of striking a balance between freedom of speech and information integrity while fostering a culture of media literacy and critical thinking.

Keywords: social media, fake news, information disorder, misinformation, media literacy, information technology

Introduction

The world's development move has taken a new face, commonly known as the Fourth Industrial Revolution. In this new era, technology is increasingly shaping the ways people live, work, and communicate (Kumar 2019). In the context of this era, developing countries like Tanzania have witnessed the rapid adoption of social media platforms which has transformed the ways in which communication,

---

John Jackson Iwata, Department of Knowledge Management, Moshi Co-operative University, P. O. Box 474, Moshi-Kilimanjaro, Tanzania, e-mail: iwata2j@gmail.com, ORCID: https://orcid.org/0000-0002-4491-650X

https://doi.org/10.1515/9783111176345-001

information sharing, and public discourse are transmitted (Mwaisaka, Makulilo, and Lusekelo 2020). In Tanzania for example, social media platforms have become a powerful force with a pervasive impact on various aspects of society (Gupta and Arora 2018). This widespread presence and adoption of social media applications by people with varied profiles and backgrounds in terms of beliefs, political affiliation, education and social status have also led to the rise of various challenges, including the rampant spread of fake news and information disorder which subsequently causes social unrest and cohesion. Sometimes, this affects societal values, changing opinions on critical issues and topics (Domenico, Sit, Ishizaka, and Nunan (2021). 2021).

Tanzania, like many other countries, especially the developing ones, faces detrimental consequences of fake news and information disorder, which are amplified by the presence and adoption of social media platforms (Mushi and Lusekelo 2022). In this context, fake news is deliberately fabricated false or misleading information presented as factual news or information, often created with the purpose of deceiving and manipulating the truth (Lugumila 2020). On the other hand, information disorder refers to the situation of sharing or spreading and using misleading information with or without the intent of harming or misleading anyone. Information disorder encompasses a broader spectrum of phenomena, including disinformation, misinformation, and malinformation (Nyanduga 2019).

Disinformation refers to the spread or dissemination of false information with the intention to deceive, mislead, manipulate, cause damage to, or misguide people, organizations, and countries (Pennycook and Rand 2019). In other words, disinformation is a strategic and calculated effort to manipulate public opinion, influence beliefs, and shape political or social discourse. On the other hands, misinformation refers to the use of inaccuracies or misleading information that have been shared unintentionally, often stemming from genuine mistakes, misinterpretations, or a lack of awareness of the truth (Pennycook and Rand 2019). However, it is important to note that even if it is unintentional, misinformation can contribute to the spread of false narratives and subsequently erode trust in reliable sources. Another category of fake news is malinformation, which refers to the circulation of genuine or truthful information with the intention of causing harm or damage to someone's reputation, status, and authority including the sharing of unauthorized information such as the sharing of personal or private information without the owner's consent or the sharing of leaked documents (Mtebe et al. 2021). These forms of information disorder pose significant challenges to the community as they distort the truth, hinder informed decision-making, and erode public trust in media and sources of information (Makulilo 2020). Furthermore, they undermine democratic processes by misinforming citizens and consequently contribute to the erosion of social cohesion and political tensions (Kapinga and Uiso 2021).

Therefore, in the context of the emerging developments and especially by focusing on Tanzania's context as the world's renowned country of peace and love, understanding the intricate relationship between social media, information disorder, and the spread of misinformation is of paramount importance for the country's tranquillity and national unity (Mwansasu, Mushi, and Mcharo 2020). The current author recognizes that the adoption of popular social media platforms like Facebook, Instagram, Telegram, Skype, Snapchat, TikTok, WeChat, Twitter, WhatsApp, and YouTube have brought positive impacts in almost all aspects of living as they have created new avenues for expression, connectivity, and engagement (Nkya, Mcharo, and Mrindoko 2021). Through these platforms, individuals now have the great opportunity of accessing and sharing content at an unprecedented speed while being enabled to voice their opinions, mobilize communities, and access a wealth of information (Birari et al. 2022). However, the lack of proper fact-checking mechanisms and the viral nature of social media news have facilitated the rapid dissemination and spread of false or information, which in turn has made it difficult for individuals to distinguish fact from fiction and vice versa (Juma et al. 2022).

The justifiable significance of exploring this intricate relationship between social media, information disorder, and the spread of misinformation extends beyond Tanzania's borders, as it sheds light on the broader challenges faced by societies in the era of advanced information technology in the Fourth Industrial Revolution (Ally and Mujuru 2021). At this juncture, it is important to note that fake news has the ability to shape public opinion, influence political narratives, and even impact election outcomes, which has far-reaching consequences for democratic processes and societal stability (Kashindye and Mboya 2020). Focusing on Tanzania's context, the author's home and motherland, valuable insights into the mechanisms through which social media platforms contribute to the spread of misinformation, its effects on public discourse and how societal trust can be gained has been highlighted (Mwambene 2022). Therefore, the chapter unravels the truth in the era of social media platforms by delving into the complex interplay between social media, fake news, and information disorder in developing countries like Tanzania. Drawing upon the discussion from scholarly research, case studies, and theoretical frameworks, the chapter analyzes specific challenges faced by societies in developing countries like Tanzania in combating the spread of misinformation through social media platforms (Mjema and Msoka 2023). Additionally, potential strategies and interventions to mitigate the negative effects of fake news and information disorder have been explored while promoting media literacy and critical thinking among individuals in developing countries (Nyerere and Kalolo 2021).

The Rise of Social Media Platforms and its Influence

The digital era of the Fourth Industrial Revolution has ushered in the rapid rise and proliferation of social media platforms, revolutionizing communication and information-sharing practices (Boyd 2012). This transformation has historical roots dating back to the Internet's early days and the emergence of online communities in the late 1990s. Platforms like Usenet in the 1980s and early blogging sites in the late 1990s laid the groundwork for virtual communication. The breakthrough moment occurred later, with the launch of key platforms including Friendster in 2002, MySpace in 2003, Facebook in 2004, YouTube in 2005, and Twitter in 2006 (Boyd and Ellison 2007), followed by Instagram in 2010.

Friendster marked the inception of contemporary social networking, MySpace allowed users to customize profiles and share multimedia content, and Facebook became a favored social media platform in Tanzania (Kaplan and Haenlein 2010). These platforms revolutionized social media by introducing features like user profiles, friend connections, and content sharing, while the advent of smartphones further accelerated their global popularity (Boyd and Ellison 2007). The subsequent establishment of YouTube provided a platform for sharing video content, preceding Twitter which introduced microblogging, enabling real-time short message sharing (tweets) and public conversations. But, more importantly Instagram revolutionized visual sharing with artistic filters and captions, offering users a unique mode of visual self-expression (Kaplan and Haenlein 2010).

Research studies conducted in Tanzania underscore the significance of these global platforms in shaping online interactions. A study by Kwayu, Lal, and Abubakre (2018) explored the impact of social media platforms on communication patterns within Tanzania's communities, emphasizing their role in fostering connectivity and information dissemination. Similar findings emerge from studies conducted elsewhere in Africa, such as that by Morris et al. (2018) which highlighted the transformative effects of these platforms on communication dynamics across the continent. The widespread use of smartphones and mobile applications has further propelled the popularity and user base of these platforms, enabling individuals to access and engage with them anytime, anywhere.

In Tanzania, as in other parts of the world, the adoption of social media platforms gained momentum in the early 2000s with the introduction of the Internet and the integration of technologies like Web 2.0 (O'Reilly 2005). Web 2.0 emphasized user participation, enabling individuals to be both consumers and producers of content. Tanzania's social media platforms embraced these principles, empowering users to create profiles, share updates, and engage dynamically with others. Hence, the collective impact of social media platforms has created a virtual space where users globally can connect, interact, and engage with one another. These platforms facilitate the sharing of personal updates, photos, videos, and opinions, fostering the sense of a virtual community (Boyd 2012).

The Impact of Social Media Platforms on Communication and Information Practices

In the ever-evolving landscape of social media, Facebook stands out as a dominant force, exerting a profound impact on communication, information sharing, and community building in Tanzania and beyond. This chapter explores the multifaceted influence of Facebook, Twitter, Instagram, and YouTube within Tanzania's context, shedding light on the distinctive roles each platform plays in shaping public discourse, cultural expression, and civic engagement. Undeniably, Facebook has emerged as the most popular social media platform in Tanzania, fostering virtual connections among its citizens (Boyd 2014). This platform has facilitated the reconnection of friends, the strengthening of family ties, and active participation in social, political, and cultural discussions. With its diverse functionalities, including news feeds, groups, events, and pages, Facebook has become a virtual nexus, allowing individuals, organisations, businesses, and communities to interact, share information, organize events, and engage in discussions, thereby influencing decision-making processes (Boyd 2014). In contrast, Twitter, with its real-time and public nature, has played a pivotal role in shaping public discourse in Tanzania. This platform has been instrumental in facilitating political discussions and social movements, empowering citizens and journalists to express opinions, engage with public figures, and participate in trending topics (Boyd 2014). Moreover, Twitter has emerged as a potent platform for activism and advocacy, employing hashtags to mobilize support for various causes and raise awareness about critical issues.

In contemporary Tanzania, Instagram has become a popular platform for visual expression, particularly among the younger generation. Known for its emphasis on visual content and fostering creativity and self-expression, Instagram has significantly influenced Tanzania's culture by promoting fashion, photography, and visual storytelling (Boyd,2014). The platform has provided a space for Tanzania's photographers, artists, and influencers to showcase their work and gain recognition. Additionally, Instagram has played a role in impacting tourism by showcasing captivating images of Tanzania's natural beauty and cultural heritage, attracting visitors and tourists globally. Elsewhere, YouTube, as a video-sharing platform, has revolutionized the creation and consumption of diverse content, spanning entertainment, education, music, and citizen journalism. Tanzanian musicians, filmmakers, and content creators have harnessed YouTube to reach a global audience, promoting Tanzania's culture and creativity (Boyd 2014). Educational content on YouTube has contributed to knowledge-sharing and learning opportunities, allowing users to subscribe to channels, share videos, and engage with creators and communities.

The collective impact of these social media platforms has democratized the ability to disseminate information, breaking down traditional barriers in media production and distribution (Bruns 2008). With these social media platforms, users can now share news, opinions, and personal experiences with a global audience, bypassing traditional gatekeepers such as mainstream media (Bruns 2008). This transformation has redefined the consumption and sharing of news during major events or crises, where social media platforms serve as primary channels for sharing first-hand accounts and eyewitness reports (Mendoza, Poblete, and Castillo 2010b). Furthermore, social media has facilitated the formation of online communities centered around shared interests, identities, or causes, transcending geographical boundaries (Boyd 2014). These communities have become platforms for civic engagement, political activism, and social movements, amplifying voices that may have previously been marginalized or overlooked. Thus, the pervasive influence of social media platforms on communication and information-sharing practices has ushered in a new era, enabling individuals to actively participate in the creation and propagation of news stories, engage in civic discourse, and build virtual communities. Understanding the distinct roles of Facebook, Twitter, Instagram, and YouTube provides valuable insights into the evolving digital landscape of Tanzania.

Usage Statistics of Social Media Platforms

The global usage statistics of social media platforms highlight its vast reach and influence on various communities. According to Statista (2021), as of 2021, Facebook had over 2.8 billion worldwide monthly active users, while Twitter had approximately 330 million users, Instagram had over 1 billion users, and YouTube had over 2 billion logged-in monthly active users. These staggering numbers of users reflect the scale at which social media platforms operate, connecting people across continents, cultures, and languages.

In Tanzania, the number of social media users has been steadily increasing, demonstrating the significant penetration and influence of these platforms in almost all of Tanzania's societies (Statista 2021). Statista (2021) shows that, as at 2021, the number of Facebook users in Tanzania had reached over 3.5 million, with Twitter and Instagram users were estimated to be around 1.4 million and 2.9 million, respectively. These statistics indicate the significant reach and impact of social media in Tanzania. The penetration of social media platforms in Tanzania's societies has been driven by the widespread availability of affordable smartphones and internet access (Tanzania Communications Regulatory Authority [TCRA], 2021). According to the TCRA, as of December 2021, the number of social media users in the country were 6.1 million, further emphasizing the significant reach of these platforms (TCRA 2021), with this number increasing by 900,000 (+20%) between 2020 and 2021. It is therefore important to note that the emergence of social media platforms has revolutionized communication and information sharing in Tanzania, facilitating connectivity, amplifying voices, and providing opportunities for individuals and communities to engage in various forms of expression.

Power of Social Media in Empowering Individuals and Communities

According to Mitu and Ndalahwa (2017), social media has recently been emerged as a powerful tool that facilitates democratization of information dissemination, empowering individuals and communities specifically for the purpose of enabling them to share their voices, engage in public discourse, and shape public opinion. In Tanzania the power of social media is evident in three key aspects such as the democratization of information dissemination, the virality and amplification of content, and the influence on public opinion. In this context, the democratization of information dissemination involves empowering individuals by enabling them to share and engage in information management processes. Social media platforms have given Tanzanians the ability to become active participants in the information ecosystem by allowing them to easily create and share content while bypassing traditional gatekeepers and intermediaries (Mitu and Ndalahwa 2017). This low barrier to entry enables marginalized voices and communities to have their stories and perspectives heard on a global scale, challenging dominant discourses and enriching public discourse (Mitu and Ndalahwa 2017). It is in this context that individuals in Tanzania can through social media engage in direct conversations with their audience, foster dialogue, and promote civic engagement in various matters (Oduor and Horst 2017). Therefore, individuals can share their experiences, participate in discussions, and advocate for social causes, amplifying their impact beyond their immediate circles and contributing to societal change (Oduor and Horst 2017).

The virality and amplification of content through social media platforms enable rapid and quick spread of such content through networks of connected individuals. Social media can quickly make such content go viral, reaching a wide audience within a short period of time irrespective of compelling or controversial content (Schoen et al. 2013). In Tanzania, social media platforms like Twitter, Instagram, and Facebook have played a significant role in amplifying important messages and mobilizing communities for collective action (Schoen et al. 2013). On Twitter, hashtags and trending topics, for example, garner widespread attention and foster discussions around specific issues or events. This ability of amplifying some content and engaging a broader audience allows the dissemination of information that may have been overlooked or suppressed by traditional media channels (Schoen et al. 2013).

More importantly, social media platforms in the country have become influential sources of news and information for many Tanzanians. Also, in terms of the marketing of some products and gaining popularity, social media have brought positive impacts. For instance, in the music and artistic industry, social media platforms have made it possible for artists to emancipate themselves from radio owners' control, and now they can use these new personalized media to broadcast their massage and activities for income generation and popularity. Social media has been shaping beliefs, attitudes, and behaviors of individuals, with the contents shared and consumed on social media platforms in Tanzania having the potential to shape public opinion and influence individuals' perspectives (Kalsnes and Larsson 2019). Exposure to diverse perspectives and information on social media can contribute to a more informed community and engaged citizenry (Kalsnes and Larsson 2019). However, the influence of social media on public opinion is not free from challenges. It is in this context that the rapid spread of misinformation, the presence of echo chambers, and filter bubbles can reinforce existing beliefs and polarize opinions (Kalsnes and Larsson 2019). This highlights the importance of media literacy and critical thinking skills to navigate the complex information landscape on social media (Kalsnes and Larsson 2019). It is crucial to critically assess and evaluate any accessed information, verify sources, and seek diverse viewpoints to ensure a more accurate and balanced understanding of the information accessed.

Fake News, Information Disorder, and its Impact in Advanced Information Technology

Tanzania, like many other countries, deals with the overwhelming influx of information facilitated by advanced information technology and digital platforms (Wardle and Derakhshan 2017). The rise of social media platforms and digital news sources has led to an exponential increase in the quantity of information accessible to individuals and subsequently the spread of fake news. The spread of fake news and misinformation has also become a pressing challenge in Tanzania's information landscape, mirroring the global trend. The simultaneous rise of misinformation, disinformation, and fake news in Tanzania has resulted in an information disorder characterized by information overload and filtering challenges (Wardle and Derakhshan 2017). Hence, the abundance of information available through social media platforms can overwhelm individuals, making it difficult to distinguish between credible and false information, which subsequently contributes to the spread of misinformation, disinformation, and fake news.

It is important to note that fake news refers to false or misleading information presented as factual and correct news, encompassing a wide range of deceptive content deliberately created in order to deceive or mislead readers, viewers, and listeners (Pennycook and Rand 2019). The constant exposure to fake news and the rapid dissemination of false information in society can lead to information fatigue and cognitive overload. In Tanzania for example, individuals may feel exhausted and overwhelmed by the sheer volume of conflicting information, resulting in apathy, disengagement, and a loss of trust in the information ecosystem (Wardle and Derakhshan 2017). Cognitively the overload of information can impair critical thinking and decision-making, and furthermore can perpetuate the spread of misinformation.

Therefore, with various criteria, fake news can take various forms including fabricated stories, manipulated images or videos, satire presented as fact, and propaganda disguised as legitimate news. Fabricated news involves entirely false information created to mislead readers, listeners, or viewers, while manipulated news involves the use of selective editing or alteration of real information for the purpose of distorting its original meaning. Although satire and parody are not intended to deceive, they can also contribute to the spread of misinformation if taken out of context. Additionally, another form of fake news is propaganda, which aims at promoting a specific political, financial, or ideological agenda and often relies on manipulative techniques (Lewandowsky et al. 2017).

Characteristics and Purpose of Fake News

The creation and dissemination of fake news in Tanzania is motivated by various factors, including political, financial, and ideological interests (Bertrand 2018). Politicians and political actors sometimes use fake news for the purpose of manipulating public opinion, discrediting opponents, or shaping election outcomes. Financial motivations involving income generation through clickbait headlines or promoting products or services is made through deceptive marketing practices. On the other hand, ideological motivations of fake news involve the spreading of false narratives with the intention of reinforcing existing beliefs, inciting division, or promoting extremist ideologies (Bertrand 2018). Therefore, fake news often exhibits certain characteristics, including emotional appeal, sensationalism, and polarisation (Berger and Milkman 2012). It targets people's emotions, seeking to evoke strong reactions and engagement. Sensationalized headlines and exaggerated claims are designed in order to grab attention and generate clicks or shares. Furthermore, fake news can contribute to the polarization of society by reinforcing existing biases, deepening divisions, and creating an "us versus them" mentality (Berger and Milkman 2012).

Consequences of Fake News in Tanzania

The proliferation of fake news in Tanzania has far-reaching implications that extend beyond the realm of information, impacting various aspects of society in the era of advanced information technology. In terms of Tanzania's democracy, fake news is normally used to undermine trust in reliable sources of information and distorts public discourse by introducing falsehoods, conspiracy theories, and divisive narratives into the public sphere (Lewandowsky et al. 2017). Fake news amplifies certain narratives, promotes misinformation, hinders informed decision-making, and hence can manipulate public opinion, influence elections, and create an environment of confusion and mistrust to some individuals, organizations or parties, and ideologies. In Tanzania, with the proliferation of social media, false information in the country is sometimes presented as factual while undermining the credibility of the media and traditional information gatekeepers. This erosion of trust can lead to widespread skepticism and cynicism among citizens, hindering individuals' ability to distinguish truth from falsehood and undermining democratic processes in the country. The proliferation of fake news threatens fundamental principles of human rights and in most cases undermines the democratic fabric in the country (Lewandowsky et al. 2017).

Additionally, fake news has socio-cultural implications, such as polarization, division, and societal unrest. Tanzania, a country known for peace, tranquillity, and brotherhood among its citizens, should be careful with false information because false information circulating through social media platforms can reinforce biases and exacerbate societal divisions (Waisbord 2018). The dissemination of false information can deepen societal divisions and create echo chambers where individuals are exposed only to information that aligns with their beliefs. This polarization can lead to increased hostility, online harassment, and offline conflicts among different social, ethnic, or political groups. Also, it hinders constructive dialogue and understanding among different segments of society and ultimately erodes social cohesion, hinders collective problem-solving, and contributes to societal unrest (Waisbord 2018).

Furthermore, the impact of fake news is not limited to the realm of politics and society but also extends to economic consequences. Fake news can lead to economic disruptions, loss of public confidence in brands, and financial losses (Bertrand 2018). In Tanzania, the spread of false information about a company's performance, product, or health can mislead customers, investors, and other stakeholders. This situation can manipulate stock prices or influence consumer behavior, leading to economic instability, financial losses, market volatility, and reputational damage. Unfortunately, the loss of trust caused by fake news can have long-term economic ramifications, affecting consumer trust, investment decisions, and business operations (Bertrand 2018).

Psychology Behind Fake News Consumption in Tanzania

There are many factors that contribute to the consumption of fake news by various people in in Tanzania. Among the factors are those psychological and, hence, understanding the psychological factors in the consumption of fake news is crucial in comprehending the spread and impact of misinformation in Tanzania. The psychology behind fake news delves into two key aspects, cognitive biases and emotional triggers, as well as echo chambers and confirmation bias. Therefore, through examining these psychological phenomena, one can gain insights into why individuals are prone to consuming and sharing fake news, leading to the perpetuation of information disorder.

Cognitive biases such as among the psychological aspects play a significant role in individuals' susceptibility to fake news and the ways in which such individuals process and interpret information (Pennycook and Rand 2019). In Tanzania, as in other parts of the world, several cognitive biases contribute to the consumption and acceptance of fake news and misinformation. For example, confirmation bias can lead individuals to seek and believe information that aligns with their pre-existing beliefs or opinions while neglecting or dismissing contradictory evidence (Nyhan and Reifler 2010). This bias can make individuals more likely to accept and share information that confirms their existing worldview, even if they know that it lacks credibility or validity. Additionally, availability bias leads individuals to rely on easily accessible information, often neglecting to fact-check or critically evaluate the accuracy of the content (Tversky and Kahneman 1973). On the other hand, emotional triggers such as fear, anger, or excitement can override critical thinking and lead individuals to share or believe fake news without careful scrutiny and integration to critical thinking. Emotional content tends to evoke strong reactions, making it more likely to be shared and amplified. Fake news stories that elicit emotions such as fear, anger, or empathy can be particularly persuasive, as they tap into individuals' emotional states and increase the likelihood of engagement and dissemination (Pennycook and Rand 2019).

Echo chambers, which in this context refer to social environments in which individuals are exposed to information that reinforces their existing beliefs and values while minimizing exposure to dissenting opinions, exacerbate the consumption of fake news. In Tanzania's digital landscape and advanced information technology, echo chambers are prevalent both online and offline through social media platforms and in offline communities. When individuals are surrounded by like-minded individuals and exposed to a limited range of perspectives, echo chambers may happen, which create a sense of validation and confirmation bias. Confirmation bias reinforces individuals' existing beliefs, as they tend to seek out and interpret information in a way that confirms their preconceived notions and understandings (Taber and Lodge 2006). When individuals encounter fake news that aligns with their beliefs and their common understanding, they are more likely to accept it without integrating such information into critical thinking and critically evaluating its credibility. This confirmation bias contributes to the perpetuation of misinformation, as individuals selectively accept and share information that supports their worldview, even if it lacks a factual basis (Pennycook and Rand 2019). The consequences of echo chambers and confirmation bias include the perpetuation of false narratives, reinforcement of polarization, and hindrance of constructive dialogue and critical thinking because individuals are exposed to a limited range of perspectives. This increases the potential for misinformation to spread and gain credibility (Pennycook and Rand 2019).

Manipulation Techniques on the Spread of Misinformation and Disinformation

In the digital age, Tanzania is not immune to the pervasive spread of misinformation and the manipulation techniques employed by malicious actors (Wardle and Derakhshan 2017). Like in other parts of the world, these actors use various techniques to manipulate and deceive online audiences. Manipulative headlines, clickbait, and emotional appeals are commonly employed tactics. Manipulative headlines aim to distort or sensationalize information in order to grab attention and generate clicks (Lewandowsky, Ecker, and Cook 2017). Clickbait, often in the form of eye-catching headlines or thumbnails, entices users to click on misleading or low-quality content while emotional appeals, such as exploiting fear, anger, or empathy, are used to manipulate users' emotions and influence their opinions. Understanding these techniques is crucial for comprehending their impact on public opinion, political discourse, and real-world consequences. Malicious actors frequently employ manipulative techniques to mislead Tanzania's social media users which include the use of manipulative headlines, sensationalist clickbait, and emotional appeals. Manipulative headlines are designed to grab attention and generate clicks, often distorting or exaggerating the underlying content. Clickbait, with its enticing and misleading nature, entices users to click on links that often lead to misleading or false information. Emotional appeals, exploiting human emotions such as fear, anger, or sympathy, aim to evoke strong reactions and override critical thinking (Lewandowsky, Ecker, and Cook 2017).

Bots and automated accounts that mimic human behavior on social media platforms play a significant role in the spread of misinformation in Tanzania. These malicious actors utilize bots and automation to amplify the reach of their content and create the illusion of widespread support or engagement. By programming these accounts to like, share, or retweet content, they increase its visibility and reach. Exploiting algorithms that prioritize popular or trending content with high engagement, these actors artificially inflate the visibility and perceived importance of misinformation, making it more challenging for users to discern between genuine and manipulated information. Bots and automation not only contribute to the virality of misinformation but also create the illusion of widespread support or consensus, further deceiving users (Wardle and Derakhshan 2017).

In Tanzania, as in many other countries, coordinated disinformation campaigns pose a significant threat to public discourse and democracy. These campaigns are orchestrated to deceive and manipulate public opinion through the strategic dissemination of false or misleading information. They involve the creation of fabricated stories, misleading images, and manipulated videos. Malicious actors may establish networks of fake accounts, employ astroturfing techniques to simulate grassroots support, or use propaganda tactics to shape public opinion. The objectives of such campaigns often include sowing discord, influencing elections, or undermining trust in institutions (Wardle and Derakhshan 2017). By targeting specific audiences, these campaigns exploit existing divisions, manipulate political discourse, and seek to influence electoral outcomes.

The manipulation techniques used to spread misinformation in Tanzania have significant implications for public opinion and democracy. Exposure to false or misleading information can influence individuals' beliefs, attitudes, and behaviors, leading to misguided decisions and actions. These techniques also extend to the political sphere, as disseminating false or misleading information allows malicious actors to shape public opinion, undermine trust in institutions, erode faith in the democratic process, target political opponents, and influence electoral outcomes. The intentional distortion of facts and spread of misinformation can disrupt the democratic process, hinder informed decision-making, and polarize society (Wardle and Derakhshan 2017).

Numerous case studies in Tanzania highlight the real-world consequences of misinformation and manipulation techniques. Instances of false information leading to social unrest, incitement of violence, or reputational damage have been observed. Furthermore, the dissemination of misleading health information can have dire consequences on public health, resulting in misguided behaviors or distrust in medical interventions. For example, the dissemination of false health information during the COVID-19 pandemic has led to harmful behaviors and the spread of misinformation about treatments and prevention measures. Additionally, political disinformation campaigns have the potential to disrupt electoral processes and erode trust in democratic institutions (Wardle and Derakhshan 2017).

Combating Fake News and Information Disorder

In this era of tremendous changes and advancements in information technology, combating fake news and information disorder has become a critical challenge (Amnesty International 2019). The fact-checking and verification of information play a significant role in promoting accuracy, accountability, and combating the spread of misinformation (Vosoughi, Roy, and Aral 2018). Fact-checkers critically assess the accuracy of information and verify claims made in news articles, social media posts, and other online sources, making them vital in combating fake news. In Tanzania, various fact-checking organizations, including Tanzania Fact Check, Jamii Forums, and Zanzibar Fact Check, have emerged to counter the spread of misinformation (Banda and Khamis 2017). These organizations employ rigorous methodologies such as source verification, cross-referencing information with credible sources, and consulting experts to assess the veracity of claims. By providing accurate and evidence-based information, fact-checkers empower individuals to make informed decisions and navigate the complex information landscape.

Addressing the challenges posed by fake news and information disorder in Tanzania requires collaborative efforts among various stakeholders, including fact-checking organizations, media outlets, tech platforms, and government agencies. Collaboration between fact-checkers, media organizations, technology companies, and civil society groups can enhance the effectiveness of fact-checking initiatives by sharing resources, expertise, and data. Partnerships with social media platforms are particularly crucial as they can provide tools and algorithms to identify and label potentially false information (Kumar and Sharan 2019). Media literacy programs and educational campaigns are also essential in promoting critical thinking skills and empowering individuals to discern reliable sources of information.

Therefore, in the country, collaborative initiatives have been established to promote accuracy and accountability. For instance, the Tanzania Media Council has partnered with fact-checking organizations to develop guidelines and train journalists in fact-checking techniques. Social media platforms like Facebook have collaborated with fact-checkers to label or reduce the visibility of false information, with these collaborative efforts aiming to enhance information integrity and foster a more informed society (Banda and Khamis 2017). Also, the Tanzania Communications Regulatory Authority (TCRA), established by the TCRA Act No. 12 of 2003, as an independent authority for the postal, broadcasting, and electronic communications industries, includes licensing and enforcing licence conditions of broadcasting, postal, and telecommunications operators, managing radio frequency spectrums, and monitoring the implementation of ICT applications. The TCRA came into being after merging the former Tanzania Communications Commission and the Tanzania Broadcasting Commission. Since its inception in 2003, it has issued a number of regulations and guidelines to manage communication industry and information sharing.

Despite these efforts, the scale and speed at which misinformation spreads on social media platforms persist and pose significant challenges and limitations for fact-checkers to keep up with the volume of false information. The sheer scale and speed of information dissemination make it difficult to address every instance of misinformation in real-time. Malicious actors continuously adapt their techniques, making it challenging for fact-checkers to keep up with evolving tactics in detecting and debunking misinformation (Vosoughi, Roy, and Aral 2018). Additionally, limited resources, including funding and staffing, constrain the capacity of fact-checking organizations to tackle the vast amount of circulating misinformation. However, concerted efforts to improve fact-checking methodologies, strengthen collaborations, and promote media literacy can enhance the effectiveness of tackling fake news and information disorder in Tanzania (Kumar and Sharan 2019). In other words, the effectiveness of fact-checking relies on the willingness of individuals to engage with verified information and the development of media literacy and critical thinking skills among the public.

Algorithmic Interventions and Platform Responsibility in Tanzania

Social media platforms, such as Facebook, Twitter, and YouTube, have become influential sources of information, and there is a growing recognition of their responsibility in curbing the spread of misinformation. According to BBC Monitoring (2017), algorithms power the content distribution on these platforms, making them influential in determining what users see and engage with. Recognizing their influence, social media platforms have taken steps to address the spread of misinformation in Tanzania, such as making changes to their algorithms to prioritize reliable sources and combat false information by implementing policies and guidelines. Content moderation policies have also been implemented to remove or label harmful or misleading content that violates community standards. However, the social media platforms face the challenge of striking a balance between preserving free speech and limiting the spread of harmful content. The decisions made in algorithmic changes and content moderation have implications for the visibility and reach of information (BBC Monitoring 2017). For example, Facebook has partnered with fact-checking organizations to flag and reduce the visibility of false content, while Twitter has introduced features to report and label misleading information. These platforms bear the responsibility of maintaining the integrity of their networks and safeguarding users from the harmful effects of misinformation.

For the purpose of addressing the concerns about the spread of misinformation, social media platforms are striving to be more transparent and accountable, providing users with tools and features to understand how algorithms work and access the information sources behind recommendations. These social media platforms publish transparency reports, outlining their content moderation efforts and actions taken against misinformation. They also engage in partnerships and consultations with governments, civil society organizations, and researchers to improve their policies and practices. Through enhancing transparency and accountability, social media platforms aim to build trust with their users and stakeholders (BBC Monitoring 2017).

The algorithms used by social media platforms, often referred to as filter bubbles, play a significant role in shaping the information landscape. These algorithms personalize content based on user preferences, engagement, and past behavior. Therefore, while personalization can enhance the user experience, it can also create filter bubbles and echo chambers, reinforcing individuals' existing beliefs and limiting exposure to new and diverse perspectives. This algorithmic curation can inadvertently contribute to the spread of fake news and the polarization of society. Both echo chambers and filter bubbles contribute to the spread of misinformation and reinforce polarization by isolating individuals from alternative perspectives (Pariser 2011).

The impact of information overload, echo chambers, and filter bubbles extends beyond individual experiences to broader societal implications. In Tanzania, these phenomena contribute to societal fragmentation and polarisation. When individuals are exposed only to information that aligns with their beliefs, it becomes increasingly challenging to engage in meaningful dialogue, find common ground, and foster empathy. This polarization can lead to heightened social tensions, divisions, and the erosion of social cohesion (Wardle and Derakhshan 2017). Furthermore, the formation of echo chambers and filter bubbles hinders individuals' ability to critically assess information, undermining democratic processes and informed decision-making. When people are exposed primarily to biased or false information, it becomes difficult to reach consensus on important issues, undermining trust in institutions and hindering collective problem-solving.

Media Literacy and Critical Thinking in Tanzania

Media literacy and critical thinking skills are vital in navigating the landscape of fake news and information disorder in the era of advanced information technology (Kumar and Sharan 2019). In Tanzania, promoting media literacy is crucial to help individuals navigate the vast amount of information available on social media platforms. Media literacy education should encompass skills such as identifying credible sources, fact-checking, understanding bias and propaganda techniques, and discerning between reliable and unreliable information. Equipping individuals with media literacy skills is the way to empower them so that they become more adept at recognizing and combating fake news and misinformation (Kumar and Sharan 2019).

Education and awareness campaigns play a pivotal role in promoting critical thinking and digital literacy in Tanzania. These campaigns are meant to educate individuals on how to fact-check information, recognize clickbait headlines, and differentiate between reliable and unreliable sources. Fostering digital literacy skills better equips individuals to discern accurate information from falsehoods and contribute to a more informed society. Nurturing a skeptical mind-set is essential in the face of misinformation (Kumar and Sharan 2019); individuals in Tanzania should be encouraged to question sources, cross-verify information from multiple reliable sources, and critically analyze the content they encounter on social media platforms. Promoting fact-checking initiatives and tools can also help individuals verify the authenticity of news articles, images, and videos before sharing them. It is through fostering a skeptical mind-set that individuals become more discerning consumers of information, reducing the spread of misinformation (Kumar and Sharan 2019).

Building Digital Resilience in Tanzania

In the face of the challenges posed by fake news and information disorder, building digital resilience is crucial (Chan, 2023). Promoting media literacy and digital citizenship plays a significant role in Tanzania's efforts to achieve digital resilience. Media literacy equips individuals with the skills to critically evaluate and analyze media content, including social media platforms. It enables individuals to assess information, recognize misinformation, and make informed judgments about the credibility and reliability of information sources. Digital citizenship, on the other hand, emphasizes responsible and ethical online behavior, encouraging individuals to engage with information critically, respect diverse perspectives, and engage in constructive dialogue (Kumar and Sharan 2019). It is through promoting media literacy and digital citizenship that individuals are better prepared to navigate the digital landscape and counter the spread of fake news.

Education and awareness programs play a significant role in building digital resilience in Tanzania. These programs can be implemented in schools, communities, and online platforms to raise awareness about the risks of fake news and information disorder. Furthermore, these programs provide individuals with the necessary tools and knowledge to critically assess information, fact-check information sources, and understand the implications of sharing misleading content. Promoting education and awareness in Tanzania can empower its citizens to become active and responsible participants in the digital sphere. Empowering individuals to identify and counter fake news is a paramount technique in building digital resilience (Kumar and Sharan 2019). Therefore, various stakeholders in Tanzania should encourage the development of fact-checking skills and provide accessible resources and platforms for verifying information. This is also a way of fostering a culture of skepticism and equipping individuals with the tools to discern accurate information from falsehoods. Individuals in Tanzania can actively contribute to combating the spread of fake news through promoting critical thinking, encouraging diverse perspectives, and facilitating constructive discussions on digital platforms.

Strengthening Regulation and Legislation in Tanzania

In order to address the challenges posed by fake news, and information disorder proliferated by social media, it is crucial to strengthen the country's regulation and legislation (Smith 2021). The establishment of a legal framework outlining ethical guidelines and responsibilities for online platforms and social media companies is crucial. These guidelines specify principles for content moderation, fact-checking, and preventing the spread of misinformation, holding platforms accountable and promoting responsible behaviour.

In Tanzania, media organizations and social media platforms should adhere to ethical principles prioritizing accuracy, fairness, and accountability in information dissemination (Smith 2021). The development and enforcement of ethical codes can combat intentional misinformation spread and encourage responsible journalism. Social media platforms should establish clear policies for content moderation, ensuring user-generated content meets accuracy and reliability standards. Balancing freedom of speech and information integrity is a delicate task in Tanzania, requiring careful consideration. Legislation can address issues like hate speech, defamation, and deliberate misinformation spread, striking a balance that protects citizens' rights while safeguarding against harmful effects.

According to Kumar and Sharan (2020), balancing freedom of speech and information integrity is a delicate task for regulators and policymakers. Upholding the right to freedom of expression is crucial, but measures must prevent malicious misinformation spread. Strengthening legislation can address issues like defamation, libel, and incitement to violence as tools for spreading false information. A careful balance is needed, ensuring regulations are fair, proportionate, and respect fundamental rights through thoughtful deliberation and consultation with stakeholders.

Conclusion

The chapter delved into the pervasive impact of social media in the digital age, with a specific focus on Tanzania. The intricate relationship between social media and the spread of misinformation has been thoroughly explored, shedding light on the challenges posed by fake news and information disorder. Throughout the chapter, key points have surfaced, emphasizing the necessity for collective action and multi-stakeholder collaboration in addressing this issue (Smith 2021). Firstly, the chapter have examined the rise of social media platforms and their influence on communication and information sharing. The emergence of platforms such as Facebook, Twitter, Instagram, and YouTube have transformed the ways people consume and share information. Social media has facilitated the democratization of information dissemination, empowering individuals to share and engage while also enabling the rapid spread of content through networks (Smith 2021). Moreover, the chapter has revealed the profound impact of fake news on society. It extends far beyond merely distorting public discourse; it threatens social and cultural harmony, fuelling polarization, division, and societal unrest. The economic consequences of fake news further disrupt markets and erode public confidence.

From recapitulating the key points discussed in the present chapter, one can learn that social media platforms have become influential sources of information, enabling individuals to share and engage in ways never seen before. However, this empowerment comes with risks, as the virality and amplification of content can lead to the rapid spread of misinformation. The influence of fake news on public opinion is significant, shaping beliefs, attitudes, and behaviors and posing threats to democracy and societal cohesion. Furthermore, the manipulation techniques employed by malicious actors, coupled with cognitive biases and emotional triggers, contribute to the spread and consumption of fake news (Smith 2021). Also, it has been observed that information disorder in the digital age poses significant challenges, including information overload and the formation of echo chambers and filter bubbles. Navigating vast amounts of information becomes increasingly difficult, while the reinforcement of pre-existing beliefs and biases can lead to fragmentation, polarization, and reduced empathy in society.

Additionally, the discussion has underscored the significance of promoting media literacy, fostering critical thinking, and enhancing digital resilience as essential measures in addressing the challenges of fake news and information disorder. Equipping individuals with the skills to navigate the vast amounts of information available and promoting a skeptical mindset has been discussed as essential in combating the spread of misinformation. Collaborative efforts, involving fact-checkers, educators, and policymakers, are crucial in promoting accuracy, accountability, and digital citizenship. Additionally, social media platforms and tech companies have a responsibility to implement algorithmic interventions, content moderation, and transparency measures to curb the dissemination of fake news.

It has also been established that the battle against fake news and information disorder is an ongoing one. As technology continues to advance, new challenges and opportunities will emerge. It is crucial for multi-stakeholder collaboration to continue, bringing together governments, civil society, technology companies, and individuals to collectively address this complex issue. Efforts should focus on striking a balance between freedom of speech and information integrity, strengthening regulation and legislation and fostering a culture of media literacy and critical thinking (Kumar and Sharan 2020). It is through embracing the collective responsibility to uphold information integrity that Tanzania's societies can strive for a more informed and resilient society.

Recommendations

In addressing the pervasive issue of fake news and information disorder on Tanzania's social media platforms, it is imperative to recognize the critical role that ethical guidelines play in shaping responsible behavior and mitigating the harmful effects of misinformation. Building upon existing legal frameworks, it is crucial for both individuals and organizations to actively engage in upholding ethical principles that prioritise accuracy, fairness, and accountability in the dissemination of information. In Tanzania, as well as globally, media organizations and social media platforms should champion the development and enforcement of robust ethical codes which serve as a foundation for responsible journalism, acting as a bulwark against the intentional spread of misinformation. Emphasizing accuracy, fairness, and accountability, ethical codes empower individuals and organizations to navigate the complex landscape of information dissemination with integrity.

Moreover, the responsibility does not solely rest on media organizations, with governments playing a pivotal role in fostering an environment that discourages the propagation of fake news. It is essential for governments to collaborate with stakeholders, including social media companies, to establish and reinforce legal frameworks that mandate adherence to ethical guidelines. Doing so does not only promote responsible behavior but also creates a deterrent against the intentional spread of misinformation. In tandem with ethical codes, social media platforms must proactively establish clear and comprehensive policies for content moderation. Therefore, setting high standards for accuracy and reliability for the stated platforms can protect the amplification of false information. Hence, information users are encouraged to contribute to a responsible information ecosystem, knowing that their content is held to stringent accuracy standards. Finally, the battle against fake news requires a multi-faceted approach involving individuals, media organizations, governments, and social media platforms.

References

Ally, M. H., & Mujuru, P. (2021). The impact of fake news on social media users in Tanzania. *International Journal of Information Science and Management*, 1(2): 112–125.

Amnesty International. (2019). Tanzania: crackdown on freedom of expression continues. https://www.amnesty.org/en/latest/news/2019/10/tanzania-crackdown-on-freedom-of-expression-continues/

Banda, F., & Khamis, S. (2017). Political and media reactions to 'fake news' in East Africa. *Journal of African Media Studies*, 9(2): 163–181.

BBC Monitoring (2017). Tanzania's fake news crackdown raises press freedom fears. https://monitoring.bbc.co.uk/product/c201l5dv

Berger, J., & Milkman, K. L. (2012). What makes online content viral? *Journal of Marketing Research*, 49(2): 192–205.

Bertrand, M. (2018). The role of fake news in the 2016 election. *Journal of Economic Perspectives*, 31(2): 211–236.

Birari, S. K., Khan, F. M., Gite, L. P., & Lakhe, R. R. (2022). Fake news identification on social media platforms using machine learning: A systematic literature review. *Journal of Information Science Theory and Practice*, 10(2): 134–152.

Boyd, D. (2012). Participating in the always-on lifestyle. In M. Mandiberg (ed.), *The Social Media Reader* (pp. 71–76). New York: New York University Press.

Boyd, D. (2014). *It's complicated: the social lives of networked teens*. Yale: Yale University Press.

Boyd, D. M., & Ellison, N. B. (2007). Social network sites: Definition, history, and scholarship. *Journal of Computer-Mediated Communication*, 13(1), 210–230.

Bruns, A. (2008). *Blogs, Wikipedia, Second Life, and Beyond: From Production to Produsage*. New York: Peter Lang Publishing.

Chan, E. (2023). Analysis of the challenge in fake news and misinformation regulation comparative in global media landscape. *SHS Web of Conferences, ICPRSS 2023, 02018*. https://doi.org/10.1051/shsconf/202317802018

Gupta, S., & Arora, N. (2018). Impact of social media on youth: A case study of India and Tanzania. *International Journal of Applied Research*, 4(3): 298–303.

Juma, S. M., Nsajigwa, S. J., Mugyabuso, J., & Mbanaso, U. M. (2022). Combating misinformation during public health crises: An assessment of Tanzanian public health agencies' social media use during the COVID-19 pandemic. *Information Development*, 38(3), 384–398.

Kalsnes, B., & Larsson, A. O. (2019). Debunking in a world of tribes: The role of evidence in online misinformation debates. *Media and Communication*, 7(2): 59–69.

Kapinga, E. F., & Uiso, C. B. (2021). The impact of fake news on democracy and social cohesion: Evidence from Tanzania. Media and Communication Studies, 1(1): 67–82.

Kaplan, A. M., & Haenlein, M. (2010). Users of the world, unite! The challenges and opportunities of social media. *Business Horizons*, 53(1): 59–68.

Kashindye, L. R., & Mboya, A. (2020). Fake news in Tanzania and its impact on democracy and political stability. *Journal of Media and Communication Studies*, 12(8), 85–99.

Kumar, R., & Sharan, S. (2019). Fake news on social media platforms: Impact on society and its prevention. In T. Choudhury, B. C. Ojha, S. Dey, & N. Kumar (Eds.), Handbook of Research on Deception, Fake News, and Misinformation Online (pp. 89–109). IGI Global.

Kumar, R., & Sharan, S. (2020). Fake news on social media platforms: Impact on society and its prevention. In T. Choudhury, B. C. Ojha, S. Dey, & N. Kumar (Eds.), *Handbook of Research on Deception, Fake News, and Misinformation Online* (pp. 89–109). IGI Global.

Kumar, V. (2019). The Fourth Industrial Revolution and education. *Journal of Education and Learning Management*, 9(2): 105–114.

Kwayu, S., Lal, B., Abubakre, M. (2018). The impact of social media on internal communications in the Tanzanian telecom industry. In: Dwivedi, Y., et al. *Emerging Markets from a Multidisciplinary Perspective. Advances in Theory and Practice of Emerging Markets*. Springer, Cham. https://doi.org/10.1007/978-3-319-75013-2_11

Lewandowsky, S., Ecker, U. K. H., & Cook, J. (2017). Beyond Misinformation: Understanding and Coping with the "Post-Truth" Era. *Journal of Applied Research in Memory and Cognition*, 6(4): 353–369.

Lugumila, C. J. (2020). Social media and fake news in Tanzania: Implications for democracy and public trust. *Journal of Media and Communication Studies*, 12(9): 115–128.

Makulilo, A. B. (2020). Misinformation, disinformation, and malinformation in the age of social media: Conceptualizing the Global South perspectives. *The Journal of Community Informatics*, 16(1): 41–60.

Mendoza, M., Poblete, B., & Castillo, C. (2010). *Twitter under Crisis: Can We Trust What We RT?* Proceedings of the First Workshop on Social Media Analytics.

Mendoza, M., Poblete, B., & Castillo, C. (2010). *Twitter Under Crisis: Can We Trust What We RT? Proceedings of the First Workshop on Social Media Analytics, 71–79*. DOI: 10.1145/1964858.1964874

Mitu, M., & Ndalahwa, M. (2017). Impact of social media on journalism practice in Tanzania. *The Journal of Media Innovations*, 4(1): 24–41.

Mjema, G., & Msoka, E. (2023). Combating fake news and misinformation on social media in Tanzania: a review of strategies and interventions. *African Journal of Information Systems*, 11(2): 48–64.

Morris, L. D., Daniels, K. J., Ganguli, B., & Louw, Q. A. (2018). An update on the prevalence of low back pain in Africa: a systematic review and meta-analyses. *BMC Musculoskeletal Disorders*, 19(1): 196. https://doi.org/10.1186/s12891-018-2075-x

Mtebe, J. S., Raphael, C., Kayanda, A., & Jumbe, A. (2021). Fake news and misinformation: Implications for digital literacy in Tanzania. *Journal of Information Literacy*, 15(2): 241–255.

Mushi, F., & Lusekelo, A. (2022). The spread of fake news and its impact on social cohesion in Tanzania. *Journal of Media and Communication Studies*, 14(3): 227–244.

Mwaisaka, F., Makulilo, A. B., & Lusekelo, A. (2020). Use of social media platforms and political participation in Tanzania: a case of youth in Dar es Salaam. *Journal of Media and Communication Studies*, 12(10): 129–144.

Mwambene, L. M. (2022). Social media and the spread of fake news in Tanzania: Challenges and opportunities. *Journal of Media and Communication Studies*, 14(4): 302–318.

Mwansasu, B. S., Mushi, F., & Mcharo, S. M. (2020). The impact of social media on political participation in Tanzania. *Journal of Media and Communication Studies*, 12(11): 145–159.

Nkya, A., Mcharo, S. M., & Mrindoko, A. M. (2021). Use of social media for civic engagement in Tanzania: A case of #WhatWouldMagufuliDo movement. *Journal of Media and Communication Studies*, 13(1): 55–70.

Nyanduga, A. C. (2019). Social media, fake news, and information disorder: A case study of Tanzania. *International Journal of Scientific and Technology Research*, 8(10): 2191–2203.

Nyerere, E., & Kalolo, R. E. (2021). Fighting fake news in Tanzania: Role of media literacy and critical thinking. *Journal of Media and Communication Studies*, 13(2): 123–137.

Nyhan, B., & Reifler, J. (2010). When corrections fail: The persistence of political misperceptions. *Political Behaviour*, 32(2): 303–330.

Oduor, E., & Horst, H. (2017). Social media and youth political participation in Tanzania: Perceptions, practices and protests. *Journal of African Media Studies*, 9(2): 163–180.

O'Reilly, T. (2005). What is Web 2.0: Design patterns and business models for the next generation of software. *Communications and Strategies*, 1(65): 17–37. http://www.oreilly.com/pub/a/web2/archive/what-is-web-20.html

Pariser, E. (2011). The filter bubble: what the internet is hiding from you. New York: Penguin.

Pennycook, G., & Rand, D. G. (2019). The implied truth effect: attaching warnings to a subset of fake news stories increases perceived accuracy of stories without warnings. *Management Science*, 67(11): 4944–4957.

Schoen, H., Gayo-Avello, D., Takis Metaxas, P., Mustafaraj, E., Strohmaier, M., & Gloor, P. (2013). The power of prediction with social media. *Internet Research*, 23(5): 528–543.

Smith, J. (2021). Regulation and legislation in the digital age. *Journal of Communication*, 25(3): 167–183.

Statista. (2021). Number of Facebook users in selected African countries as of January 2021 (in millions). https://www.statista.com/statistics/872660/number-of-facebook-users-in-selected-african-countries/

Taber, C. S., & Lodge, M. (2006). Motivated skepticism in the evaluation of political beliefs. *American Journal of Political Science*, 50(3): 755–769.

Tanzania Communications Regulatory Authority (2021). Communications statistics: quarter 2 – 2021/2022. Dar es Salaam: Mamlaka ya Mawasiliano Tanzania.

Tversky, A., & Kahneman, D. (1973). Availability: a heuristic for judging frequency and probability. *Cognitive Psychology*, 5(2): 207–232.

Vosoughi, S., Roy, D., & Aral, S. (2018). The spread of true and false news online. *Science*, 359(6380): 1146–1151.

Waisbord, S. (2018). Fake news: a typology of scholarly definitions. *Journalism Studies*, 19(13): 1–18.

Domenico, G., Sit, J., Ishizaka, A., and Nunan, D. (2021). Fake news, social media and marketing: a systematic review. *Journal of Business Research*,124:329–341.

Wardle, C., & Derakhshan, H. (2017). Information disorder: toward an interdisciplinary framework for research and policymaking. Strasbourg: Council of Europe Report.

Oluwabunmi D. Bakare-Fatungase and Nosakhare Okuonghae

Information Disorder Ecosystem: The Place of Information Professionals

Abstract: Advancements in technology and a lack of information literacy skills among various groups of society have been identified as the cause of the information disorder ecosystem. Amid the Covid-19 pandemic, information disorder was widely regarded as an infordemic that affected the fight against the pandemic. In addition, the issue of misinformation, disinformation, and malinformation have dominated most political and social discussions around the world. It is based on this premise that this chapter discusses the roles of information professionals in a polluted information space. This study identifies challenges encountered by information professionals in addressing the issue of information pollution, and equally notes that librarians are more proficient in undertaking the role of training information users on how to evaluate digital information sources hinged on the principles of the Big 6 Information Literacy Model. Therefore, the study recommends that library schools must provide librarian trainees with adequate training to function effectively in this regard, while libraries can also partner with other stakeholders to promote information literacy. Finally, it is also emphasized that individuals and organizations can help combat and trustworthy information are dissemination though proper channels.

Keywords: information disorder, freedom of information, misinformation, false information, Covid-19 pandemic, social media, information literacy, information professionals, librarians, Nigeria

Introduction

The emergence of the internet and social technology has engineered an avenue for unrestricted access to information, which has had a direct influence on the way information is produced, searched, shared, or communicated. This is in tan-

---

Oluwabunmi D. Bakare-Fatungase, Department of Information Management, Faculty of Communication & Information Sciences, Lead City University,
e-mail:Bakare.oluwabunmi@lcu.edu.ng, ORCID ID: 0000-0002-4665-3969
Nosakhare Okuonghae, Law Librarian, Samson Folawunmi Adedoyin Library, Glorious Vision University, Ogwa, Edo State, Nigeria, e-mail: nosakhareokuons@gmail.com,
ORCID ID: 0000-0002-2281-6217

https://doi.org/10.1515/9783111176345-002

dem with the principles of librarianship in which Information Professionals (IPs) also known as librarians are saddled with the responsibility of ensuring that users get the right information, through the appropriate channel, and use it rightly in making informed decisions. This is well enunciated in the Five Laws of Ranganathan which are "books are for use; every person his/her book; every book its reader; saves the time of the library user, and the library is a growing organism." Bringing this to bear in the digital space in which information in the form of content such as text, videos, audio, animation, or images are churned out almost every second reveals that "Information is for use; every person his/her information; every information its user; save the time of the information seeker and the information ecosystem is a growing organism." All of this is based on the fact that the user is digitally literate enough to embrace netiquette and to decipher the rightness or wrongness of information at any point in time.

Given the availability of smart devices, an individual with an Internet connection who has joined the throng of information producers to share information which at times does not have real truth, is producing what some scholars call fake news, misinformation, or false information. In this study, these terms will be used interchangeably; although technically they are different, jointly they represent what is now commonly referred to as Information Disorder Ecosystem (IDE). According to Wardle and Derakhshan (2017), the term information disorder can be understood as any of the following phenomena: misinformation, disinformation, and malinformation. The information landscape is constantly changing and growing with billions of people interacting across different social media platforms such as Facebook, Instagram, WhatsApp, TikTok, Twitter, etc., allowing users to share information seamlessly without adhering to the principles of netiquette (Soler-Costa *et al.* 2021).

Szabo (2021) opined that within the digital information environments, fake news travels faster and gets more attention than real news, leaving behind adverse consequences to the unsuspecting digitally illiterate audience. The idea that information should be freely accessible to all is rooted in the concept of freedom of expression which negates the principle of censorship. Freedom of expression is one of the major pillars entrenched in the pursuit of true democracy. Thus, the right to express one's thoughts and ideas is keenly protected by the constitutions of modern societies, especially those which aspire to attain the status of a democratic state. Leading the free world in this regard, is the United States of America (USA or simply U.S).

The First Amendment of the U.S. Constitution prohibits governments from tempering the individual right to freedom of speech. Article 10 of the European Convention on Human Rights gives an individual the liberty to hold opinions and receive and impart information and ideas, even when others do not share the same opinion on the subject. Similarly, section 39 of the Constitution of the Federal Republic of Nigeria 1999 (as amended) added that *"every person shall be enti-*

tled to own, establish and operate any medium for the dissemination of information, idea and opinion." Essentially, these provisions permit individuals to read and publish any information they may wish (Garner 2019). However, it must be stressed that this right will not protect individuals who consciously publish slanderous or libellous information about a person or institution in a civil action. To avoid this pitfall, ImaobongNdifreke, Bantai, and Jonah (2020) highlight the obligation of Information Professionals (IPs) in inculcating ethical standards of information behaviour into the entire spectrum of the society since the library as a social system cuts across all given constituencies.

Media giants have taken reasonable steps to moderate content on digital platforms with features that discourage the spread of false information or potentially harmful materials, however, this move has been fiercely criticised because it gives few people the power to determine what is "harmful" or "truthful," and as truth at times is relative. However, despite this steps being taken, misinformation is on the increase, thus increasing the versatility of the information disorder ecosystem, which is the result of two factors; unethical use of digital platforms and the lack of digital and information literacy skill, making internet users susceptible to deception (Szabo 2021). Information plays a significant role in affecting the decisions, attitudes, and emotions of the audience on a given subject and therefore information professionals have a crucial role to play in ensuring individuals are armed with the right skills to distinguish fact from fiction, to depopulate the information disorder ecosystem.

Reflecting on the Concept of Information Disorder Ecosystem (IDE)

Information Disorder is a term that represents all phenomena that happen in the information universe. Wardle and Derakhshan (2017) divided it into three categories, based on the degree of harm and falseness: misinformation, disinformation, and malinformation. Misinformation according to Merriam-Webster (n.d.) is incorrect or misleading information. Disinformation, on the other hand, is the conscious attempt to deceive people by spreading false information (Ireton & Posett 2018). Bad actors spread disinformation via text, memes, GIFs, photos, audio, and video, with dis-informers often effective because they exploit the psychological vulnerabilities of their audience. Disinformation can have a far-reaching effect on the public and make it difficult for people to distinguish between facts and fiction, which makes it harder for members of the society to know the truth about certain political, social, academic, and security issues in their local environment or around the globe. The negative intention of the peddler distinguishes it from misinformation which, although false by all standards, is not deliberate. Malinformation on the other hand is factual information disseminated to cause harm, for instance revenge porn which, although factual, causes psychological harm to the victim who may not have intended it to be circulated. Thus, understanding the subtle differences between these terms makes the information user a deliberate, intentional, and savvier media consumer.

Considering the aforementioned, the Information Disorder Ecosystem (IDE) is an information space characterised by the sharing of false or factual information with the intent to cause harm (although, sometimes, it may not be intentional due to negligence or ignorance). The concept of information disorder is not a new occurrence, however, digital innovation has exacerbated the situation to the point that rumormongers and peddlers of false information now see it as an opportunity to exploit those seeking instant fame or a way to draw traffic to a website, often being reckless in the nature of the information they publish. Wardle and Derakhshan (2017) noted that emotional contents travel very fast because most social media platforms are configured in such a way that allows for a call to action through the functionality of features of "like," "comment," and "share" which tend to give the contents wide visibility and facilitate the diffusion of the information on the cyberspace.

The desire for information is enormous, with information being proven as the most important factor for human development across different epochs of human history. Information is rightly conceived as a tool for realising individual and corporate goals (Ayaowei, Memory, & Mercy 2021). Schiller (2024) affirmed that information is a critical resource that can enhance an individual's ability to make quality decisions and improve job productivity. Elsewhere, Bakare and Okuonghae (2022) submitted that information is the common thread that runs through the 17 Sustainable Development Goals (SDGs). In this context, information should be neutral, factual, free from bias, and should enlighten the holder. However, information disorder does the opposite. In recent times information has been weaponized as an instrument of chaos, fear, and violence, which can hinder meaningful development in any society and increase the panic level in tension situations. Such was seen with the COVID-19 pandemic, where the World Health Organization (2020) condemned the level of misinformation circulating in the social media space by non-health practitioners about the pandemic. Furthermore Johnson and Bakare (2021, 8) asserted that "In the current global pandemic, there are new ways to deliver information both real and fake; it is left for librarians to sort out the real information for their users to avoid misinformation." In response to this development, the agency established its own social media team to tackle the issue of misinformation and reduced its impact to the barest minimum.

The usage of information on social media platforms has indeed been instrumental to many social changes. An example is the End SARS protest in Nigeria; while it seems hard to provide specific details of the events which transpired at the Lekki toll gate on October 20, 2020, that led to the termination of the protest, it is easy to recall how they started: with a video showing how a young man was brutalised and shut down by an operative of the SARS- Special Anti-Robbery Squad which trended on social media and led to a nationwide protest.

Apart from the pandemic, another event that brought the issue of misinformation to bear was the circumstances surrounding the US election of 2016. It was at this time that the phrase "fake news" gained prominence. According to Wardle and Derakhshan (2017), the term fake news is inadequate for the menace of information pollution as it fails to capture all aspects of the concept of information disorder. Christopoulou (2018), while in agreement, noted that the term has assumed a different meaning within the political circle; he further noted that politicians and propagandists are now in the habit of appropriating every news item disagreeable as fake. This is nothing but a mere conflict of opinions which is a far cry from the subject of misinformation (Szabo 2021).

Efforts have been made by digital giants to curb the spread of misinformation. It is common for users to be given a warning regarding the nature of their post, and at times such a post is deleted by the company, for instance when a Twitter account shares information contents that are not true. Other social media platforms like YouTube give the audience room to report and block harmful channels promoting misleading information contents. Wikipedia, on the other hand, allows anyone (volunteers) to contribute to its content which has, however, been criticised by stakeholders, especially academics who argue that the fact that anyone is allowed to make an edit makes it an unreliable source of information. On the contrary, according to Wikipedia policy, only facts are allowed on the website and such factual statements have to be supported by credible sources regardless of the nature of the information. Thus, any information (especially contentious content) not backed by a credible source is removed by the site's administrators. These checks represent one way misinformation can be curtailed, with opinion statements distinguished from factual statements.

Global Perspectives and Impact of Information Disordered Ecosystem (IDE)

One major factor that impeded the fight against the global COVID-19 Pandemic was the chaos in the information chain brought about by misleading health information circulating in the media space. Some conspiracy theories portrayed the virus as a biological weapon developed in China. Others claimed it was an invention by world elites to reduce the world's population (Igbinovia, Okuonghae, & Adebayo 2020). This information. which had since been debunked due to lack of credible evidence, nonetheless gave the wrong impression of the origin/nature of the virus (Kaszeta 2020). While commenting on the danger of conspiracy theories and misinformation, Tedros Adhanom Ghebreyesus, the Director General of the World Health Organization made a profound remark during the fight against Covid-19: *"We are not just fighting a pandemic; we are fighting an infodemic"* (World Health Organization 2020). This statement reflects the view of misinformation and its attendant consequences, wherein it is likened to a pandemic that requires urgent attention from relevant stakeholders.

In sub-Saharan Africa, misleading information about the cure for the virus gained wide publicity despite the apparent lack of scientific verification. This fact was corroborated by Igbinovia, Okuonghae, and Adebayo (2021) who analysed misinformation regarding COVID-19 in different areas due to factors such as its origin, causes, prevention, and how to control its spread. In Nigeria, it was commonly reported on social media that salt water, vinegar, and warm water reduces the likelihood of contracting the virus. Elsewhere, Knight (2020) noted that similar medical advice on the use of alkaline water to prevent virus infection had been debunked. It was also widely reported that herbal medicine (*"Agbo"*) could cure COVID-19 infection, with the herbal concoction subsequently widely used in the hope that it could cure a COVID-19-infected patient. However, the usage of local herbal medicine in the use of COVID-19 is yet to be verified. Although in Nigeria a study by Ogaugwu *et al.* (2020) revealed that high temperatures can reduce the spread of the virus but do not stop it, the researchers also noted that contrary to popular belief, that there was no correlation between high humidity and mortality rate.

In Nigeria, social influencers were known to have been involved in making wide speculations on issues outside their expertise thereby spreading false information. A case in point was the COVID-19 Pandemic; while some claimed that it was a sham and did not exist, others like a prominent cleric claimed that 5G technology was responsible for the outbreak of the pandemic. However, the claim had been refuted by another cleric who asserted that the claim was an extreme interpretation of the pandemic; he in turn observed that the COVID-19 pandemic was similar to the 1981 influenza pandemic commonly referred to as the Spanish flu which was also caused by virus H1N1 influenza A virus (Aso Rock Mirror News 2020).

In India, it was also noted that influential public personalities were responsible for the unfortunate dissemination of fake news stories during the pandemic, such as several unverifiable medical remedies falsely promoted on social media such as the use of cow urine and cow dung (Shimizu 2020). In contrast, misinformation and the spread of fake news were successfully contained in Peru during the COVID-19 pandemic. Alvarez-Risco *et al.* (2020) reported that the Ministry of Information was of the opinion that those who misinform others with fake news do so to obtain benefit or agitate public peace. Thus, a subsequent national law decreed that anyone who creates and shares false information to benefit themselves or cause public panic will, if found guilty, be sentenced to two to four years of imprisonment. This development is believed to have flattened the curve in the fight against the infodemic in Peru.

In summary, studies have highlighted various causes of the rising spate of information disorders. Avarez-Risco *et al.* (2020) identify a low literacy rate as responsible for the rise in information pollution. Adeeb and Mirhoseini (2023) assert that the spread of fake news could be attributed to the increase in the use of social media platforms for information sharing. Kabir (2020) and Batchelor (2017) aver that the information explosion makes it difficult for people to identify and authenticate information sources. Meanwhile Johnson and Bakare (2021, 8) assert that *"information explosion is causing information to be chunked out on a daily basis, and the onus lies on librarians to collect, organise, store and disseminate the appropriate information for the consumption of the users."* Shimizu (2020) blames the spread of fake misinformation on the role of influential personalities on social media, while Apuke and Bahiyah-Omar (2020) linked the spread of misinformation to poor information literacy.

The Mediating Role of Information Professionals (IPs) in the Information Disorder Ecosystem (IDE)

In their report, Wardle and Derakhshan's (2017) report on information disorder advocated for an interdisciplinary approach. Consequently, it is expected that relevant experts in the information chain have a significant role to play in curtailing the carnage of the information disorder ecosystem. Information professionals are people who are principally concerned with the handling of information and the information user. According to IFLA (2011, as cited in ImaobongNdifreke *et al.* 2020, 1), defined information professionals are saddled with the responsibilities of "facilitating access to information for all for personal development, education, cultural enrichment, economic activity and informed participation in and enhancement of democracy." They are professionals who act as intermediaries between information sources, information systems, and information. They may include librarians, archivists, record managers, information scientists, information brokers, Journalists, Data managers, editors, and information consultants (Feather 1998; Aina 2004)

In this study, we would like to streamline our discussion on librarians and library staff. A librarian is a person who works professionally in a library and provides information services to the user (Iwuchukwu & Emuchay 2022). Librarians have the task of satisfying the information need of the user by using varied information resources, be they physical or digital. Library professionals have embraced emerging technologies in rendering services to the users in contemporary times which goes a long way to show the significant role they play in solving any societal problem with the arsenal of information products at their disposal (Bakare, 2023). Essentially, a librarian's role involves the collection, storage, processing, and dissemination of recorded information/knowledge in a library, be it analogue or digital. They also carry out other tasks such as providing personal assistance to library users in finding information; engaging in user education in inculcating the requisite information literacy skill-set needed to navigate the digital landscape; engaging in the training of library users to become self-dependent learners; providing answers to users' queries either in person, by telephone, or over the internet; and most importantly educating library users in an academic environment on information usage, particularly within the digital information environment.

Several studies have shown that information users in an academic environment are not immune from the problem of misinformation. In fact, El Rayes *et al.* (2018) surveyed undergraduates of Notre Dame University on the issue of fake news, with results indicating that the students lack the required skills and competency to evaluate the authenticity and accuracy of information sources. This study is particularly relevant because it shows that being educated alone is not sufficient and that there is a need to train people on how to evaluate information before using or circulating such information. However, Khan and Idris (2019) contend that income, level of education, and internet information verification skills are relevant predictors of self-efficacy to detect misinformation in cyberspace. They further revealed the increased use of social media may have contributed to the menace of misinformation and that any plausible solution must involve the empowerment of individuals to recognize and halt the spread of false information. Therefore as gatekeepers, Information Professionals (IP) are equipped with the professional competencies to know that there is a need for an individual to be information literate which is hinged on having the ability to verify the source, authorship, authenticity, accuracy, quality, and currency of information.

Furthermore Abba and Kamba (2022) noted that information literacy is the spirit behind lifelong learning that empowers people from all walks of life to seek, evaluate, use, and create information effectively. Similarly, Leetaru (2019) and Apuke and Bahiyah Omar (2020) linked the spread of misinformation to poor information literacy. In essence, being literate is different from being educated as such, with information literacy significant in the current information age due to the importance of not falling prey to the information-disordered sphere or being a perpetrator of the inhumanness that laces this landscape. This is in tandem with the work of Jones-Jang, Mortensen, and Liu (2019) whose investigation was based on a sample of 1,299 ordinary US citizens which revealed that information literacy shields the audience from being victims of misleading information. The study further revealed that information literacy, not other literacies such as media literacy, news literacy, and digital literacy, is key to identifying false information.

From the foregoing, it is indisputable that empowering people with information literacy is the most crucial factor in the fight against misinformation. Kimani and Onyancha (2015) as cited by Abba and Kamba (2022) observed that information literacy is a field of study that is synonymous with library practice. Library professionals, in meeting the information needs of their users, attempt to study how people seek and use information, which makes library professionals most suitable for this task. However, Israel and Nsibiwa (2018) observed that the field of Library and Information Science (LIS) is yet to expand the application of the concept of information literacy to electronic information resources, with the authors suggesting that libraries should collaborate with faculty members to support the integration of information literacy courses into the school curriculum. Elsewhere, Hudson *et al.* (2020) reported that the promotion of health literacy in schools resulted in enhanced basic cancer literacy among middle and high school students.

The roles of library professionals in this regard stem from two perspectives; first, they usually interact with information users on a daily basis; and second, they are often skilled in evaluating information sources across different media. As noted above, undergraduates and everyday people are not immune to misinformation unless they are further trained to authenticate and use information accurately. In other words, through user education, library professionals can impart users with the requisite information literacy skills which are imperative not only for accessing information but also evaluating information resources, taking into account their sources. According to Lau, Bonilla, and Garate (2014), information literacy involves the acquisition of lifelong learning skills through user education, library orientation, and user training for the development of information fluency, competencies, and skills.

Ojedokun and Lumade (2005) observed that information literacy, among other things, enables individuals to make effective decisions. Furthermore, they submitted that information literacy is the capacity of individuals to critically evaluate information sources. Similarly, Ayaowei, Memory and Mercy (2021) noted that information literacy entails the ability to identify information needs, understand the organisation of information, identify appropriate information sources, locate these sources, critically evaluate the sources, and disseminate such information appropriately. In agreement, the Young African Leaders Initiative course on understanding and countering disinformation by Watcher (2022) proposes nine ways to counter misinformation in the social media space:

1. Make it a habit to carefully examine article headlines. Is the information shocking or sensational? The chances are that it may not be true or might not mean what you first thought it did.
2. Check the number of likes and comments on a post. Is it consistent with the number of followers of the account?
3. Take a closer look at the account that posted the content. Does it have a profile picture? What sort of content has this account posted and shared in the past and how frequently do they post? What type of accounts does this user interact with? Do those seem reputable?
4. Ask yourself if this account is posing as someone else. Government officials and large news publications will always have verified accounts on social media platforms. The verification symbol can act as a guiding compass for users to determine which accounts are authentic and which are not.
5. Do not interact with content you believe to be misinformation or a deepfake.
6. Report suspicious and imposter content to the social media platform, which can reduce the visibility of the content to other users. See an example below:

   "Hello, Andrew Okuonghae,
   Thank you for your video report on March 30, 2023. We'll take a careful look at this content.
   What happens next
   After our review, we may:
   - **Remove the content, if it violates our Community Guidelines**
   - **Place restrictions on it, if it's not suitable for some audiences**
   - **Keep the content up, if it doesn't violate our Community Guidelines"**

7. Promote accurate sources and content from verified accounts or backed by clear citations on your social media platforms.
8. Ask for sources when information is shared with you and consistently scrutinise content that arises during in-person conversations. For example:
9. Educate your friends, family, and community.

In addition, try to confirm the story from mainstream media and avoid using blog posts because they are often not credible and merely represent the opinions of individuals who may have ulterior motives.

Moreover, with the increase in technology usage and digital information, it is appropriate for everyone to be information compliant, which implies being able to evaluate information appropriately. It is, therefore, logical and appropriate for everyone to be information literate. For librarians to effectively assume the role of educators, they have to expand the user education curriculum and collaborate with other information stakeholders, such as Nigeria Library Association (NLA), the Nigerian Universities Commission (NUC), and the Librarian Registration Council (LRCN), to promote information literacy programs (Bakare & Bakare, 2024).

The Big6 Information Literacy Model developed by Mike Eisenberg and Bob Berkowitz in 1990 explains the role of information is to maintain order in the information disorder ecosystem, with researchers considering this model a process for information problem–solving. This framework derives its name from the proposed six steps and processes required to determine an individual's information literacy ability: task definition, information-seeking strategy, location and access, problem formulation, use of information synthesis, and evaluation. This process model integrates information searching and uses skills along with technological tools in a systematic process to find, use, apply, and evaluate information for specific needs and tasks. The six elements under the systematic steps may be highlighted as follows:

1. **Task definition**: this entails defining the problem and identifying the information needed.
2. **Information-seeking strategy**: this implies determining the range of possible resources and then selecting the best resources.
3. **Location and Access**: this locates resources and finds information in resources.
4. **Use of information within resources**: this includes reading and viewing information, as well as deciding which parts you want to use (and leave out).
5. **Synthesis**: this involves organising resources or information from multiple sources and presenting the information.
6. **Evaluation**: this last level involves judging the product (effectiveness) and problem-solving process (efficiency).

The suitability of this model is premised on the fact that it encapsulates the basic sub-construct of the information literacy skill which includes the ability to identify information needs, access and evaluate the information sources, and more importantly use the information. The model is relevant to all information users who may have information needs but also seems unequipped because of the diverse information resources available.

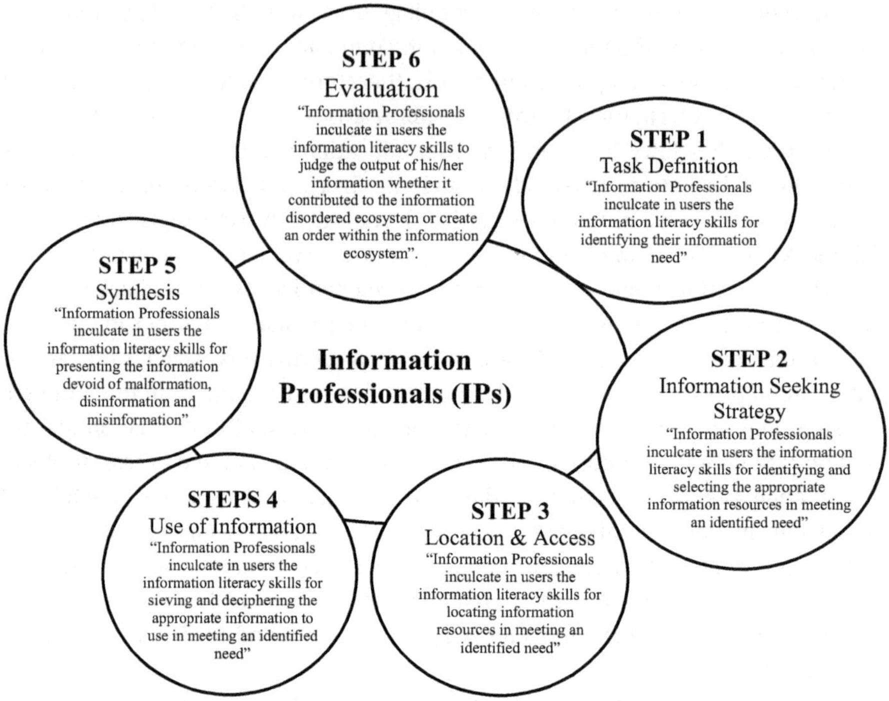

Figure 1: Conceptual model showcasing the roles of information Professionals (IPs) in the Information Disorder Ecosystem (Bakare-Fatungase & Okuonghae, 2025).

In view of the conceptual model in Figure 1, information professionals need the six steps of information literacy skills to be able to guide the information disorder ecosystem. Through these steps, they are able to bring about order. For instance, with information literacy skills, the information users will be able to conceptualise their information need which is task definition. This is operationalised as "Information Professionals inculcate in users the information literacy skills for identifying their information need." In Step two, Information Seeking Strategies, the individual will need to possess information literacy skills to be able to decide which information-seeking strategy they will need to employ, which is operationalised as "Information Professionals inculcate in users the information literacy skills for identifying and selecting the appropriate information resources in meeting an identified need." Furthermore, Step 3 of the framework, Location & Access, emphasises the ability to locate and access information. The importance of this step cannot be overemphasised, as information users must be skilled in accessing information in whatever format especially digitally because the sophisticated nature of dis/misinformation has been exacerbated by digital technology. This is operationalised as "Information Professionals inculcate in users the information literacy skills for locating information resources in meeting an identified need." In Step 4, the model explains the ability to use information, which is vital but lacking in many information users. This skill is needed to navigate effectively in the current complicated digital space, with information users needing to be able to apply information received, and operationalised as "Information Professionals inculcate in users the information literacy skills for sieving and deciphering the appropriate information to use in meeting an identified need." Meanwhile, Steps 5 and 6 dwell on the users' ability to synthesise and evaluate information. Synthesis is operationalised as "Information Professionals inculcate in users the information literacy skills for presenting the information devoid of malformation, disinformation and misinformation," while Evaluation is operationalised as "Information Professionals inculcate in users the information literacy skills to judge the output of his/her information whether it contributed to the information disordered ecosystem or create an order within the information ecosystem." This skill strengthens the individual's ability to distinguish facts from fiction which fosters healthy information dissemination.

Finally, it is advisable for institutions, agencies, and government parastatals to maintain credible information systems to disseminate information, with information dissemination viewed as an anecdote to speculation, disinformation, and misinformation (Ayaowei, Memory, & Mercy 2021). This position seems to suggest that misinformation thrives when a society is uninformed, which is why most government agencies and corporate bodies now own and operate their official social media handles on Facebook, Twitter, and YouTube channels to communicate with their audiences. In addition to social media handles, government agencies should ensure that they maintain an updated website for providing accurate information to members of the public. Members of society who are not sure whether the information is from a trusted source can verify this on the parastatal's website. Users should know how to spot genuine government official websites; for instance, in Nigeria, official government websites use *".gov" or ".ng"* domain extensions. However, a secure government website uses the likes of *"https://"* which is an indication of a secured connection to a government official website. The website will also have an official contact email, as it is against standard practices for government official websites to list private emails such as Gmail or Yahoo, which is an important indicator that an information user should be aware of.

Challenges Encountered by Information Professionals in Addressing Issues of Information Disorder in the Digital Age

The impact of the information disordered environment is innumerable as it causes irreparable damage to the victim. It has been observed that information content shared on social media has the capacity to inflame people's emotions and force them to make irrational decisions. Schafer and Schadauer (2018) observed that fake news poses a serious threat to global well-being and global perspective since it is often used to promote violence, hate crimes, war, racism, and other social ills. However, librarians' roles in dealing with this issue are challenging because of the general decline in the use of library space and resources. More so, Dube and Jacobs (2023) reported that library practitioners may be unable to provide quality services because some are lacking in Information Communications Technology (ICT) skills and competencies, and searching skills, as well as suffering from power failures, etc. Within the Nigerian space, some of the challenges are:

1. Library and Information Science (LIS) Curriculum: The LIS curriculum within the Nigerian space does not cater to the professional competencies needed for solving pertinent issues that are peculiar to this age. Bakare and Bakare (2024, 1) asserted that "the discipline of librarianship is supposed to be hinged on a curriculum that fits into the contemporary practice of sustainable environmental transformation," however, this is not the case.
2. The library as a social institution is saddled with the responsibilities of solving all societal issues and is not given the recognition it deserves within the Nigerian space.
3. There is a lack of training of librarians and library staff for the upgrading of their professional skillset in meeting the contemporary needs of users.
4. There is a lack of funding for the library as a focal information institution in Nigeria in advocating against the information disorder ecosystem.
5. There is a complacent attitude and mindset of librarians in realising the significant roles they play in bringing orderliness to the chaos of the information disorder ecosystem.

Summary

Information is a resource that enables users to make informed decisions, however, a disordered information ecosystem does the opposite. The term information disorder is globally regarded as an epidemic, a plague that can do more damage to society than any disease. Unguarded false information may spiral out of control and cause chaos, fear, and violence, especially in a volatile environment. Essentially, misinformation is caused by various factors, in particular the lack of information literacy among members of the society. On the other hand, it is equally true that information disorder is driven by the active use of social media platforms by influential public personalities who are called "influencers." These platforms allow individuals to create and share audio-visual information, regardless of their authenticity. In a democratic society, this behaviour is deemed justifiable by the notion of freedom of expression, however, there is a need to curtail the spread of false information. Library professionals have been noted to be prime agents for flattening the curve through their role in promoting information literacy among diverse information users within or outside the library through information outreaches.

Conclusion

The term information disorder is not a new concept but recent development in Information Technology has impacted who shares information and how. "Who" refers to the personality involved in the creation and dissemination of information, and it is crucial to note that apart from the traditional information producers, private individuals are now actively involved in determining the kind of information that is circulated on a global scale. "How" refers to the medium of information distribution. Social media has been identified as a catalyst for the advancement of polluted information; as noted above, studies have revealed that effective information dissemination is key to inhibiting the spread of false information. However, it must be emphasised that there is a need to equip citizens with the right information literacy skills to identify, locate, and evaluate information sources before forming opinions about them. Having established that Information Professionals' roles within the discourse of the Information Disorder Ecosystem cannot be taken for granted, it is imperative for the library as a social institution to be actively involved in empowering its users with these critical skills.

Recommendations

This study revealed the danger of a polluted information ecosystem, the role of library professionals in curtailing the spread of misinformation, the benefit of an organised channel of information dissemination as well as empowering users with the requisite information literacy skills. Therefore, we recommend the following:

1. Library schools should ensure that students are adequately taught information literacy skills: it is costly to assume that information literacy is naturally associated with Library and Information Science, as not all librarians can effectively navigate the digital space. This is important because discussion on information evaluation by most curricula centres on books and other materials in hard form.
2. Organise literacy programs: libraries and librarians who are information gatekeepers can inculcate information literacy skills as part of their numerous soft skill campaigns to empower individuals who are not only affected by false information being spread but are also information conduits. Such a program will help strengthen the dwindling relevance of library practitioners in society.
3. Legislations: there is a need for Nigeria to revisit the issue of regulating the social media space. Suitable legislation like that in Peru will ensure that people who knowingly create and share misleading information are justifiably penalised.

Impact of the Study

The study is impactful to:

1. **Individuals:** the study has emphasised the significant role of individuals as information conduits in the twenty-first century. It argued that individuals as content creators must be acknowledged as important stakeholders in the fight against the spread of false information in society.
2. **Library Professional as Educator:** librarians and library staff should provide many services to the user, with regard to the pursuit of an organised information ecosystem. These practitioners should relish the opportunity to empower users with information literacy skills that are needed to navigate the information landscape seamlessly.
3. **Policy:** given the roles of librarians in developing a literate society, policymakers must ensure that librarians are duly recognized for this role, and the library as a social institution must be given the recognition it deserves in society.
4. **Government and corporate bodies:** these organisations should note that effective information dissemination is important in curbing misinformation.

Therefore, a proper channel of communication must be developed, deployed, and improved to ensure timely and accurate information is disseminated to counter misleading information.

5. **Theory:** the theoretical contribution is based on the conceptual model hinged on the Big 6 model of Information Literacy. This was operationalized to situate the significant roles information professionals play within the discourse of the information disordered ecosystem.

References

Abba, M. A. & Kamba, M. A. (2022). Repositioning the Impact of Information Literacy Skills in Library and Information Practice among academic libraries in Kano State. Nigerian Library Association Compendium of Conference Papers NLA @ 60, Abuja, Nigeria. 324–336

Adeeb R. A. & Mirhoseini, M (2023). The Impact of Affect on the Perception of Fake News on Social Media: A Systematic Review. *Social Sciences* 12(12), 674; https://doi.org/10.3390/socsci12120674

Aina, L. O. (2004). Library and Information Science Text for Africa. Third World Information Services Limited, Ibadan.

Alvarez-Risco, A., Mejia, C. R., Delgado-Zegarra, J., Del-Aguila-Arcentales, S., Arce-Esquivel, A. A., Valladares-Garrido, M. J., Rosas del Portal, M., Villegas, L. F.,Curioso, W. H., Sekar, M.C. & Yañez, J. A. (2020). *Am. J. Trop. Med. Hyg.*, 103(2), 583–586

Apuke, O. D & Bahiyah Omar, B. (2020), Fake news proliferation in Nigeria: consequences,
Motivations, and prevention through awareness strategies", *Humanities and Social Sciences Reviews*, 8(2), 318–327

Ayaowei, I. T; Memory D. & Mercy, U. (2021). Information Dissemination Strategies in Rural Development Authorities in Bayelsa State. Lagos Journal of Library & Information Science 10(1/2), 125–146

Bakare, O. D. (2023). Emerging Technologies as a Panacea for Sustainable Provision of Library Services in Nigeria. In *Global Perspectives on Sustainable Library Practices* (pp. 1–21). IGI Global.

Bakare, O. D., & Bakare, B. M. (2024). Redesigning library and information science curriculum for sustainable environmental transformation among students of the Department of Library and Information Science, Lead City University. *Journal of Education for Library and Information Science*, 20210016, 1–21.

Bakare O. & Okuonghae, N. (2022). Information Managers as Change Agents in Achieving
Sustainable Development in the 21st Century. *Journal of Environmental Science and Economics 1(2), 58–66*

Batchelor, O. (2017). Getting out the Truth: the Role of Libraries in the Fight Against Fake News. *Reference Service Review. Doi*: 10.1108/RSR-03-2017-0006.

Aso RockMirrorNews (2020). 5G: Pastor Sam Adeyemi Warns Oyakhilome, Other Leaders http://asorockmirrornews.com/2020/04/09/5g-pastor-sam-adeyemi-warns-oyakhilome-other-leaders/

Constitution of the Federal Republic of Nigeria 1999 (as amended)

Christopoulou, A. (2018). The Information Disorder Ecosystem: A Study on the Role of social media, the Initiatives to tackle disinformation and a Systematic Literature Review of False Information Taxonomies. (Masters thesis), International Hellenic University

Dube, T. V., & Jacobs, L. (2023). Academic library services extension during the COVID-19 pandemic: considerations in higher education institutions in the Gauteng Province, South Africa. *Library Management*, *44*(1/2), 17–39.

El-Rayes, M. Chebl, C. Mohanna, J. and Hage, R. (2018). The Case of Undergraduate Students at Notre Dame University-Louaize, Lebanon.*Reference Service Review*, 46(1) 146–149

Feather, J. (1998) *Information Society: A Study of Continuity and* Change. 2nd Edition. London: Library Association.

Garner, B. A. (2019). Black's Law Dictionary,(Black's Law Dictionary (Standard Edition)). *Thomson Reuters*.

Hudson, L., Samons K.M., Dicken H.E., Prichard, C., Weiss, L.T, Edward, J., Vanderpool, R.C., Vanderford, N.L. (2020). A brief educational intervention enhances basic cancer literacy among Kentucky middle and high school students. J Cancer Educ doi:10.1007/ s13187-020-01696-3.

IFLA. (2011). Code of ethics for librarians and other information workers. Retrieved from https://cdn.ifla.org/files/assets/faife/nationalcodeofethics/nigeria-code-of-ethics-and-professional-conduct-for-librarians.pdf

ImaobongNdifreke, O. B. O. T., Bantai, R. C., & Jonah, E. A.(2020). The Library Profession in Nigeria. Retrieved from https://www.globalacademicstar.com/download/article/the-library-profession-in-nigeria-35929.pdf

Israel, R.O. & Nsibiwa, Z. (2018). Information Literacy Skills in using electronic information resources. *Library Philosophy and Practice*. Available at: https://digitalcommons.unl.edu/libphilprac/1947

Iwuchukwu, C.O. & Emuchay, B. N, (2022). Development and Future of academic library and Information Profession in Nigeria: Analysis from 1962–2022. Nigerian Library Association Compendium of Conference Papers NLA @ 60, Abuja, Nigeria. 244–251

Ireton, C. & Posseti, J. (2018). Journalism, Fake News & Disinformation: Handbook for journalism education and Training UNESCO: https://unesdoc.unesco.org/ark:/48223/pfoooo265552

Johnson, A. B., & Bakare, O. (2021). Covid-19 Effects on Libraries Goes Beyond Books: Lead City University in Perspective. *Library Philosophy and Practice (e-journal), 6610*.

Jones-Jang, S. M., Mortensen, T., and Liu, J. (2019). Does Media Literacy help Identification of Fake News? Information Literacy helps, but other Literacy Don't, *American Behavioral Scientist*, 65(2), 1–18

Igbinovia, M. O., Okuonghae, O., & Adebayo, J. O. (2021). Information literacy competence in curtailing fake news about the COVID-19 pandemic among undergraduates in Nigeria. *Reference Services Review*, *49*(1), 3–18.

Kabir, A. (2020). Coronavirus:10 fake news you should ignore, premium times. Available at: www.premiumtimesng.com/entertainment/naija-fashion/383639-coronavirus-10-fake-news-you-should-ignore.html

Knight, V. (2020). Covid-19: Beware of Online Tests and Cures, Experts Say", the Guardian, Available at: www.theguardian.com/world/2020/mar/31/coronavirus-covid-19-test-cures

Kaszeta, D. (2020). No, the coronavirus is not a biological weapon, The Washington Post, Available at www.washingtonpost.com/outlook/2020/04/26/no-coronavirus-is-not-biological-weapon

Khan, M. L; & Idris, I. K (2019) Recognize Misinformation and Verify Before Sharing: A Reasoned Action and Information Literacy Perspective. *Behaviour and Information Technology*38(12), 1–50

Lau, J., Bonilla, j. L., Garate, A. (2014). Diving into deep water: Development of an information literacy rubric for undergraduate course syllabi. In S. Karbanoglu, S. Spiranec, E. Grassian, D. Mizrachi. & R. Catt (Eds.), Information Literacy. Lifelong learning and digital citizenship in the 21st century. ECIL 2914. Communications in Computer and Information Science, 492. Springer. http://doi.org/10.1007/978-3-319-14136-760

Leetaru, K. (2019). A reminder that fake news is an literacy problem- Not a technology problem. Available at: www.forbes.com/sites/kalevleetaru/2019/07/07/a-reminder-that-fake-news-is-an-information-literacy-problem-not-a-technology-problem/6e3e1cl6a6f.

Merriam-Webster (n.d.). Misinformation. In *Merriam Webster.com dictionary*. Retrieved August 17, 2024, from https://www.merriam-webster.com/dictionary/misinformation

Ogaugwu, C. Mogaji, H., Ogaugwu, E., Nebo,U., Oko, H., Agbo, S & Agbon, A. (2020). Effect of Weather on COVID-19 Transmission and mortality in Lagos Nigeria. *Scientifica* (Cairo) Doi: 10.1155/2020/2562641.

Ojedokun, M., &Lumade, E. (2005) User education: the quintessence of quality readers services for teacher education libraries in Nigeria. *Nigerbiblios* 11(4)

Sabo, D. (2021) Digital information Literacy competencies in the world of contradictory information. *Folia Turuniesia* 105–116

Schafer, C. & Schadauer, A. (2018). Online fake news, hateful posts against refugees, and a surge in xenophobia and hate crime in Austria. *Refugee News, Refugee Politics*, 109–116.

Schiller, D. (2024). How to Think About Information. University of Illinois Press

Shimizu, K. (2020). CoV, Fake news, and racism. Lancet, 395(10225), 685–686.

Soler-Costa, R., Lafarga-Ostáriz, P., Mauri-Medrano, M., & Moreno-Guerrero, A. J. (2021). Netiquette: Ethic, education, and behavior on internet—a systematic literature review. *International journal of environmental research and public health*, *18*(3), 1212.

Szabo, L. V. (2021). Mass Media, Social Media and Technological Evolution Today: A Theoretical Approach. *Journal of Media Research-Revista de Studii Media*, *14*(41), 95–105.

VGU Research guide (2017). Big 6 information literacy model. Retrieved from http://vgulibguide.Wordpress.com/inf-literacy-skills/big6-model

Watcher, S.J. (2022 November). Understanding and Countering Disinformation. Young African Initiative Video (YALI) Lesson 3 [Video]. YouTube https://www.youtube.com/watch?v=a9m3IK_2iI8&t=54s

Wardle, C. & Derakhshan, H. (2017). Information Disorder: Toward an interdisciplinary framework for research &policymaking. *Council of Europe*, Pdf

World Health Organization (WHO) (2020), "Coronavirus", Available at www.who.int/health-topics/coronavirus=tab_3(accessed 1 May 2023).

Anne Sam Etuk

Unraveling the Web of Deception: Exploring the Role of Technology in the Proliferation of Online Misinformation

Abstract: Misinformation is false information that is shared with no intention of harm. It has become a common part of our social media environments and compromises the ability of our people to form erudite opinions. With the advent of sophisticated technology, there is an apprehension about misinformation and how it should be regulated, however, technology as a major tool for the propagation of false information also offers ways to scrutinize its real impact. In this chapter, the trend of misinformation from a technological point of view will be analysed and the role of technology in tackling the spread of misinformation will be looked into. The study will also look at how technologies enable and support the creation and spread of misinformation and disinformation, as well as what approaches could be used to mitigate these issues. The study recommends, majorly, that technology companies should try and review the vast amount of information and misinformation circulating on their platform.

Keywords: Misinformation, Disinformation, Fake news, Online Deception

Introduction

Today, about half of the world's population has access to the Internet through the use of technology, where they can create, propagate, and consume information instantly and globally. Social media through the use of diverse technology allow populaces to spread information rapidly without confirmation of truth, and to paraphrase this information to fit their intentions and precise beliefs. However, the internet is only the latest communications technology used to spread misinformation. Though misinformation is a common problem in all media, it is intensified in digital social media due to the speed and ease in which posts are spread, and the difficulty of providing countervailing corrective information. At the same time, unconstrained misinformation usually leads to social disorder and the prevalence of negative emotions, ultimately having a huge impact on society.

Anne Sam Etuk, Nigerian Institute of Advanced Legal Studies, University of Lagos Campus, Akoka, Yaba – Lagos, Nigeria

https://doi.org/10.1515/9783111176345-003

Technology has witnessed an impressive evolution in the past few decades, which has in turn transformed every aspect of our lives and helped us evolve with it. It has contributed more than anything to help humankind live a life of luxury and convenience. It is also because of technology that we know our world and outer space better. Every field owes its advancement to technology, and this clearly indicates the importance of technology in every aspect of our lives. It allows a small number of individuals to influence and manipulate the opinions of a larger audience. In addition, the targeting and crowd dynamics created by social media allow for ideas to spread faster than ever before. Technology affects communication by making it possible to work closely with clients who may even be on the other side of the planet. It is possible to maintain communications with clients from a remote setting, using apps such as Zoom, Microsoft Teams, and Google Meet, to mention but few (Bakir and McStay 2017).

Misinformation is false, inaccurate, or misleading information that is communicated regardless of an intention to deceive (Merriam-Webster Dictionary 2020). It is false information that is shared with no intention of harm. Though there is misconception in the use of misinformation and disinformation, disinformation on the other hand is false information that is shared with the aim of deceiving the consumer. Misinformation and disinformation both refer to fake or imprecise information and a key dissimilarity between them lies in the intention; information is deliberately created to mislead, as disinformation commonly refers to, whereas misinformation is inadvertent (Saurwein and Spencer-Smith 2020). However, misinformation in this perspective is the use of general terms to include false or inaccurate information that is spread in social media that can cause distress and diverse kinds of negative effects, particularly when prompt intervention is absent. Misinformation is incorrect or misleading; it's different from disinformation, which is deliberately deceptive. Rumors are information not attributed to any particular source, and so are unreliable and often unverified, meaning they can turn out to be either true or false (Wardle and Derakhshan 2017).

In recent years, the spread of misinformation on social media platforms has caused public concern, not only because misinformation can easily confuse people, causing them to make wrong decisions, resulting in economic and material losses, but also because misinformation engenders misperceptions, which have affected many domains, including economy, health, climate change, foreign policy, etc. It is now a common part of our social media environments and is compromising the ability of our societies to form informed opinions. The reckless spread of misinformation on social media usually occurs from users sharing posts from friends or mutually-followed pages, with these posts often shared from someone the sharer believes they can trust. Other misinformation is created and spread with malicious intent, sometimes to cause anxiety, other times to deceive audiences. There are times when rumors are created with malicious intent, but shared by unknowing users, with several motivations for misinformation like making money, polarizing people, or shifting blame, while, on the other hand, it may happen because of ignorance, individual egos, or misguided intention (Menczer 2016).

Classification of Misinformation

Misinformation is classified into the following categories:

Unintentionally-Spread Misinformation: Some misinformation is unintentional to deceive its recipients. Consistent and caring users may contribute to the dissemination simply due to their confidence of information sources, such as their friends, family, colleagues, or influential users in the social network. They usually try to inform their social network of friends of a certain issue or situation and not to mislead them. The cases of Ebola and Covid-19 are clear instances of widespread misinformation.

Intentionally-Spread Misinformation: Often some misinformation is deliberately spread to mislead its recipients. This is usually by writers and unified sets of propagators who have a clear aspiration and plan to cause misinformation. Conspiracies, rumors, and fake news that were trending during Covid-19 are clear examples of intentionally-spread misinformation.

Urban Legend: This is intentionally-spread misinformation that is associated to imaginary stories about local events. The intent can often be amusement.

Fake News: Fake news is intentionally-spread misinformation in the format of news. Fake news can be used as misinformation and become viral through social media.

Unverified Information: This is information that has not been checked or proven to be true. It lacks confirmation from trustworthy sources, making its accuracy doubtful. Unverified information may cause fear, hatred and astonishment, just like other types of misinformation.

Rumor: A rumor is unverified information that can be true. An example of a true rumor is about the deaths of several ducks in Guangxi, China, which was claimed to be caused by avian influenza; this was a true rumor until it was verified by the government. A similar example of avian influenza, which turned out to be false, was that some people had been infected through eating well cooked chicken.

Spam: Spam is unsolicited information that unfairly overwhelms its receivers. It has been found in instant messaging, email, and social media.

Troll: The troll's intention is to cause distraction and disagreement to a certain group of individuals. This is different from other kinds of misinformation that try to persuade recipients, with trolling aiming to increase the tension between ideas and ultimately deepen hatred. For example, the probability for a median voter to vote for a certain candidate can be aroused by being trolled. In 2016, the troll army that was claimed to be controlled by the Russian government was accused of trolling at key moments during the US election.

Hate speech: Hate speech refers to abusive content on social media that targets certain groups of people, expressing prejudice and threats. A dynamic interplay was found between the 2016 presidential election and hate speech against some protected groups, with the peak of hate speech reached on the day of the election.

Cyberbullying: Cyberbullying is a form of bullying which happens online, usually on social media, that may consist of any form of misinformation such as rumor and hate speech.

Levels of misinformation

Misinformation occurs at different social layers such as individual, group, and socio-structural levels.

The Individual Root Level: At this level, efforts are sought to focus on the citizen's individual ability to recognize disinformation or misinformation and thus correct their views based on what they received. The concern is that the inability to identify misinformation leads to postulation that all populaces are misinformed and therefore incapable to distinguish and reasonably assess information that surface from social media. Other factors that influence misinformation at the individual level are motivations and emotions that influence motivated reasoning processes.

The Group Level: The second root is at the group level. People's social networks have truly changed as the social media environment has evolved, which thus allows a different web of social networks to persist in allowing individuals to selectively disclose information which unfortunately is in a biased format. The problem with the homogenous social groups is that this raises a misinformation mindset, allowing for deception to be accepted since it emerges as a social norm due to the decrease in contradictory information. These social networks produce a clustering effect which can end up being specific rumor variations.

The Societal Level: The third level of misinformation is the societal level which is influenced by both the individual and group levels. The common figures associated with misinformation include politicians as well as other political actors who attempt to shape the public opinion in their favor. The role of the mass media is to be a corrective agent to prevent misinformation to citizens. Objectivity has been a common thread that media such as in America has lacked, which has thus been a contributor to the plague of misinformation. Print media has evolved into radio, television, and now the Internet which go hand in hand with paid commercial actors to generate tailored content to attract viewers. The intent is to reach target audiences which have dramatically shifted, such as Facebook who utilize their sources to have data collection as well as profiling tools that track each user's preferences for products and allow for ads that are targeted for that viewer (Scheufele and Krause 2019).

Understanding Misinformation from a technological point of view

It is becoming ever clearer that the increase of technology has served to catalyze the conception, spreading, and consumption of misinformation at scale. Hence, from a technology perspective, the major factors that have appended to the present situation are:

Democratization of content creation: Anyone can start a website, blog, Facebook page, Twitter account, or similar and begin creating and sharing content. There has been a makeover from a highly centralized system with a few major content producers to a decentralized system with many content producers. This overflow of information makes it awkward to differentiate between true and false information. Additionally, tools to create particular fake content have become accessible to non-technical actors, lowering the barrier to entry in terms of time and resources for people to create realistic fake content.

Wide and immediate reach and interactivity: Content created on one side of the planet can be viewed on the other side of the planet nearly instantaneously. Moreover, these content creators receive immediate feedback on how their campaigns are performing and can easily iterate an A/B test based on the number of likes, shares, clicks, comments, and other reactions on social media. For instance,

Russian disinformation campaigns, such as the false claim that the U.S. created AIDS in the 1980s and Pizzagate in 2016 U.S. election, show how false stories can spread slowly at first, then rapidly go viral, underscoring the ongoing danger of disinformation.

Speedy news cycle and monetary incentives: The more clicks a story gets, the more money gets made through ad revenue. Given the rapid news sequence and superfluity of information sources, getting more clicks requires producing content that catches consumers' attention, for instance, by appealing to their emotions or playing into their other reasoning prejudices. In addition, social media platforms that provide free accounts to users make money via targeted advertising. This allows a variety of entities, with intentions ranging from helpful to harmful, to get their messages in front of users.

Organic and intentionally created filter bubbles: In today's web experience, individuals get to choose what content they do or do not want to see, for example, by customizing their social media feeds or visiting specific websites for news. The resulting echo chambers are often called filter bubbles and can help misinformation spread to receptive people who consume it within a limited broader context. A study by Flaxman, Goel, and Rao (2016) submitted that filter bubbles may not be as extreme as feared; however, cognitive biases like the backfire effect suggest that exposure to alternate views is not necessarily effective in helping people change their minds. Furthermore, entities wishing to spread misinformation can intentionally create filter bubbles by directly and accurately targeting potentially vulnerable individuals.

Scale and obscurity in online accounts: People wishing to spread misinformation can influence the weak identity and account managing contexts of numerous online platforms to create a large number of accounts in order to artificially create the impression that particular content is popular.

Algorithmic curation and lack of transparency: Content curation algorithms have become increasing complex and fine-tuned. For instance, the algorithm that determines which posts, in what order, are displayed in a user's Facebook news feed, or YouTube's recommender algorithms, have been frequently criticized for pushing people towards more extreme content. At the same time, these algorithms are not transparent for end users. This complexity can make it challenging for users to understand why they are having specific viewing experiences online and what the real source of information is (Tufekci 2018).

Misinformation Detection

Misinformation detection seems to be a classification problem, which has the same setting as text categorization. In detecting misinformation, focusing on the text content will be appropriate, with the methods categorized as:

- **Content-based misinformation detection:** directly detecting misinformation based on its content, such as text, images, and video.
- **Context-based misinformation detection:** detecting misinformation based on the contextual information available in social media, such as locations and time.
- **Propagation-based misinformation detection:** detecting misinformation based on the propagation patterns, i.e., how misinformation circulates among users.
- **Early detection of misinformation:** detecting misinformation at an early stage before it becomes viral, usually without adequate data or accurate labels.

Tools to detect misinformation

To detect and display the diffusion of misinformation, tools such as Truthy, Rumor Lens, and Twitter trails can be used. These tools are based on a semi-automatic approach where users can explore the propagation of a rumor with an interactive dashboard. However, they do not monitor the social media stream automatically to detect misinformation, but require the user to input a specific rumor to investigate. In trying to address this issue Shao and colleagues developed Hoaxy, a platform that automatically monitors the social stream and detects and analyzes online misinformation. Following this trend Facebook has recently released new tools to help fight the spread of fabricated news stories. As opposed to Hoaxy, Facebook tools not only use a combination of content and network analysis but also include user feedback to accurately identify fake news, with this system under continuous development and testing. However, current efforts to combat misinformation have been criticized because they fall short on preventing misuse of the platform. Google's proposal to tackle misinformation also includes asking users for feedback by providing a link at the bottom of the snippet box (Chengcheng et al. 2016).

Technologies That Aid the Spread of Misinformation

The overwhelming appearance of technology has increased access to diverse information and its spread as well. Hence, some of these technologies that aid the spread of misinformation on social media are:

False Video: False media has existed for as long as there has been media to falsify, such as forgers faking documents or works of art and teenagers have faking driver's licenses. With the emergence of digital media, the problem has been intensified and tools like Photoshop make it easy for relatively unskilled actors to perform sophisticated alterations on photographs. More recently, developments in Artificial Intelligence have extended that capability to permit the creation of realistic false video, with the most popular technique to produce photorealistic false video being the reanimation of faces. In a typical circumstance, a target video, such as an announcement by a politician, is reanimated to show the same sequence of facial expressions as a source video, such as a political opponent sitting in front of a webcam in their living room (Kim et al. 2018).

Algorithms: Algorithms are designed to determine how information is propagated on both search engines and social media platforms. Humans program algorithms to make choices about what information is amplified, propagated, and recommended to other users, therefore having the power to decide which values or goals their algorithms optimize, with little oversight. As social media platforms profit from clicks and engagements, platforms have been criticized for profiting off of misinformation and disinformation, as well as creating areas of homogenous thinking and contributing to bias. Search engines like Google also use algorithms to prioritize certain sources of information over others, also without oversight or transparency. Adding to the transparency challenge that comes with algorithms is that they are not available to the public or to researchers, potentially due to commercial concerns. Social media sites have their own usage policies and moderation, by which they are able to remove posts and accounts they determine to have broken their terms of service. This moderation is largely private and without oversight from independent bodies, standardized regulations, or codes of conduct from governing bodies. In addition, moderation is increasingly being automated using artificial intelligence or machine learning, potentially exacerbating current moderation concerns (Gorwa, Binns, and Katzenbach 2020). There are concerns about possible biases in moderation, as well as a lack of transparency in decisions and a lack of recourse for when posts are removed.

Tracking and Targeting: These entail sending messaging and advertising to a specific audience across platforms, in order to reach the audience when and where they are ready to engage with the brand. The economic pattern of today's web is based on advertising, where users receive content for free in exchange for seeing advertisements. Advertisements are gradually targeted at specific clients, which are facilitated by extensive and erudite tracking of user browser behaviors. These technologies of tracking and targeted can be leveraged not only by legitimate traditional advertisers but also by entities wishing to spread misinformation for monetary gain as well as directly malicious purposes. Media tracking is the automated process of scanning media for the keywords and topics specified as being relevant to public relations campaign (Englehardt and Narayanan 2016). The main goal of data tracking is organizing and analyzing the data to derive insights. It helps users make better decisions by providing them with insights from data analysis.

Though there are a number of anti-tracking browsers, such as in browser extensions, there exists no fully reliable way for users to avoid tracking. Tracking can be and often is sooner or later influenced to target specific content at a user, via advertisements on Facebook or rooted in other websites. Researchers have found that even when advertising platforms like Facebook provide some transparency for users about why certain ads are targeted at them, the information provided is overly broad and vague, failing to improve users' understanding of why they are seeing something or what the platform knows about them. The act of targeting itself can raise privacy concerns, because it can allow potentially spiteful advertisers to extract private information about users stored in advertising platforms like Facebook, and can also be used to intentionally or unintentionally target individuals based on sensitive attributes like health issues. These very ecosystems, which underlie the economic model of today's web and thus cannot be easily displaced, can be directly exploited by disinformation campaigns to target specific vulnerable users and to create and leverage filter bubbles (Lerner et al. 2016).

Social Bots: A social bot is an automatic program that stimulates human behavior on social networks. A common tactic seen in recent disinformation campaigns is the use of social media bots to amplify low-credibility content to give the false impression of significant support or interest, or to fan the flames of partisan debate. Indeed, bot accounts are common on social media platforms. Twitter itself recently suspended bot accounts, resulting in a cumulative 6% drop of followers across the platform. These bots slowly build trust and credible profiles within online communities and then either programmatically or manually disseminate false news and extremist perspectives in order to sow distrust and division among human users. Social bots have been shown to be effective at swaying online discussion as well as inflating the amount of support for various political campaigns and social movements (Varol et al. 2017).

Technological Role in Tackling Misinformation

Facebook and Google are extremely important internet companies, with both owning more than one product with over a billion users. Twitter is a slightly different case, with just one product and one third of a billion monthly users, but with a particular relevance to journalism, news, and current events. Google's major products include search, Drive, Gmail, YouTube, Android, and Chrome, with each having different competitors and a different relationship to the misinformation debate. Facebook owns Facebook itself, Messenger, WhatsApp, and Instagram. These three companies rightly focus on tackling misinformation and protecting their users from misinformation. Other important internet companies include Amazon, Apple, and Microsoft. In the field of digital assistants with voice interfaces, these correspond to Alexa, Siri, and Cortana (Thomas, Munoz and Miller 2019). Digital assistants such as these and Google's will be making increasingly important information choices for their users, providing one answer, not lists of ten options. Furthermore, there are many other companies whose products and services have a role to play, not just internet companies.

However, technology companies, especially digital media platforms, play a key role in tackling misinformation. As their platforms are often used to launch and spread misinformation, they are well placed to quickly identify potential misinformation, fact-check, and verify it. However, the way in which a platform acts upon suspected misinformation has significant implications for human rights. In order to tackle misinformation, digital media companies should address the associated business model. In the digital space, one of the main sources of revenue is digital advertising and host websites make their pages as captivating as possible to induce users to click and spend time on them. Misinformation is often characterized by news containing shocking and catchy headlines, thus potentially feeding into the business model of digital advertising. Digital media companies have taken different approaches to tackling misinformation, which have ranged from flagging the content, removal of the content, suspension or removal of accounts, to demoting and de-ranking content (Hui-Fei and Chi-Hua 2017).

Automated fact-checking algorithms can be used to identify suspect content and assess whether it constitutes misinformation. Due to the high volume of online content on social media platforms, the platforms have introduced automated processes to screen and assess whether a piece of information constitutes misinformation. However, automated fact-checkers could bring new forms of harm to individuals when they are not operating under adequate human supervision that corrects eventual biases and mistakes. The data on which the algorithms rely may be incomplete and potentially biased, thus resulting in incorrect assessments that could seriously threaten a user's freedom of expression. Several digital media platforms employ human supervision on their automated fact-checkers or a mixed human automated fact-checking system.

Moreover, the possibility for users to challenge the removal or de-ranking of their content, also for the lack of effective grievance process outside of the company, is limited. Companies' internal processes for reviewing user complaints can be generally opaque. In order to provide an effective protection of users' basic rights, it is fundamental for digital media platforms to be fully accountable for their actions and to provide users with accessible and transparent grievance mechanisms that include effective remedial processes. The initiative of some platforms to provide users with a comparison of misinformation with an authoritative source rather than removing or demoting the content could be a viable solution to limit the adverse impact on freedom of expression (Hameleers 2020).

Technological Advancements

Fact-checking websites are one of the most used technologies to debunk fake news in this digital era. Due to the hike in fake news and misinformation, there is a substantial increase in the number of fact-checking websites available online, both independent and professional. The main objective of a fact-checking website is to collect and analyze all the misinformation or disinformation circulating on online platforms and to prove them wrong with a considerable number of facts. The digital world has also threat from social bots, which can be defined as a social media user account that appears to be handled by a human but, in fact, is equipped with automated software that runs the account. These types of accounts are extensively used in the diffusion of misinformation (Schmuck and von Sikorski 2020) and to counter this threat, technologies that can use artificial intelligence should be developed.

Technological advancements in the form of fact-checkers, automated fake news detection, and omission software have helped to curb the misinformation crisis to an extent on social media platforms. Many platforms use technology to enable users to make better judgments about the content which appears on them. For example, Facebook has delivered a context button to links which indicates the Wikipedia data about the author, the region where the hyperlink has been shared in addition to the range of instances of sharing. Elsewhere, YouTube has a gadget that automatically highlights news videos from authoritative sources as some important news activities occur (Saurwein and Spencer-Smith 2020). Automated fact-checking also lets Google comprehend and highlight fact-checked articles and, additionally, prepare a database of fact checks that may be used for voice search engines including Amazon Echo.

Conclusion

The spread of misinformation is made possible largely through social networks and social messaging. Misinformation is a story fueled by technology. Powerful new technology makes the manipulation and fabrication of content simple, and social networks dramatically amplify falsehoods peddled by the public. Through the use of technology, it is possible to prevent and control misinformation in advance by understanding the diffusion rules of misinformation, to reduce the spread of misinformation on social media more effectively, and thereby reduce the harm caused by misinformation from the source. As an important channel of information dissemination, technology has not only become an important platform for the communication of true information but has also contributed to the spread of misinformation in social networks. Hence, the spread of online misinformation on social media misleads the public and prompts them to make wrong decisions, while also bringing about great threats to the public's physical and mental health as well as economic properties. It is therefore recommended that technology companies should try and review the vast amount of information and misinformation circulating on their platform. Social media and other media companies involved in moderating online content should ensure greater transparency regarding content moderation practices and policies, as well as the algorithmic systems underpinning their platforms.

References

Adam, M. (2017). Facebook. "Working to Stop Misinformation and False News." Last accessed on 13 June 2017, https://newsroom.fb.com/news/2017/04/working-to-stop-misinformation-and-false-news/.

Adrien, F., Lada, A. A., Dean E., and Justin, C. (2014). Rumor Cascades. *In Proceedings of the 2014 International Conference on Web and Social Media.*

Allcott, H., & Gentzkow, M. (2018). Trends in the Diffusion of Misinformation on Social Media. Sage Journals. https://journals.sagepub.com

Amazeen, M. A., & Bucy, E. P. (2019). Conferring resistance to digital disinformation: The inoculating influence of procedural news knowledge. *Journal of Broadcasting & Electronic Media*, *63*(3), 415–432.

Bakir, V., & McStay, A. (2017). Fake News and the Economy of Emotions. *Digital Journalism*, *6*(2), 154–175.

Berinsky, A. (2017). Rumors and Health Care Reform: Experiments in Political Misinformation. *The British Journal of Political Science*.

Bowden, I., (2016). How Google is tackling fake news, and why it should not do it alone.[Online] Available at: http://searchengineland.com/google-tackling-fake-news-not-alone-264058

Burkhardt, J. M. (2017). Combating Fake News in the Digital Age. Library Technology Reports, 53, 8, p. 5–33.

Castillo, C., Mendoza, M. and Poblete, B. (2013). Predicting information credibility in time-sensitive social media. Internet Research, 23(5):560–588, 2013.

Chen, A., (2019). 'Limiting Message Forwarding on WhatsApp Helped Slow Disinformation', MIT Technology Review.

Chengcheng, S., Giovanni L., Ciampaglia, A. F., and Filippo M. (2016). Hoaxy: A platform for tracking online misinformation. *In Proceedings of the 25th International Conference Companion on World Wide Web. International World Wide Web Conferences Steering Committee*, 745–750

Chu, Z., Gianvecchio, S., Wang, H. and Jajodia, S. (2010). Who is tweeting on twitter: Human, bot, or cyborg? *In Proceedings of the 26th annual computer security applications conference, pages 21–30.*

Cubbon, S. (2020). Identifying 'data deficits' can pre-empt the spread of disinformation.First DraftFootnotes.https://medium.com/1st-draft/identifying-data-deficits-can-pre-empt-thespread-of-disinformation-93bd6f680a4e

Eunice, Y. and Everett, R. (2017). Chinese social media giant opens up on 'fake news' and charges of stifling dissent. Last accessed on 25 May 2017, http://www.cnbc.com/2017/03/23/sina-ceo-charles-chao-on-fake-newsand-charges-of-stifling-dissent.html.

Ferrara, E., Varol, O., Davis, C. A., Menczer, F., and Flammini, A. (2016). The rise of social bots. Communications of the ACM 59 (2016), 96–104.

Flaxman, S., Goel, S., and Rao, J. M. (2016). Filter bubbles, echo chambers, and online news consumption. Public Opinion Quarterly 80, S1 (2016), 298–320

Friesem, Y., and Gutsche, R. E. (2019). Teaching truth, lies, and accuracy in the digital age: Media literacy as project-based learning. *Journalism & Mass Communication Educator*, *74*(2), 185–198.

Ghenai, A. and Mejova, Y. (2018). Fake cures: user-centric modeling of health misinformation in social media. *Proceedings of the ACM on human-computer interaction, vol. 2, pp. 1–20.*

Glaser, A., (2017). Google is rolling out a fact-check feature in its search and news results.

Gorwa, R., Binns, R., Katzenbach, C., (2020). Algorithmic content moderation: Technical and political challenges in the automation of platform governance.

Guacho, G. B., Abdali, S., Shah, N. and E. E. Papalexakis (2018). Semi-supervised content-based detection of misinformation via tensor embeddings. *In 2018 IEEE/ACM International Conference on Advances in Social Networks Analysis and Mining (ASONAM), pages 322–325.*

Gupta, A., and P. Kumaraguru (2012). Credibility ranking of tweets during high impact events. *In Proceedings of the 1st workshop on privacy and security in online social media, page 2.*

Halliday, J., (2017). *Facebook and Twitter should do more to combat fake news, says GCHQ*. [Online] Available at: https://www.theguardian.com/media/2017/mar/14/facebooktwitter-gchq-combat-fake-news.

Hameleers, M. (2020). Separating truth from lies: comparing the effects of news media literacy interventions and fact-checkers in response to political misinformation in the US and Netherlands. Information, Communication & Society, 1–17.

Hanselowski, A., Avineshp.V., S., Schiller, B., Caspelherr, F., Chaudhuri, D., Meyer, C. M., and Gurevych, I. (2018). A retrospective analysis of the fake news challenge stance-detection task. In COLING

Heath, A., (2016). *Facebook is going to use Snopes and other fact-checkers to combat and bury fake news.* [Online] Available at: http://www.businessinsider.com/facebook-willfact-check-label-fake-news-in-news-feed-2016-12

Hui-Fei, L., and Chi-Hua, C. (2017). Combining the Technology Acceptance Model and Uses and Gratifications Theory to examine the usage behavior of an Augmented Reality Tour-sharing Application, Symmetry, Vol. 9, Iss 7, p. 113.

James, V. (2019). Deepfake Detection Algorithms Will Never Be Enough, The Verge, June 27, 2019, https://www.theverge.com/2019/6/27/18715235/deepfake-detection-ai-algorithms-accuracy-will-they-ever-work

Kim, H., Garrido, P., Tewari, A., Xu, W., Thies, J., Niessner, M., Perez, P., Richardt, C., Zollhofer, M., And Theobalt, C. (2018). Deep Video Portraits. In ACM; Siggraph.

Lerner, A., Simpson, A. K., Kohno, T., and Roes Ner, F. (2016). Internet Jones and the Raiders of the Lost Trackers: An Archaeological Study of Web Tracking from 1996 to 2016. In USENIX Security Symposium.

Lim, S. S., & Tan, K. R. (2020). Front liners fighting fake news: global perspectives on mobilising young people as media literacy advocates. Journal of Children and Media, 14(4), 529–535. https://doi.org/10.1080/17482798.2020.1827817

McDowell, Z. J., & Vetter, M. A. (2020). It Takes a Village to Combat a Fake News Army: Wikipedia's Community and Policies for Information Literacy. Social Media + Society, 6(3), 205630512093730.https://doi.org/10.1177/20563051209373 09

Menczer, F. (2016). The spread of misinformation in social media. *In Proceedings of the 25th International Conference Companion on World Wide Web (pp. 717–717).*

Merriam-Webster Dictionary. (2020). Misinformation. Retrieved 25 May 2021 from https://www.merriam-webster.com/dictionary/misinformation.

Nekmat, E. (2020). Nudge effect of fact-check alerts: Source influence and media skepticism on sharing of news misinformation in social media. *Social Media + Society*, *6*(1).

Pennycook, G & Rand, D. G. (2019). Fighting Misinformation on Social Media Using Crowd sourced Judgments of News Source Quality. *Proceedings of the National Academy of Science (PNAS), 116(7), 2521–2526.*

Saurwein, F., & Spencer-Smith, C. (2020). Combating Disinformation on Social Media: Multilevel Governance and Distributed Accountability in Europe. *Digital Journalism, 8(6), 820–841.*

Scheufele, D. and Krause, N. (2019). "Science audiences, misinformation, and fake news". *Proceedings of the National Academy of Sciences. 116 (16):7662–7669.*

Schmuck, D., & von Sikorski, C. (2020). Perceived threats from social bots: The media's role in supporting literacy. *Computers in Human Behavior, 113, 106507.*

Seo, H., Blomberg, M., Altschwager, D., & Vu, H. T. (2020). Vulnerable populations and misinformation: A mixed-methods approach to underserved older adults' online information assessment. New Media & Society, 146144482092504.

Shao, C., Ciampaglia, G. L., Flammini, A. &Menczer, F., (2016). Hoaxy: A platform for tracking online misinformation. *National University of Defense Technology, China; Indiana University Network Science Institute; School of Informatics and Computing Indiana University, Bloomington*, pp. 745–750.

Shao, C., Ciampaglia, G. L., Flammini, A., and Menczer, F. (2016). Hoaxy: A platform for tracking online misinformation. *In Workshop on Social News on the Web (SNOW).*

Sharma, K., Qian, F., Jiang, H., Ruchansky, N., Zhang, M. and Liu, Y. (2019). Combating fake news: A survey on identification and mitigation techniques:1901.06437.

Shu, K., Sliva, A., Wang, S., Tang, J. and Liu H.(2017). Fake news detection on social media: A data mining perspective. ACM SIGKDD Explorations Newsletter, 19(1):22–36, 2017.

Thomas M. Brill, Laura Munoz & Richard J. Miller (2019): Siri, Alexa, and other digital assistants: a study of customer satisfaction with artificial intelligence applications, Journal of Marketing Management, doi: 10.1080/0267257X.2019.1687571

Tristan, H. (2019). Our Brains Are No Match for Technology, The New York Times, sec. Opinion, December 5, 2019. https://www.nytimes.com/2019/12/05/opinion/digital-technology-brain.html.

Tufekci, Z. (2018). YouTube, the great radicalizer. The New York Times Opinion, Mar. 2018. https://www.nytimes.com/2018/03/10/opinion/sunday/youtube-politicsradical.html

Varol, O., Ferrara E, Davis C. A., Menczer, F., and Flammini, A., (2017) Online Human-Bot Interactions: Detection, Estimation, and Characterization.

Wardle, C., and Derakhshan, H. (2017). Information disorder: Toward an interdisciplinary framework for research and policymaking. Tech. rep., Council of Europe, Sept. 2017.

Wu, L., Morstatter F., Carley K. M. and Liu, H. (2019). Misinformation in Social Media: Definition, Manipulation, and Detection. ACM SIGKDD Explorations Newsletter 21 (2). ACM New York, NY, USA:80–90.

Oluwole O. Durodolu and Stella C. Nduka

Collaborative Approach for Successful Information Literacy Skills: A Panacea for Information Hygiene

Abstract: Information pollution poses a grave danger to the quality of information and diminishes public knowledge. With this in mind, the attention of experts has shifted to the importance of acquiring hygienic information practices in order to stem the tide of fake and adulterated news which has affected the ability of citizens to access authentic information to assist them in combatting the scourge of the coronavirus (COVID 19) that is still ravaging the world. This study investigates a collaborative approach for bringing about successful information literacy skills and solutions for information hygiene. It adopts an interpretive research paradigm in which extant documents, journals, policy documents, conference papers, and books were studied systematically and cautiously evaluated relative to the subject matters of information literacy, information hygiene, fake news, and misinformation. The study affirms that good information hygiene is premised on having a relentless commitment to fact finding activities, engagement with divergent points of view, substantiating information, and not magnifying or disseminating misinformation, which is the bedrock of information literacy.

Keywords: information literacy, information hygiene, fake news, information pollution and misinformation

Introduction and Background to the Study

As humanity continues to progress with the evolution of digital age, the manner of information consumption, adaptation and transmission is also changing, with humans consumed with information overflow. This also been referred to as infobesity by Rogers, Puryear and Root (2013), infoxication and infosaturation by Dias (2014), information anxiety by Wurman (2012), and information explosion by various authors, with these different descriptions revealing the overwhelming chal-

Oluwole O. Durodolu, Department of Information Science, University of South Africa, Pretoria, South Africa, e-mail: woledurodolu@gmail.com, https://orcid.org/0000-0003-2734-8165
Stella C. Nduka, Library and Information Science Programme, Department of Mass Communication, University of Lagos, Lagos, Nigeria, e-mail: stellacnduka@gmail.com

https://doi.org/10.1515/9783111176345-004

lenges associated with the extreme quantity of regular information that an average individual has to deal with and the resulting problem of processing information to separate facts from fiction. This is further compounded with the easy access to technology that can be utilized to promote various ideologies regardless of whether sources are genuine or from well informed researchers, meaning ability is required to effectively deal with this prodigious information because it has the tendency to diminish the value of one's decisions. Valuable information increases the prospect for better decisions, however, the deluge and torrent of the opposite type of information is impacts detrimentally on the ability to make good decisions (Rogers, Puryear, and Root 2013). With the abundance of valuable information side by side with enormous quantities of adulterated data, the overconsumption of information driven by the internet has created a new concept known as the "post-truth" society where people consume information that confirms or reaffirm their pre-conceived ideas, views, opinions, and ideologies rather than making efforts to identify the truth independently. As the online setting becomes more difficult to circumnavigate, distrust and ambiguity continues to surge, with people becoming comfortable with familiar facts rather than trustworthy ones (Paora and Heravib 2020).

In a world struggling with information pollution, where the intellectual landscape is inherently littered with contamination of information, Durodolu, and Adekanye (2017) are of the opinion that knowledge of information literacy is significant in navigating through this modern predicament, viewing information literacy as a veritable compass to survive misinformation that now constitutes a threat to knowledge. Accidental access to fake news is the consequence of a lack of information literacy skill and critical thinking, as an information literate citizen is equipped with the ability to ascertain the truthfulness and dependability of information by applying useful evaluative benchmarks which will help them to reduce the threat posed by fake news (Durodolu, Nduka and Tinyiko (2022). The coronavirus pandemic has further complicated the management of fake and adulterated information which has compromised the ability to access lifesaving information, which has put people's life at risk. Durodolu, Ibenne, and Dube (2021) assert that, in a post-truth period, where deceptive information has skyrocketed, it is essential to acquire new knowledge in dealing with misinformation because of the danger it poses to the quality of knowledge. To acquire such knowledge academic institutions must emphasize several information literacy strategies so as for citizens to have the aptitude to appraise information and detect fake news. In view of this, no nation is immune to the disastrous effects of the coronavirus (COVID-19) pandemic and infodemic, the epidemic of information, which continues to inflict disaster on the socio-political and economic stability of countries. Therefore, the oxymoronic nature of misinformation and fake news increases anxiety regarding issues of validity and authenticity because the conception of news is reinforced by verifiability (Durodolu, Chisita, and Dube 2021).

The emergence of the COVID-19 pandemic has brought about a lot of fake news and misinformation surrounding the virus. Many individuals have spread unconfirmed information regarding the pandemic, thereby, creating many fears or myths in society. The spread of such information has continued to generate many challenges for information experts who have the enormous task of providing unbiased information regarding the coronavirus pandemic. One way of providing such information is to practise good information hygiene, that is, to regularly fact-check any information before disseminating or forwarding the content to others.

Information hygiene, according to Collins Dictionary (2020), is a careful evaluation of the information that an individual consumes and broadcasts. For, example, fake information surrounding the coronavirus pandemic has flooded the internet, which has led to specialists calling on the entire populace to practise information hygiene. Ilyin (2013) posited that information hygiene as a concept appears rather absurd but that the phrase is not new and the problem is acute. This is because we live in a world where there is excess of information and people have to process huge amounts of data on a daily basis.

For people to have good information hygiene information literacy they must be equipped with information literacy skills. This is needed more than ever, in part because the urge to consume information is growing daily. Therefore, the importance of information literacy skills in information hygiene cannot be overemphasized.

Information literacy, as observed by Cobban and Seale (2003), depicts the individual need to develop proficiency in locating and evaluating necessary information from a variety of sources. Due to the proliferation of fake information relating to crises such as the COVID-19 pandemic, a lot of misinformation has been circulating among many people in society due to unrestricted access to social media tools. Librarians are thus tasked with coordinating efforts to educate the citizens in information literacy so that they become educated consumers of information.

Information literacy can be defined as the ability to recognize when information is needed as well the ability to locate, evaluate, and effectively use the needed information (American Library Association 1989 in Godbey 2013). Information literacy is one of the highest priorities in education and thus many organizations, especially in the education sector, have turned to collaborative approaches to reduce misinformation or bad information hygiene in society. Information literacy is thus the panacea for good information hygiene among individuals in the society. To achieve information literacy, the individual must be given repeated opportunities to work with the same information, which will help them in life.

Research Questions

What are the collaborative approaches for successful information literacy skills and the solution for information hygiene?

Significance of the Study

This research is essential because of its contribution to the body of knowledge in a variety of ways. The relevant literature in the area of information literacy and information hygiene was revised in an effort to reveal something unique, detect a gap in research, and scrutinize and appraise the prevailing research to uncover strengths and weaknesses of the subject matter. The reason for the literature review was to improve and understand existing research and debates relevant to the topic under review, and to present that knowledge in the form of a written report. The study is also significant because it will add value to the study of Library and Information Studies (LIS) in relation to information hygiene, fake news, and information literacy.

Methodology

This research is channelled by the interpretive paradigm since it deliberates on relevant and contemporary concepts related to topical issues and in which opinion is supported by extant literature to bring out new knowledge. The interpretivist paradigm holds that reality is multi-layered and complex as a single occurrence can have various interpretations and analysis. In investigating a phenomenon, research techniques are used to help understand how people understand and interact with their social environment (Hsieh and Shannon 2005). In this study, previous documents, journals, policy documents, conference papers, and books were studied systematically and cautiously evaluated relative to information literacy, information hygiene, fake news, and misinformation. The motivation for this study is to identify a gap in current research and give birth to new knowledge.

Information Hygiene

According to Bojanic (2021), information hygiene emerged as a construct in the late 1990s. The concept was coined by Linda Stone,(2009) continuous partial attention to explain the condition where the human mind is always searching for information so as not to miss out, regardless of the type of information. Bojanic (2021 further identified three steps to practising information hygiene:

- Choose what not to say
- Be real
- Offer hope

The coronavirus pandemic has brought about much discrepancy and misinformation relating to the virus. Consequently, there is the need to focus on information hygiene in order to level out the path of dangerous fake news online by taking adequate and proactive measures to curb its spread (The News Literacy Project 2021). In addition, fake information that is "liked or shared" many times may have a negative impact on people, thereby, escalating a crisis regarding the misinformation. One such crisis is fake news or misinformation regarding the COVID-19 pandemic plaguing the world, which has created much panic among people because of the confusion about the virus. The easy access to information through social media tools has generated a lot of bad information hygiene, however, many social media organizations are making great efforts to reduce fake information or misinformation relating to the-19 pandemic, with many structures for combating misinformation created.

Twenty-eight countries participated in a survey conducted by Edelman Trust Barometer (2021) between October 19 and November 18 2020 on the Covid-19 pandemic. The survey results indicated that there is an outbreak of misinformation and widespread mistrust among societal institutions and leaders around the world. In addition, the survey equally submitted that the increase of misinformation is threatening information regarding Covid-19 as well as its recovery because there are many people who practice poor information hygiene, that is, they do not check their sources and ensure credible and factual information is shared to others. To an extent, information hygiene involves information security due to the fact that some information can be harmful for consumption.

The Edelman Trust Barometer (2021) report revealed that good information hygiene means: having regular engagement with news; considering contradictory points of view; authenticating information; and not spreading fake information. This report has lots of good information on information hygiene, which was measured across four facets:

- Regular engagement with news
- Engagement with differing points of view

- Information verification, that is, not assuming information is true simply because it supports one's point of view and
- Avoid spreading fake information, which means to check information reliability before forwarding such information to people (King and Ockels 2021).

Furthermore, King and Ockels (2021) submitted that bad information hygiene is a significant societal problem which is obviously not easy to rectify and recommended people practicing good information hygiene so as not to spread misinformation.

Most sites with misinformation about certain issues that can be entertaining for human consumption can increase the number of visitors to the site within a short period of time because many people, regardless, are interested in the type of information available on the site. Individuals must thus be aware of the basics of information hygiene in order for them to be protected from misinformation also known as fake information, especially information surrounding major issues in the society, such as the COVID-19 pandemic. As posited by Bojanic (2021), a well planned approach to information hygiene would have a great role in shaping perceptions and establishing loyalty that would outlive the current coronavirus predicament. Therefore, everyone needs to focus on information hygiene and challenge dangerous fake news online by taking positive measures to reduce its spread in society. This is because the decisions about what type of information to believe and share can have a great impact on other people.

The News Literacy Project (2021) identified five steps that can help to practice good information hygiene:

- Recognize the effects of your information decisions: one's decisions and actions can unintentionally spread fake information regarding a crisis, e.g., misinformation on the coronavirus influences other people and can have consequences for society.
- Take 20 seconds to practice good information hygiene: scan the text for fact checks, conducting a quick search for the specific assertion and then looking for reliable sources so as to not spread any unconfirmed news.
- Filtering information: as cited by the World Health Organization, there is an excess amount of information which is the cause of misinformation regarding COVID-19, and it is hence recommended that an individual focus his/her attention on credible sources.
- Learn to identify misinformation patterns: one should be able to spot misinformation outlines in any information so as not to be deceived by "copy-and-paste" unfounded information.
- Assist in the sanitizing of social media feeds: individuals should help to flag misinformation when posted on social media because refusing to do so would leave behind fake information that will influence those who read it after them.

Information Literacy

The Association of College and Research Libraries' Information Literacy Competency Standards for Higher Education (2000) defines an information-literate person as someone who is able to: determine the extent of information needed; effectively and efficiently access the required information; critically evaluate the information and its sources; incorporate selected information into their knowledge base; effectively use information to accomplish a desired goal; understand the economic, legal, and social issues surrounding the use of information; and access and use information ethically and legally (ACRL Competency Standards 2000).

Some aspects of information literacy include using information technologies, such as personal computers, e-mail, software programs, and the Internet, as well as evaluation of the information obtained using the Internet and online electronic resources, while other components of information literacy regard the ethical use of information and information technologies. Some primary goals of information literacy are to get people thinking critically about the information they encounter and building people's ability to use that information to create new knowledge.

Ranaweera (2008) defines information literacy as the set of skills required to find, retrieve, analyze, and use information. This is the reason why the twenty-first century is regarded as the information era, due to the explosion of information and the information sources that must be navigated. A person may not achieve good information hygiene without acquiring and practicing exceptional information literacy skills, which highlights how such skills can empower people with the critical skills needed to explore the information space (Ranaweera 2008). The acquisition of the necessary information literacy skills helps a person to apply their knowledge from the familiar environment to the unfamiliar, that is, people are able to critically evaluate and identify the source of any information they see before sharing such information to others. Ranaweera (2008) highlighted how this information literacy concept evolved from several basic library concepts such as library instructions, bibliographic education, user education, and information literacy programs.

Philadelphia University (2006) provided an institutional definition of information literacy as it concerned their students, which can also be adapted into the current study. The information literacy definitions are highlighted as the ability to:
- articulate one's information need
- identify, locate, and access appropriate sources of information to meet the information need
- effectively use information resources, no matter the format
- critically and ethically apply the information
- determine if the need has been adequately met

In addition, Bundy (2004) described information literacy under three main components:

i. Generic skills, which includes: problem solving; collaboration; teamwork; communication; and critical thinking.
ii. Information skills, which includes: information seeking; information use; and information technology fluency.
iii. Values and beliefs, which includes: using information wisely and ethically; social responsibility; and community participation.

Figure 1: Information literacy framework developed by (American Library Association (ALA) 2016).
Source: Volunteer State Community College http://libguides.volstate.edu/instruction/infolit.
The Framework for Information Literacy, developed by the Association of College & Research Libraries (ACRL), comprises of six interconnected concepts that outline how individuals interrelate with information. At its core is Information Literacy, surrounded by key principles:
Authority is Constructed & Contextual – Information credibility varies by context.
Information Creation as Process – Information undergoes structured development.
Information Has Value – Information holds legal, ethical, and social significance.
Research as Inquiry – Research is an evolving, iterative process.
Scholarship as Conversation – Knowledge grows through academic dialogue.
Searching as Strategic Exploration – Effective searching requires adaptable strategies.
This framework promotes critical thinking, ethical use of information, and lifelong learning.

The framework above was used to describe information literacy, showing the research process of IL in a more holistic and process-oriented process than the more linear set of standards used previously. The American Library Association

(ALA) in 2016 changed the standard of information literacy to the framework illustrated above, which shows that is a direct relationship between information literacy and the other variables. For instance:

Authority is constructed and contextual: this indicates that information resources reflect their inventors' knowledge and credibility and is assessed based on the information need and the context in which the information will be utilized.

Information creation as process: in this case, information in any format can be created to convey a message which is shared via a selected delivery method. The recurring process of research, creating, revising, and disseminating information varies, and the resulting product reflects this variation.

Information has value: this explains that information has many aspects of significance, such as being a commodity, a means of education, a means to influence, and a means of negotiating and understanding the world. In addition, legal and socioeconomic security influences the information production and dissemination; for example, information on the COVID-19 pandemic has a lot value to the citizenry.

Research as inquiry: research is a recurring process which relies upon an increasing number of innovative queries whose responses in turn develop additional questions or lines of inquiry in any disciplines. This implies that individuals should be able to ask questions regarding any information they can not substantiate so as to reduce misinformation.

Scholarship as conversation: individuals must cite the contributing work of others in their own information production, as this will help people know the original source of any information shared.

Searching as strategic exploration: searching for information is often unpredictable and repetitive, requiring the appraisal of a variety of information sources and the mental ability to change and follow other avenues as new understanding progresses.

In order other to practice good information hygiene, one has to develop very effective and efficient information literacy skills. An information literate person should also simultaneously develop the following attributes:
- awareness of how one engages with the digital world
- awareness of how to find meaning in the information one discovers
- awareness of how to articulate what kind of information an individual needs
- awareness of how to use information ethically

- awareness of the role an individual can play in communication in his/her career and
- understanding of how to evaluate information for credibility and authority (Madison College Libraries 2020).

The diagram below can be used to illustrate the keys areas of information literacy any individuals should be aware of:

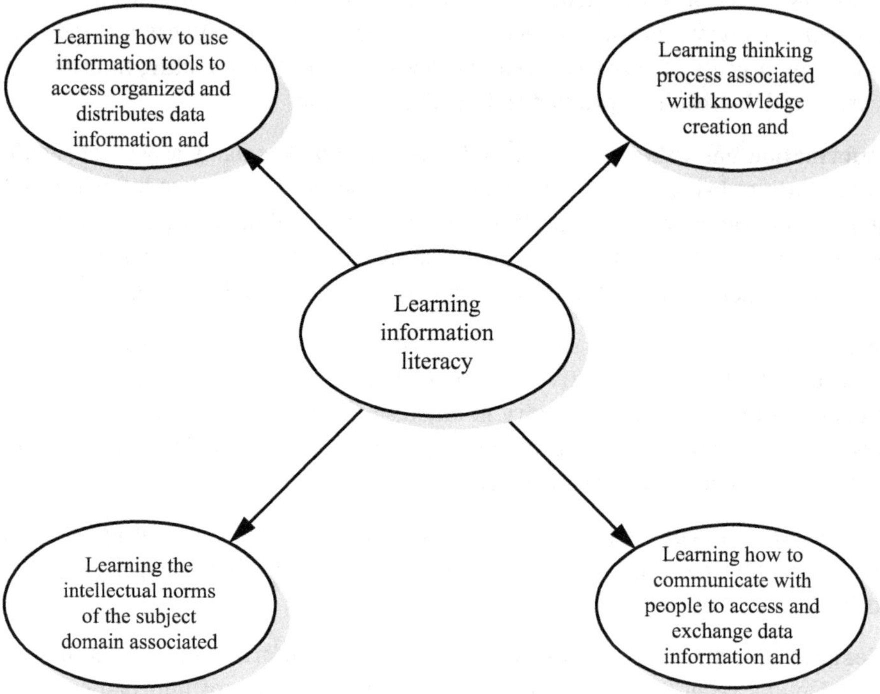

Figure 2: Key areas of information literacy (Hepworth 2000).

From the foregoing, it can be established that information literacy skills would go a long way to reduce the wide spread of fake news or misinformation regarding any particular issue affecting society. The implication is that, if possessing the required information literacy skill, individuals will be able to identify or confirm any information before "liking" or "sharing" it with other people. Therefore, information literacy skills may be considered as a panacea for practicing good information hygiene in the society. This summation can be represented using the information literacy core competencies developed by Grand Valley State University Libraries (2021).

Conclusion

Availability of information from countless sources is the hallmark of the twenty-first century, but access to information does not automatically translate to quality information. Information overload may pose a significant problem to those who lack the processing capability, therefore, when information available outweighs the information processing ability information overload occurs, which has the possibility of reducing the quality of decisions. In view of this, information pollution is the adulteration of the information landscape with inappropriate, fake, and misinformation with the aim to deceive and cause commotion as well as influence public debate and diminish the quality of knowledge. This can lead to poor decision making by those not skilled in information literacy skills who can become victims. The research affirms that good information hygiene and information literacy skills remain the solution to the misuse of information that that has overwhelmed the information space.

References

American Library Association (ALA) (2016) Framework for information literacy for higher education. American Library Association. Retrieved from www.ala.org/acrl/standards/ilframework.

Association of College and Research Libraries (ACRL) (2000) Information Literacy Competency Standards for Higher Education. Retrieved from http://www.ala.org/ala/acrl/acrlstandards/informationliteracycompetency.htm

Bojanic, I (2021) Information Hygiene: How Not to Contribute to a Culture of Clutter. Gogos. Retrieved from https://gongos.com/thinking-post/information-hygiene-how-ot-to-contribute-to-a-culture-of-clutter/

Bundy, A. (ed.) (2004) Australian and New Zealand Information Literacy Framework principles, standards and practice, 2nd ed. Adelaid: Australian and New Zealand Institute Information Literacy. Retrieved from https://adbu.fr/wp-content/uploads/2013/02/Infolit-2nd-edition.pdf

Canisius College (2021). The Framework for Information Literacy. Retrieved from https://libguides.canisius.edu/faculty/infolitresources

Cobban, S.J., and Seale, L.N. (2003). A collaborative approach for improving information literacy skills of dental hygiene students. *International Journal of Dental Hygiene*. 1(1):49–56. Doi: 10.1034/j.1601-5037.2003.00005.x. Retrieved from https://pubmed.ncbi.nlm.nih.gov/16451546/

Collins dictionary (2020) definition of information hygiene. Retrieved from https://www.collinsdictionary.com/submission/22205/information+hygiene

Dias, P. (2014). From 'infoxication' to 'infosaturation': a theoretical overview of the cognitive and social effects of digital immersion. Retrieved on 29/08/2021, from https://repositorio.ucp.pt/handle/10400.14/14939

Durodolu, O.O & Adekanye, E. A. (2017). The Perception of Information Literacy Skill among the Librarians: A Survey of University of Lagos Library. International Journal of Processing and Communication. Vol. 5,No.1 & 2. Pp. 148–159

Durodolu, O. O., Nduka, S. C. and Tinyiko, T.V (2022). Media Literacy and Fake News: Evaluating the Roles and Responsibilities of Radio Stations in Combating Fake News in the COVID-19 Era. International Journal of Information Sciences Management (IJIMS). Vol. 6 (1), pp. 32–49

Durodolu, O.O, Ibenne, S.K and Dube, TV (2021). New mental costumes: Leveraging ACRL's value of information frame in dealing with infodemics in a post-truth era. College & Research Libraries News (College & Research Libraries News). Vol. 82 (4). Pp. 175

Edelman Trust Barometer (2021). Daniel J. Edelman Holdings, Inc. Retrieved from https://www.edelman.com/trust/2021-trust-barometer

Godbey, S. (2013). Collaboration as an essential tool in information literacy education 9–16: Context, qualities and implications. *Student Research Journal*, 22: 1–16 Retrieved from http://scholarworks.sjsu.edu/slissrj/vol2/iss2/3.

Grand valley state University Libraries, (2021). The Information Literacy Core Competencies (ILCCs). Retrieved from https://www.gvsu.edu/library/information-literacy-core-competencies-59.htm

Hepworth, M. (2000) 'Approaches to providing information literacy training in higher education: challenges for librarians. *The New Review of Academic Librarianship*, *6*(1)21–34.

Hsieh, H. F., & Shannon, S. E. (2005). Three Approaches to Qualitative Content Analysis. Qualitative Health Research, 15(9), 1277–1288. doi:10.1177/1049732305276687 PMID:16204405

Ilyin, Y. (2013) Information hygiene at work. Retrieved from https://www.kaspersky.com/blog/information-hygiene-at-work/14899/

King, S., and Ockels, C. (2021) what is Good Information Hygiene? Small Business Lab. Retrieved from https://www.smallbizlabs.com/2021/01/what-is-good-information-hygiene.html

Madison College Libraries (2020) Information Literacy: Guide for Students: What is Information Literacy? Retrieved from https://libguides.madisoncollege.edu/InfoLitStudents

Philadelphia University (2006) what is Information Literacy? Retrieved on 29/08/2021, from https://www.philau.edu/infolit/what_students.htm

Ranaweera, P. (2008). Importance of information literacy skills for an information literate society. In NACLIS 2008, Colombo (Sri Lanka), 24th June 2008. [Conference paper] Retrieved from http://eprints.rclis.org/11956/1/Microsoft_Word_-_Prasanna_2.pdf

Stone, L. (2009). Information hygiene. Retrieved 29/08/2021 from https://lindastone.net/2009/11/30/beyond-simple-multi-tasking-continous-partial-attention

The News Literacy Project (2021) Practice information hygiene. Retrieved from https://newslit.org/educators/sift/the-sift-practicing-information-hygiene-coronavirus-misinformation-covid-19-journalism/

The News Literacy Project, (2021). Classroom Connection: Practicing information hygiene. Retrieved from https://newslit.org/updates/classroom-connection-practicing-information-hygiene/

Rogers, P. Puryear, R and Root, J. (2013). Infobesity: The enemy of good decisions, how to attack information overload. Retrieved on 29/08/2021, from https://www.bain.com/insights/infobesity-the-enemy-of-good-decisions

Wurman, R. S. (2012). Information Anxiety: Towards Understanding. Retrieved on 29/08/2021, from https://scenariojournal.com/article/richard-wurman/

Paora, S. and Heravib, B. (2020). Information literacy and fake news: How the field of librarianship can help combat the epidemic of fake news, The Journal of Academic Librarianship. Volume 46, Issue 5.

Joshua Onaade Ojo and Taiwo Ajayi

Information Literacy as a Panacea for Curbing Information Disorder in the Information Age

Abstract: The imperativeness of information and communication technologies in strengthening information disorder in the digital age through social media comes into focus because of the complex evolvement that has to be deciphered. The information age represents the period of information services enhancement which came into being as a veritable institution charged with the sole responsibility of information generation, organization, dissemination, preservation and conservation. Information literacy is the process of evaluating, accessing, interpreting and analyzing different forms of information before it is used. With information disorder as a widespread global problem it has become necessary to combat the various challenges that information disorder it disorder it has caused with all its antecedents and attributes such as fake news, mis-information, disinformation, mal-information, and infonomics. All these terms were coined to convey the underlying value of information in terms of its production, market demand, and economic impact in the information age. The study examines the following objectives: identifying the role of information literacy in curbing information disorder; identifying the panacea for curbing information disorder finding out the features of information disorder in the information age; examining some of the constraints caused by information disorder in society; and ascertaining the role of information professionals in curbing information disorder in the information age. The use of desktop publishing and secondary sources of information were the methodology used for the conduct of this study, while information sources from both online and public access information resources were also used. The findings from the study revealed the use of fake news social media accounts, the proliferation of scam websites, hashtags, rumors, memes, fake content development, and social feeds. All these have become worrisome and disturbing to the generality of people in society. Notwithstanding, the research suggested some appropriate solutions to ameliorate the menace of information disorder.

Joshua Onaade Ojo, University of Lagos, University Main Library, Research and Bibliographic Department, Gandhi Library, Akoka, Yaba
Taiwo Ajayi, Lagos State University of Science and Technology, University Library Complex, Readers' Services Department, Sagamu Road, Ikorodu

https://doi.org/10.1515/9783111176345-005

Keywords: information literacy, information disorder, information age, fake news, panacea, mis-information

Introduction

Information is knowledge, information is power, and information is security. This is a slogan or mantra of a media house to remind people of the importance of information. Apparently, the information age is referred to as the computer age, information society, new economy, or knowledge-based economy, among other nomenclatures with different unique features as indicated above. The Information Age is also synonymous with the digital age as it is being experienced in the twenty-first century characterized by the rapid shift from traditional industry that brought about an industrial revolution through industrialization to a knowledge economy based on information computerization and technology (Okai and Ujoumunna 2021).

The information age allows affordable and instant access to information, resulting in the rampant spread of disinformation, misinformation, and malinformation. Hate speech, defamatory remarks, and rumors by actors like trolls, cyber troops, fake news syndicates, bot networks, hardline religious groups, political parties, and even some governments are some of the characteristics of information disorder (Towhid 2020). The abuse or interaction of any of the abovementioned features of information in the society will lead to information disorder in knowledge based economy or information society.

However, information disorder with its antecedent or forbearer such as fake news, misinformation, disinformation, and mal-information reflected the negative side of man's quest for information for active participation in a knowledge economy or digital age which is significantly gaining popularity nowadays. Fake news assumes a position as any posted or printed article that is demonstrably false, regardless of opinion on the matter (Nazari 2016). This is a complex phenomenon that came into being with the introduction of information and communication technology in the twenty-first century as founded by some schools of thought. However, some researchers may have a hard time separating fact from fiction; they should but do not know that librarians are information literacy experts, essentially the "information gatekeepers" who can direct individuals to dependable resources like peer-reviewed, vetted content in full-text databases (Umahi and Mmejim 2021). The role of information literacy as a panacea for curbing information disorder is the notion of a developing information society. It is from this perspective that fake news, misinformation, disinformation and malinformation are central to this discourse of information disorder as the statement of the problem (Durodolu and Ngoaketsi 2019).

Information disorder has its consequences and is prevalent in our daily activities where information is being passed from one person to another, most especially through social media. The Internet to some extent has changed the political campaign climates of nations where propaganda, one of the elements of fake speech and fake news, is used when targeting political enemies, which has no truth and is simply used to disparage one's opponents. Information disorder has different characteristics such as satire or parody, namely: depicting no intention to cause harm but with the potential to deceive; false content, when headlines are shared with false contextual information; misleading content to frame on issue or individual; false connection, when headlines, visuals, or captions don't support the content; impostor content, when genuine content is shared with false contextual information and others are manipulated; propaganda, false information for damaging people's characters; sponsored content, maligned information to damage people's characters; and fabricated content, which is designed to deceive and do harm (Dias 2017).

According to Durodolu and Ibenne (2020), with growing dependency on social media for reportage, coupled with media errors with the potential to threaten the boundaries of knowledge and reliable information, attention is now being drawn to the credibility of using social media and other media outlets. This increasing attention is because of apparent disorderliness in the information milieu as a result of powerlessness to regulate activities on social media, coupled with the dilemma of tampering with the fundamental rights of the individual to free speech.

Information disorder is also a threat to knowledge. Unlike the traditional media houses with specific addresses and locations, identifying the whereabouts of the promoters of fake news is challenging as information can be manufactured in remote locality but with a global impact. Tracking down individuals peddling fake news for charges of defamation or libel is difficult reflects the intercontinental nature of social networks.

Various forms of information disorder have also added to the crisis of public communication, leading to fake news and hate speech. Fake news and hate speech affect polity and threaten democratic governance and result in fear and tension which jeopardizes public peace and security. People and ethnic groups have entered into conflict against each other because of fake news peddled through platforms by people with different motives. In the course of the phenomenon of information disorder, the likes of misinformation and disinformation have false and harmful impacts such as in Nigeria during the End SARS protest which was a disaster in the annals of Nigeria, when about 80 BRT buses worth 3.9 billion Naira were destroyed by hoodlums (Dias 2017).

Tired of these abnormalities, the Federal Government of Nigeria (FGN) shut down Twitter near the end of 2021 in a bid to stop the fake content that stemmed hate speech and aggravated the defamation of individuals, associations, and organizations and, ultimately, violence or disorderliness.

Objectives of the Study

1. To identify the role of information literacy in curbing information disorder;
2. To identify the various forms of information disorder in an information age;
3. To identify the various actors and entities involved in spreading information disorder;
4. To identify the role of information professionals as information literacy experts in combating information disorder;
5. To examine some of the constraints caused by information disorder in society.

Significance of Study

Information is strength and power but it is very difficult to differentiate authentic information from fake news in the political milieu, which is due to the emergence of the digital revolution, political transition, and economic developments. With the meteoric rise of the internet, social media, and Internet-enabled personal devices in the twenty-first century, the Information age has become the most significant period in the history of communication, which in turn has formed the fundamental problem associated with the rise in information disorder.

Concept of Information Age

Information has emerged as the prime element in the twenty-first century, with the increasing relevance of computers and subsequent development of information technology as its in today's society. With information explosion and technology innovation, information is no longer confined in printed materials, accessible in a single and, physical location. Information and communication technology (ICT) refer to the technologies that are used to collect, store, edit and communicate information in various formats (Raji 2018). Indeed, the term information age is used to denote the present time which is witnessing a revolution in the use, flow, and control of information. It is also referred to as the digital age, computer age, information society, new economy, or knowledge-based economy, among other nomenclatures, although with different unique features.

The imperativeness of information and communication technologies in strengthening information disorder in the digital age through social media has come into focus because of the evolving complexity that has to be deciphered. The information age represents the period of information services enhancement which came into being as a veritable institution charged with the sole responsibility of information generation, organization, dissemination, preservation, and conservation. Information literacy is the process of evaluating, accessing, interpreting, and analyzing different forms of information before it is used.

The information age is a historic period in human history in the twenty-first century characterized by the rapid shift from the traditional industry that the Industrial Revolution brought through industrialization to an economy based on information computerization and technology. It refers to the post-industrial society rich in information and communication technologies. The information age has remained a period where digital technology is prevalent and of common use throughout the world (Ujoumunna 2021). The information age began in earnest with the widespread use of the Internet and has moved into the current era where social, economic, and political activities are dependent on information and communication technologies. Accordingly, personal computers and other subsequent technologies have been introduced to give users the ability to easily and rapidly transfer information from one location to the other in different formats.

Likewise, the outset of the information age is associated with the digital revolution, just as the Industrial Revolution marked the onset of the Industrial Age. As the information age progresses, the phenomenon is that the digital industry creates a knowledge-based society surrounded by a high-tech global economy that spans its influence over how the likes of the manufacturing and service sectors operate in an efficient and convenient way. John (2016), while comparing the features of a commercialized society with a knowledge-based economy, notes that in a commercialized society, the information industry is able to allow individuals to explore their personalized needs, therefore simplifying the procedure of making decisions for transactions and significantly lowering costs for both the producers and buyers while in a knowledge-based economy, with economic activities based on the purpose of information efficacy. Moreover, the term "Information Age" is used to denote the present time which is witnessing a revolution in the use, flow, and control of information. This new invention or phenomenon is extensively used to highlight the information-based world where the speed, technology, and dramatic decreases in the cost of creating, processing, transmitting, and searching for information empower a wide range of actors. This chapter focuses on these aspects of the Information Communications Technology dominated world and its impacts on society by focusing on the developments in the past two decades.

Concept of Information Disorder

Information disorder in this discourse can be traced to the emergence of the digital revolution through political transition and economic developments, aided by the spectacular rise of the internet, social media, and internet-enabled personal devices from the middle of the twentieth century to the twenty-first century. The Information Age has become the most significant period in the history of communication, making available the highest volume of information ever accessible to people (Born 2017). It broke the traditional hierarchy and democratized the production and consumption of news and information.

Towhid (2020) posited that significant development of the media landscape has led to the deregulation of media houses in some countries. The multi-channel television grew in influence, the number of newspapers rose, first-generation internet access via dial-up modems started to spread, and the number of options available to audiences and advertisers continued to expand between the 1980s and 1990s. This change of media landscape accelerated in the first decade of the twenty-first century as the pattern of content creation, distribution, and consumption witnessed even more radical shifts.

The rise of digital technologies has made a profound contribution and changes to the way people communicate, interact, and learn about the world. There has been a dominant growth in the information explosion through the incessant use of search engine, due to the explosive growth of social media websites and the spread of mobile web access, which in turn has led to a decline in printed newspaper circulation and the decreasing popularity of television as the key source of information and entertainment. Through this changing in environment, online media has become the fulcrum of information propagation. People increasingly depend on this online medium platform for their means of communication and entertainment needs (Dias 2017).

This could be due to many factors such as the penetration of internet facilities to both rural and urban areas of the country, access to low cost phone devices, and affordable mobile data packages, with mobile phone no longer not available to the poor. On-demand access and the receiving of content as it happens contribute to information disorder. Invariably, this has led to the democratization of media system by removing barriers to publication, with anyone able to produce content and share it with people anywhere in the world. However, online platforms do not take into account how informed users are, either as disseminators or consumers of information (Banerjee 2020).

Towhid (2020) remarked that until very recently there was no differentiation of who the actors were and their motivations for generating information. Thus, any kind of content could reach a global audience without undergoing the filters of traditional media houses. Also, nothing on the internet is lost, so these kinds of contents remain in the information ecosystem and reappear in front of audiences. The volume of such contents becomes so vast that it often dilutes credible voices and changes people's perception of news and information, which ultimately results in a chaotic online environment which has been popularly termed as "information disorder."

If information is actually knowledge, power, and security, the level of its abuse leading to other features of its conceptualization such as fake news, misinformation, malinformation and disinformation etc., has eroded the growing dependency on social media, along with the potential of threatening the boundaries of knowledge and credible and reliable information. The information milieu, as a result, is often powerless to regulate activities on social media coupled with the dilemma of tampering with fundamental rights of free speech. Unlike traditional media, identifying the whereabouts of promoters of fake news is challenging as information can come from anywhere in the world (Durodolu and Ibenne 2020).

Concept of Information Literacy

Information literacy is the term for continuity of learning in an information society or knowledge based economy that does not depend on one's educational attainment. The term literacy does not have an acceptable universal definition as it is viewed from different perspectives which are dependent on the context in which it is being defined. Mokhtar and Majid (2008, n.p.) detail that "Literacy has been seen as the ability to read and write implying tests of literacy and numeracy; on the other hand, is seen as entity where by books and users are brought together into direct contact" Elsewhere, Imeremba (2015), cited in Osedo (2021), states that information literacy through having the skills to search for current and credible information sources that will empower people to make the right decisions is critical Finally, Doyle (1992), sees the concept of information literacy as the ability to access, evaluate, and use information from a variety of sources, and asserts that to be information literate, a person must be able to recognize when information is needed and have the ability to locate, evaluate, and effectively use the needed information (ALA 2000). These descriptions imply that those who are information literate are those who understand their information needs and where to seek information, and are able to locate desired information in order to satisfy these information needs.

With the use of information literacy as the panacea for curbing information disorder in an information society, two appropriate definitions might include Lawal et al. (2014) cited in Kuhlthau (2014), who posited that information literacy is a way of learning, not a discrete set of skills and attributes, and Kuhlthau (2014), who identified six stages of the process, namely, task initiation, topic selection, pre-focus exploration, focus formulation, information collection, and search closure. More relevant in the context of this discourse is the definition of Doyle (1992) which differs from that of IFLA (2017) with her inclusion of ten specific attributes that represent tangible steps in the progression of information needs satisfaction through accessing, evaluating, and using information: (a) recognition of an information need; (b) recognition and concession of the importance of reliable information to intelligent decision-making; (c) formulation of questions based on information need; (d) identification of information sources; (e) development of search strategies; (f) accessing information sources; (g) evaluating information; (i) organization of information for application and integration of new information into the existing knowledge base; and (j) information use for critical thinking and problem solving. These attributes are in-exhaustive but the most appropriate to guide against information disorder in an information society.

Role of Actors in Information Disorder

Most of the targeted actors in information disorder are those who lack information literacy, due to the fact that some of the features of information literacy are supposed to guide information that is released to the public. Such major actors are journalists, politicians, propagandas, and blackmailers who strive on the pretense of being the custodian of information but who in fact hide under the influence of social media which is their major tool of propagating information disorder. They have forgotten that information literacy is the ability of one to conquer their environment using information as a tool (Idoko and Osadebe 2016) and rather are embedded in fabricated or impostor content, all in the name of creating information disorder.

Beyond this, some of the other popular terms associated with information disorder include junk news, misinformation, propaganda, rumors, conspiracy theories, hate speech, and defamatory speech, with different stakeholders like think-tanks, media practitioners, digital intelligence community, social media platforms, and governments using different definitions to describe these elements (Towhid 2021). Some of these things happened during election periods when there were threats made by individuals towards their opponents, often to discourage them from contesting and inflict public disgrace on such politicians in order to tarnish their image. Such instruments of destruction are embedded in false claims, leading to misleading information, with the potent aim of harming another by an impostor whose aim is to manipulate and fabricate the truth.

Such stories may be malformation, which is described as, "genuine information that is shared with intent to cause harm." Such information is "based on reality, but used to inflict harm on a person, organization or country" as defined by Wardle and Derakhshan (2018). Subsequently, these are messages with some truth but are created, produced, or distributed by actors who intend to harm rather than serve the public interest. The role of social media has made information readily available to society at any time and it connects vast groups of people along with their information (Aral 2020). The effect of information communication technologies has impacted the way people communicate information and the way misinformation is spread. Misinformation has an impact on societies' ability to receive information which then influences the likes of communities, politics, and medical fields (Lewandowsky et al. 2012).

Actors Mode of Operations in Information Disorder

Most of the actors operate at the higher level of interactions using various modi of operandi, such as hostile information campaigns, influence, information, soft power, and interference. An idea developed by the Centre of Excellence for National Security (CENS) of the S. Rajaratnam School of International Studies (RSIS) was the term "Hostile Information Campaigns" which includes the spreading of disinformation or biased information in the defender state, such as spreading narratives by traditional media (i.e. newspapers) through proxies, or under convert identities, or carrying out these activities using automated social media accounts (trolls) to create coordinated campaigns, often disguised as local opinions (Rahman et al. 2020).

The essence of this idea or theory as propounded by the Centre of Excellence for National Security in (1834) is to propagate hate speech, harassment, religious defamation, rumor, attacks, and vigilantism through the association of actors. Such actors use information for media syndication through media ownership by open advertising devoid of editorial manipulation and issuing of public statements. Most of the actors buy the influence of their editorial board through luxurious contracts and money laundering so as to have their adverts published without any hindrances.

Another major modus operandi use by information disorder practitioners is the use of soft power, which is prevalent among nations through their public diplomacy, strategic communication, and foreign assistance in the form of foreign investment. For instance, the global phenomenon about "climate change has been one of the most debated topics in the past few decades. Even though the majority of nations now agree on the basic ideas related to climate change and participate together in international initiatives, such as the Kyoto Protocol and the Paris Agreement, comprehensive and effective steps are yet to be taken. The political process has been stagnant and plagued by claims of national interest" (Mavrodieva et al. 2019, n.p.). The recent few decades have seen a rise in public figures and pop culture products focused on raising awareness on climate change through effective campaigns, using the advances of technology. At the same time, social media platforms—such as Instagram, Twitter, and Facebook—have provided an opportunity for the general public to discuss and share opinions instantly with vast cross-border networks.

This phenomenon is considered to be a new form of soft power which can provide input into the discussion on climate change and possibly affect the current international political mechanisms (Fernandez et al. 2016). According to Anderson (2017), there is need for cautioning as this can also lead to opinion silos and opinion reinforcement, which can be both positive and negative for the issue of climate change (Nye 1990). The soft power concept was introduced by American political scientist Joseph Nye in the late 1980s as a neoliberal theory which was defined as the ability to influence the behavior of others to get the outcomes you want, which uses co-option rather than hard power enforcement. Soft power can be used by nations to influence other nations but also by a number of non-state actors, such as international organizations and institutions. This aims to shape the preferences of others employing assets such as attractive cultural or political values, or even through personalities revered by public (Nye 2004).

Interference also plays another crucial role as an actor or agent in information disorder influences government decisions through policy documents with politicians and with academics such as in educational institutions. They are present in non-government organizations through their activities so they can affect their targeted audience and operate through the likes of cyber-attacks, hacking, and leaking information. For instance, Wardie and Derakhshan (2018, n.p.), in their Facebook published report by three members of their Security team, entitled "Information Operations and Facebook," outlined the use of the platform by state actors, defining information operations as "actions taken by organized actors (governments, or non-state actors) to distort domestic or foreign political sentiment, most frequently to achieve a strategic and /or geopolitical outcome. These operations can use a combination of methods, such as fake news, disinformation or networks of fake accounts aimed at manipulating public opinion (which is refer to as 'false amplifiers').

Additionally, in South Africa, an email leak in May 2017 exposed large-scale disinformation efforts by the powerful Gumpta family to distract attention from its business dealings with the government. These efforts included paying Twitter users to abuse journalists, spreading disinformation, and the use of bots to amplify fabricated stories. In contrast to the above-mentioned, unofficial actors are those who work alone or with loose networks of citizens to create false content to harm, make money, or entertain other like-minded people (Elisee 2017). During the US 2016 elections, journalists tracked down one of these "unofficial" agents who in an interview admitted that his "whole idea from the start was to build a site that could infiltrate the echo chambers of the alt-right, publish blatantly [false] or fictional stories and then publicly denounce those stories and point out the fact that they were fiction." Later people were amazed at how quickly fake news could spread and how easily people believe it (Sydell 2017).

Tool for Curbing Information Disorder

Various attempts by stakeholders have been and are being made to curb information disorder, but these are yet to make any impact, thus necessitating this type of opinionated paper. The best approach is to educate practitioners about the need to equip themselves with various formats of information literacy: media literacy for journalists, information literacy for academia, digital information literacy, and other forms of literacy that can contribute and ameliorate information disorder, misinformation, mal-information, fake news, hate speech, and all forms of abusive use of information. Adjin-Tettey (2021, n.p.) posited that "Concerns have been raised about fake news, disinformation, and misinformation. One such thing is that false information could pollute the public sphere and damage democracy." Elsewhere, Pogue (2017, n.p.) considered "fake news as one of the greatest threats to democracy, journalism, and freedom of expression." For media literacy among journalists and media personnel, media and information literacy (MIL) has been conceptualized as consisting of "the knowledge, the attitudes, and the sum of the skills needed to know when and what information is needed; where and how to obtain that information; how to evaluate it critically [to make educated judgments about information] and organize it once it is found; and how to use it in an ethical way" (International Federation of Library Associations and Institutions 2011,).

Furthermore, UNESCO posits that media information literacy provides answers to: how to contribute both online and offline content wisely; the rights of online and offline information consumers and creators; and "how to engage with media and ICTs to promote equality, intercultural and interreligious dialogue, peace, freedom of expression, and access to information" (UNESCO n.d.) cited in Adjin-Tettey 2021). The most cited definition is that of Aufderheide (1993) which sees it as the ability of a citizen to access, analyze, and produce information for specific outcomes. Congruently, UNESCO (2017) asserts that media information literacy considers all forms of media and other information providers such as libraries, archives, museums, and the Internet, irrespective of the technologies used.

However, UNESCO (2013) defines media information literacy as the skills and competencies that empower citizens to access, retrieve, use, understand, evaluate, create, as well as share information and media content in all formats, using various tools in critical, ethical, and effective ways, in order to participate and engage in personal, professional, and societal activities. The essence of these definitions and concepts is to serve as the guiding principles in public speech and social media interactions in order to avoid information disorder, mal-information, misinformation, and fake news in the information age.

Solutions for Curbing Information Disorder

To curb information disorder in all ramifications, stakeholders must be involved in providing solutions since the challenges come from them. From the foregoing, therefore, media information literacy certainly recognizes the primary role of information and media in people's everyday lives. It lies at the core of freedom of expression and information since it empowers citizens to understand the functions of media and other information providers, to critically evaluate their content, and to make informed decisions as users and producers of information and media content (UNESCO 2017). It is on the basis of some of the aforementioned concepts that Aduloju (2021, n.p.) recommended "the need to empower people to interact and engage critically with media in local and global socio-political, economic and cultural contexts; that will enable people, through the use of the media, to become active citizens, able to exercise their democratic rights and responsibilities. She went further to say that people should be engaged to analyze media messages concerning their institutions, codes and conventions, audiences, technologies and ideologies."

It is also germane that government and stakeholders: provide ways for people to make informed choices; inspire the creative and democratic use of the media to express and communicate ideas, information, and opinions; entertain and promote participation in the creation of content; encourage the use of information and communication technologies (ICTs) to shape and share alternative content, media products, and services to meet the individual, community, and regional needs and interests; promote interpersonal communication to build positive relationships, family life, and community; and finally, enhance the dignity of the individual and the potential for ethical/moral and spiritual development, in solidarity with the poor, and with respect for cultural heritages (Aduloju 2021). These are the ways the pertinent issues of information disorder in the society can be reduced.

Conclusion

Information literacy as a panacea for curbing information disorder in the information age has presented an open discussion for further research with clarity of various concepts. Disinformation, fake news, and misinformation might appear in diverse forms, from various sources on social media platforms, and make many different claims in order to cause fear, panic, disunity among people and in most cases to build followers or likes. There is a need in to counter the scourge of fake news around various issues that is germane for democratic governance of a nation.

References

Adjin-Tettey, T.D. (2021). Combating fake news, disinformation, and misinformation: Experimental evidence for media literacy education. *Cogents Arts and Humanities*, *9*(1). https://www.com/doi10.1080/233111983.2022.2037229.

Aduloju, T.E. (2021). Media and information literacy: A critical response to the challenge of informdemic in the Covid-19 pandemic era in Nigeria. International Journal of Research and Innocative in Social Sciences, *5*(6).

African Library & Information Associations (2019). Weaponisation of misinformation the role of librarians in curbing the epidemic. https://www.web.aflia.net/weaponisation-of-inormation-on-the role of-librarians-in-curbing-the-epidemic.

Anderson, A.A.(2017). Effects of Social Media Use on Climate Change Opinion, Knowledge, and Behavior. Clim. Sci. 2017, doi:10.1093/acrefore/9780190228620.013.369.

Aral, Sinan (2020). The hype machine: how social media disrupts our elections, our economy, and our health–and how we must adapt (First ed.). New York.

Aufderheide, P. (1993). Media literacy. A report of the National Leadership Conference on Media Literacy. Washington, DC: Aspen Institute, Communications and Society Program.

Banerjee, R. (2020). Information Crises: The EU's Response and the South Asian Digital Media Landscape, Brussels: South Asia Democratic Forum.

Born, K. (2017) The future of truth: Can philanthropy help mitigate mis-information? Hewlett Foundation. Available at: http://www.hewlett.org/future-truth-can-philanthropy-help-mitigatemis-information.

Centre of Excellence for National Security (1834). National Counter Terrorism Centre Opening Speech National Security Adviser during the dedication of new building at Abuja.

Dias, N. (2017), Reporting on a new age of digital sastroturfing, First Draft News, https://firstdraftnews.com/digital-astroturfing/

Doyle, C. S. (1992). Outcomes measurement for information literacy within the National Education Goals of 1990, Final Report to the National Forum on Information Literacy, Summary of Findings (Syracuse, NY: ERIC Clearinghouse on Information Resources, p. 2 ERIC, ED 351033.

Durodolu, O.O. & Ibenne, S.K. (2020). The fake news infordemic vs information literacy. *Library Hi Terch News*. Emerald Publishing Limited, DOI: 10.1108/LHTN-03-2020-0020.

Durodolu, O. and Joseph M. Ngoaketsi (2019). Multilateral relationship between Information Literacy, Self-concept and Metacognitive ability. In Metacognition in Learning; Edited by Nosisi Feza, London: Intech Open Limited ISBN 978-1-78985-114-4. https://www.intechopen.com/chapters/76617

Elisee, V. (2017). The Gumpta scandal how a British PR firm came unstuck in South Africa. The new stateman. http://www.newstateman.com/culture/observation/2017/07/gupta-scandal-how-british-pr-firm-cam-unstuck-SouthAfrica

Fernandez, M.; Piccolo, L.S.G.; Maynard, D.; Wippoo, M.; Meili, C.; Alani, H. (2016). Talking Climate Change via Social Media: Communication, Engagement and Behavior. In WebSci '16 Proceedings of the 8th ACM Conference on Web Science; Conference Paper; ACM: New York, NY, USA, 2016; pp. 85–94, doi:10.1145/2908131.2908167.

John, C. (2016). Characteristics of digital age. Retrieved August 27, 22 from: http://thecomplexgroup4.blogspot.com/2022/07/characteristics-of-digital-age.html.

Idoko, N.A. and Osadebe, N.E. (2016). Information literacy: A panacea for unhealthy youth development in 21st century Nigeria. *Journal of Applied Information Science and Technology*, 7(2).

IFLA (2017). Open Educational Resources and Libraries: A briefing. Retreived July 2022 from: https://www.ifla.org/files/assests/clm/nes/oerr_ifla_brief.pdf.

Imeremba D. U, (2015) Value of Information Essentials of Information Systems and Networking. A Perspective on Nigerian libraries and information resources. Keny and Brothers ventures; 3 Owerrri Road Asaba Enugu

Lawal, V., Stilwell, C., Kuhn, R., & Underwood, P.G. (2014). Information literacy re-lated practices in the legal workplace: The applicability of Kuhlthau's model to the legal profession. *Journal of Librarianship and Information Science*, 46(4), pp: 326–346.

Lewandowsky, S.; Ecker, U. K. H.; Seifert, C. M.; Schwarz, N. Cook, J. (2012). "Misinformation and Its Correction: Continued Influence and Successful Debiasing". *Psychological Science in Public Interest*, 13(3):106–131. doi:10.1177/1529100612451018.

Mavrodieva, A.L. Rachman, O.K., Harahap, V.B.,& Shaw, R (2019). Role of Social Media as a Soft Power Tool in Raising Public Awareness and Engagement in Addressing Climate Change. license (http://creativecommons.org/licenses/by/4.0/)

Mokhtar, I.A. & Majid, S. (2008). Information literacy standards, guidelines ans their implementation: An analysis. DESISOC, *Journal of Library and Information Technology*, *28*(2), 5–12.

Nazari, M. (2016). Actuality of determining information need in GIS: A context-to-concept approach, Library and Information Science Research, 38, 2: 133–147.
Nye, J. (1990). Bound to Lead: The Changing Nature of American Power; Basic Books: London, UK, 1990
Nye, J. (2004). Soft Power: The Means to Success in World Politics; Public Affairs: New York, NY, USA, 2004.
Okai, E. J. & Ujoumunna, J. C. (2021). Libraries in the digital age. In S.K. Ibenne & C.P. Uzuėgbu (Eds.), Libraries and Society. Lagos, Nigeria, Zeh Communications Limited. (pp: 97–118).
Osedo, O.A. (2021). Libraries, information literacy and sustainable development. In S.K. Ibenne & C.P. Uzuegbu (Eds.), Library and Society. Lagos, Nigeria, Zeh Communications Limited. (pp: 239–273).
Pogue, D. (2017). How to stamp out fake news. *Scientific American*, *316*(2), 24–24. https://doi.org/10.1038/scientificamerican0217-24.
Rahman, M.F.A, Haciyakupoglu, G., Hui, J.Y., Leong, D., Ling, T., and Ang, B. (2020). Countermeasures against Foreign Interference, Singapore: S. Rajaratnam School of International Studies, Nanyang Technological University, p. 6.
Raji, S.K. (2010). The role of ICT as a panacea for national development. Retrieved from www.nacoss.or,ng.
Sydell, L. (2017). We tracked down fake news creator in the suburbs. Here's we learned, NPR. http://www.npr.org/sections/alltechconsidered/2016/11/503146770/npr-finds-the-head-of-a-convert-fake-news-operations-in-the-surbubs.
Towhid, A.B. (2020). Information disorder in the information age: actors, tactics, and impacts in south and southeast. Bangladesh Institute of International and Strategies Studies, *Bliis Papers*, 28.
Towhid, A. B. (2021). Information disorder in the information age, tactis and impacts in South and South East Asia. Bangladesh Institute of International and Strategic Studies (BIISS).
Umahi, B. O. & Mmejim, I.C. (2021). Information and communication technologies for digital developments in libraries. In J.O. Akidi, K.N. Igwe & J.C. Ujoumunna (Eds.), Libraries in the Era of Digital Technologies. Lagos, Nigeria, Zeh Communications Limited. (pp: 161–179).
UNECSO, (2017). Media and information literacy. Retrieved from http://www.unesco.org/new/en/communication-andinformation/media-development/media-literacy/mil-as-compositeconcept/#.
UNESCO, (2013). Global media information and literacy: Assessment framework, country readiness and competencies. Paris, France: United Nations Educational, Scientific and Cultural Organization (UNESCO).
Wardle, C and Derakhshan, H. (2018) "Thinking about 'Information disorder': formats of misinformation, disinformation, and mal-information", in Cherilyn Ireton and Julie Posetti (eds.), Journalism, 'Fake News' & Disinformation, Paris: UNESCO, p. 46.

Adeyinka Tella

The Impact of the Disease Infodemic and Infopandemic on Individuals with Disabilities

Abstract: The chapter examined the impact of a disease infodemic and infopandemic on individuals with disabilities. To achieve this objective, the chapter featured on overview of the global phenomenon of the rise of infopandemic, its global scope and dynamics, its impact on public perception, the challenge of balancing the need for transparency with the potential for information overload, information accessibility for persons with disabilities, psychosocial implications, health communication strategies, socio-economic disparities, intersectionality and diverse experiences, policy and intervention frameworks, building resilience in public health, and the impact of an infopandemic on persons with disabilities. Based on the discussions, a conclusion was drawn, with appropriate and relevance recommendations were made.

Keywords: outbreak of disease, pandemic, infodemic, infopandemic, misinformation, individuals with disabilities

Introduction

In recent times, the global landscape has witnessed not only the surge of diseases but also the parallel proliferation of information surrounding them, giving rise to what is now termed a "disease infodemic" and "infopandemic." This phenomenon, characterized by the rapid dissemination of both accurate and misleading information related to health crises, has profound implications on various segments of the population (Caceres et al. 2022). Among those significantly affected are individuals with disabilities, whose unique challenges and vulnerabilities are often exacerbated by the abundance of information, both reliable and misleading. As underscored by Goyal et al. (2023), understanding the intricate interplay between the outbreak of the disease infodemic and infopandemic and its impact on peoplewith disabilities is crucial for developing effective strategies to mitigate the

Adeyinka Tella, Department of Library and Information Science, University of Ilorin, Nigeria, Department of Information Science, University of South Africa, e-mail: tella.a@unilorin.edu.mg, https://orcid.org/0000-0002-5382-4471

https://doi.org/10.1515/9783111176345-006

adverse consequences and promote inclusive public health measures. In this context, it becomes imperative to delve into the specific challenges faced by individuals with disabilities during these times of heightened information flow.

The accessibility and reliability of information, the adaptability of health communication strategies, and the overall inclusivity of public health measures are crucial aspects that demand careful consideration. It is frequently difficult for people with disabilities to obtain timely and correct information, which makes it more difficult for them to make well-informed decisions on their health and well-being (Centre for Disease Control and Prevention 2020). Moreover, the proliferation of misinformation and disinformation during a health crisis can contribute to heightened anxiety and confusion among individuals with disabilities. The potential consequences of relying on inaccurate information can be particularly severe for this demographic, as their unique healthcare needs may not be adequately addressed or may be misrepresented. Therefore, a nuanced examination of the intersection between the disease infodemic, infopandemic, and the experiences of persons with disabilities is essential for fostering a more inclusive and resilient public health response (Rohwerder et al. 2022).

By comprehensively exploring these dynamics, insights that pave the way for targeted interventions, policy adjustments, and community support mechanisms can be uncovered. Addressing the specific challenges faced by individuals with disabilities in the context of the disease infodemic and infopandemic is not only a matter of equity but also a strategic imperative for building a more robust and responsive public health framework that prioritizes the well-being of all members of society (Gréauxet al. 2023).

Within the broader framework of the disease infodemic and infopandemic, individuals with disabilities confront a myriad of barriers that impact their ability to access and process information effectively. Physical barriers, such as the lack of accessible formats in official health communications, pose significant challenges. For those with visual or hearing impairments, traditional communication channels may not be sufficient, thereby necessitating the incorporation of alternative formats such as braille, sign language interpreters, or audio descriptions to ensure equal access to critical health information (Karnay, Furqan, and Saeni 2023).

Cognitive and intellectual disabilities further compound these challenges. The complexity of health-related content, coupled with the fast-paced nature of information dissemination during a crisis, may overwhelm individuals with cognitive impairments (Rockville2018. Tailoring information to various cognitive abilities and employing plain language strategies becomes crucial for ensuring comprehension and informed decision-making. Additionally, the socio-economic disparities often faced by persons with disabilities can exacerbate their vulnerability during health crises. Limited access to technology, healthcare resources, and support systems can further isolate this population, hindering their ability to stay informed and actively engage in preventive measures. Furthermore, the intersectionality of disability with other factors, such as age, gender, and ethnicity, adds layers of complexity to the experiences of individuals within this community (Rockville2018. Recognizing and addressing these intersectional dimensions is vital for crafting targeted and culturally sensitive communication strategies that resonate with diverse populations, ensuring that no one is left behind in the quest for accurate health information.

As we navigate the intricate landscape of the disease infodemic and infopandemic, it becomes evident that a one-size-fits-all approach to communication and public health interventions is insufficient. Tailoring strategies to accommodate the diverse needs of individuals with disabilities is not only an ethical imperative but also a pragmatic necessity for building a resilient and inclusive public health ecosystem capable of withstanding the challenges posed by infodemics and pandemics alike.

Overview of the Global Phenomenon

According to the WHO (2023b), an outbreak of a disease is defined as the number of cases above the normal expectation. The disease-causing agent, as well as the extent and kind of prior and current exposure to the agent, influence the number of cases. Epidemics are typically brought on by infections that are spread by direct contact between humans and animals, through the environment, or through other media. Additionally, exposure to chemicals or radioactive materials may result in outbreaks. For instance, mercury exposure is the cause of Minamata disease. However, even after a careful study, the reason of an outbreak may remain unknown.

The transmission of infectious diseases is influenced by environmental variables, including but not limited to food, water, sanitation, and air quality (WHO 2023). Epidemic-prone communicable illnesses are influenced by a multitude of environmental factors, with the most crucial beingthe climate, food, sanitation, and water supply.

Diarrheal illnesses can be brought on by contaminated food, unsanitary living conditions, inadequate excreta disposal facilities, a lack of safe water, and poor hygiene. In an emergency, these illnesses are a leading source of pain and fatalities. There are several ways that climate might impact the spread of illness: local climate can have a significant impact on disease vector population size and spread; heavy rains can cause flooding, which can lead to sewage overflow and widespread water poisoning; and certain data indicate that diseases may be transported by wind or air currents from one area to another (WHO 2023).

The global phenomenon of the disease infodemic and infopandemic represents an unprecedented challenge in the information age, where the rapid dissemination of information can both empower and overwhelm societies (Pian, Chi, and Ma2021). This sub-section aims to provide a comprehensive introduction to these interconnected phenomena, with Table 1 providing a comprehensive chronological table with country-specific outbreak details and death tolls for each epidemic and pandemic event from 1950 to 2023:

Table 1: Specific outbreaks of disease in the world, 1950–2023.

Year	Event	Pathogen	Country of Outbreak	Death Toll
1957–58	Asian Flu Pandemic	Influenza A (H2N2)	Worldwide	1–2 million
1968	Hong Kong Flu Pandemic	Influenza A (H3N2)	Worldwide	1–4 million
1981	HIV/AIDS Pandemic Emerges	HIV/AIDS	Global	32 million (as of 2022)
2002–03	SARS Outbreak	SARS-CoV	Asia, Worldwide	774 deaths
2009	H1N1 Influenza Pandemic	H1N1 Influenza Virus	Global	151,700–575,400
2014–16	West African Ebola Virus Outbreak	Ebola Virus	West Africa	11,325 deaths
2015	Zika Virus Outbreak	Zika Virus	Americas, SE Asia	Limited fatalities
2019–20	COVID-19 Pandemic	SARS-CoV-2 (Coronavirus)	Global	5.4 million (as of 2022)
2021	Marburg Virus Outbreak in Guinea	Marburg Virus	Guinea	Limited fatalities

Self-generated from the literature

It is important to note that death tolls can vary and are subject to updates and revisions. Additionally, the impact of each outbreak on different countries can be complex, affected by factors including population density, public health measures, and healthcare infrastructure (Alam and Sultana 2021; OECD 2020).One thing noticeable during the outbreak of global disease is that different kinds of information, both correct and fake, circulates, which the world witnessed during the Covid-19 outbreak. This gave birth to the concepts of the concepts of infopandemic and infodemic.

Concept

Infodemic and infopandemic share a relationship. The COVID-19 outbreak and reaction, according to Zielinski (2021) and the WHO (2023b), have been followed by a major infodemic—an abundance of information, some of which is correct and some of which is not—that makes it difficult for people to obtain dependable sources and assistance when they need it.

An infodemic is defined as a significant increase in the amount of knowledge related to a particular issue that might grow rapidly owing to a specific incidence, such as the Covid-19 epidemic. In this case, there are rumors and false information spreading, as well as information that has been manipulated with questionable motives. This issue is exacerbated in the digital age by social networks, which allow it to spread like a virus more quickly and widely (PAHO, 2020) An excess of information, both online and offline, is referred to as an infodemic. It involves intentional attempts to spread false information in order to thwart the public health response and further the goals of other organizations or people, with examples including fictitious treatments like gargling with salt water, consuming oregano, or even ingesting bleach. According to Pulido et al. (2020), conspiracy theories like the idea that the virus was created by the US or that eating "bat soup" was the source of the infection are examples of other false information that spreads misconceptions.

Person with disabilities

The UN Convention on the Rights of Persons with Disabilities (Fina, Cera, and Palmisano 2017) defines persons with disabilities (PWDs) as those who have long-term physical, mental, intellectual, or sensory impairments that, when combined with other barriers, may prevent them from fully and equally participating in society (World Bank 2021; America Disability Act 2021). "Those who have long-term physical, mental, intellectual, or sensory impairments which, in combination with various barriers, may hinder their full and effective participation in society on an equal basis with others" is how the United Nations Office of the High Commissioner (2006) defines persons with disabilities.

WHO (2021, n.p.) defined disability as "an umbrella term covering impairments, activity limitations, and participation restrictions" in the publication *International Classification of Functioning, Disability, and Health.* Approximately 15% of the world's population, or over 1 billion people, live with a disability. Between 110 and 190 million adults over the age of 15 struggle with major life issues. In addition, the aging population and the quick spread of chronic illnesses are factors in the rising rates of disability (World Bank 2021).

The Rise of Infopandemic

The United Nations Educational, Scientific, and Cultural Organization (UNESCO) created the term "disinfodemic" to emphasize the negative information component of an infodemic, even though the World Health Organization (WHO) first used the term "infodemic" (Posetti and Bontcheva 2020). Another type of false information that has been recognized by the Council of Europe is called "malinformation," which is defined as "information that is based on reality and used to inflict harm on a person, social group, organization, or country" (Wardle and Derakhshan 2017, n.p.). Put differently, malinformation is a type of deception that is intentionally created with the intent to harm people rather than to advance a particular political or ideological cause (Media Defence 2020). It is critical to discern between communications that are true and those that are false, as well as between true messages that are manufactured, spread, or developed by "agents" with the intention of doing more harm than good to the public interest. Consequently, malicious software, or "malware," is related to misinformation and intended to harm users. Since disinformation also spreads over analog communication networks, concerns about misinformation date back to the early days of the World Wide Web (Zielinski 2021; APA 2023). In essence, there are three ways to get things wrong: misinformation, disinformation, and malinformation. The other two are deliberate, but the first is an accident. These divisions, of course, are not permanent. Individuals who take in false information and then disseminate it merely because they think it is true (instead of advocating for a specific political viewpoint or ideology) are disseminating false information because they are distributing what they genuinely think is true but that someone else falsified for their own gain (College of Staten Island Library 2023). The pandemic experience seems to have reinforced the WHO's emphasis on the misinformation aspect of the issue, with Tedros Adhanom Ghebreyesus, the director general of the organization, stating, "We're not just fighting the virus." Indeed, the fight is also against conspiracy theorists and trolls that spread false information and impede the outbreak response (Ireton and Posetti 2018). Lastly, to round up this overview of false information, the area of agnotology in information management is another that has recently gained importance. Ignorance is the subject of agnotology. Proctor and Boal proposed the term in 1995 (WHO 2020). According to these scholars, ignorance can result from political and cultural conflicts in addition to simply the lack of knowledge. There is a political geography and a history to ignorance.

The rise of the infodemic marks a distinctive feature of the modern information age, particularly evident during health crises and pandemics (WHO 2021; Zielinski 2021). There are various factors that contribute to the surge of an infodemic (Pian, Chi, and Ma 2021):

Digital Connectivity: the ubiquity of digital platforms and the widespread access to the internet facilitate the rapid dissemination of information, both accurate and misleading. Social media, online forums, and instant messaging platforms enable information to reach a global audience within seconds.

Speed of Information Sharing: information can spread exponentially through social networks, reaching a vast number of individuals in a short period. The real-time nature of digital communication accelerates the amplification of information, making it challenging to discern reliable sources from misinformation.

Diverse Information Sources: the democratization of information allows a multitude of sources to contribute to the narrative, ranging from official health organizations to individual opinions and unverified claims. The abundance of information sources makes it difficult for individuals to filter and verify the accuracy of the information they encounter.

Viral Nature of Content: controversial or sensational content often gains traction more quickly than factual information, leading to the viral spread of misinformation. Emotional or fear-inducing messages tend to capture attention and circulate widely, contributing to the amplification of the infodemic.

Amplification through Echo Chambers: echo chambers are produced by social media algorithms and tailored content streams, which expose users to information that confirms their preexisting opinions. This selective exposure can reinforce and amplify misinformation within specific communities.

Challenges in Content Moderation: platforms face challenges in effectively moderating and curating content, leading to the proliferation of false or misleading information (Kaushal et al. 2023).The sheer volume of user-generated content makes it difficult to implement real-time content moderation at scale.

Lack of Information Literacy: insufficient media literacy and critical thinking skills among the general population contribute to the susceptibility to misinformation. Individuals may struggle to discern credible sources, leading to the inadvertent spread of inaccurate information.

The rise of the infodemic underscores the need for proactive measures to promote media literacy, enhance information verification processes, and establish clear communication channels during health crises. Understanding the dynamics of the infodemic is essential for developing strategies that mitigate its impact on public perception and decision-making.

The Global Scope and Dynamics

The infodemic and infopandemic are phenomena characterized by the rapid spread and dissemination of information, both accurate and inaccurate, across various communication channels during a health crisis. Unlike traditional epidemics that may be geographically confined, the infodemic and infopandemic have a global reach, impacting diverse communities worldwide (Damme et al. 2020; Mailman School of Public Health 2021).

The interconnectedness of the digital world allows information to transcend geographic boundaries. Social media, online news platforms, and digital communication tools enable the rapid dissemination of information to a global audience (Grey Group International 2024).Information, whether accurate or misleading, can spread almost instantly across the globe. Real-time communication platforms contribute to the rapid transmission of messages, influencing public perceptions and behaviors worldwide. The infodemic affects diverse communities with varying cultural backgrounds. Information is interpreted and shared within the context of different cultural norms, beliefs, and values, influencing how individuals perceive and respond to health-related information.

Information is disseminated in multiple languages, ensuring that individuals from diverse linguistic backgrounds are both contributors to and consumers of information (Canestrino, Magliocca, and Li2022). Language accessibility contributes to the global impact of the infodemic. Health crises, such as pandemics, are inherently global in nature. The infodemic accompanying these crises addresses common health concerns that resonate across borders, affecting individuals regardless of their geographic location.

The digital nature of the infodemic minimizes traditional geographic barriers to information access. However, challenges related to the digital divide persist, impacting certain communities' ability to access and critically evaluate online information. Misinformation, fuelled by the infodemic, can contribute to a global sense of fear and uncertainty (Rocha et al. 2021). Misleading narratives can transcend borders, influencing public attitudes and responses on a global scale. Efforts to counter the infodemic and infopandemic require global collaboration. International organizations, governments, and communities worldwide must work together to address the challenges posed by misinformation and promote accurate health communication. Understanding the global reach of the infodemic is essential for developing strategies that transcend regional boundaries. Efforts to mitigate the impact of misinformation should be inclusive, culturally sensitive, and tailored to the diverse perspectives and needs of communities around the world (Gupta et al. 2023).The global scope and dynamic of the infodemic highlight its pervasive influence across diverse regions and communities worldwide. Some of the key aspects that characterize the global nature of the infodemic identified by Pian, Chi, and Ma (2021) and the WHO (2021) are discussed as follows.

Cross-Border Impact: the infodemic transcends national borders, affecting populations across continents. Information spreads seamlessly across the interconnected global landscape, influencing public perceptions and responses on a massive scale.

Multilingual Challenges: the global nature of the infodemic introduces multilingual challenges, as information circulates in various languages. Language barriers may impede accurate understanding and contribute to the spread of misinformation in different linguistic communities.

Cultural Variability: cultural differences play a significant role in shaping how information is received and interpreted. The infodemic manifests differently in diverse cultural contexts, and effective communication strategies must consider these variations to ensure relevance and resonance.

Technological Influence: the dynamic role of technology, especially social media and digital platforms, amplifies the global reach of the infodemic. Information can be disseminated rapidly, reaching individuals across the world in real-time.

Dissemination through Global Media Outlets: traditional and digital media outlets contribute to the global spread of information. International news agencies, websites, and online platforms play a crucial role in shaping the narrative of the infodemic, influencing perceptions on a global scale.

Global Health Organizations and Authorities: global health organizations, such as the World Health Organization (WHO), are central in disseminating accurate information globally. However, they also face challenges in countering the rapid spread of misinformation and ensuring that accurate information reaches diverse populations.

Collaborative Efforts and Information Sharing: international collaborations and information sharing are essential in addressing the infodemic. Governments, health organizations, and experts from different countries must coordinate efforts to counter misinformation and provide consistent, evidence-based information.

Vulnerability of Global Communities: vulnerable populations in various parts of the world are particularly susceptible to the impact of the infodemic. Disparities in access to information, healthcare resources, and technology contribute to the challenges faced by these communities.

Understanding the global scope and dynamic of the infodemic emphasizes the need for a coordinated, cross-cultural approach to information dissemination and response efforts. Collaborative strategies that account for cultural nuances and regional differences are essential to effectively address the challenges posed by the infodemic on a global scale.

Impact on Public Perception

This section discusses how the infodemic and infopandemic can shape public perceptions, attitudes, and behaviors during a health crisis, influencing individuals' responses to the situation (Dutta et al. 2022). The following detail how these phenomena can impact different aspects.

Formation of Perceptions, Information Overload: the abundance of information, both accurate and misleading, can overwhelm individuals. Sorting through vast amounts of data may lead to confusion, making it challenging for individuals to form coherent and well-informed perceptions of the health crisis.

Misinformation Impact: exposure to misinformation can distort perceptions. False or exaggerated information may create fear, anxiety, or a sense of complacency, depending on the nature of the misleading content.

Trust in Institutions: the infodemic can influence trust in public health institutions and authorities. Individuals may develop attitudes based on the perceived reliability of information sources, impacting their willingness to follow recommended guidelines and interventions.

Emotional Responses: misleading information with emotional content can evoke strong emotional responses. Fear-inducing or sensationalized messages may lead to heightened anxiety, panic, or skepticism about official health guidance.

Adherence to Health Measures: the accuracy and clarity of information influence individuals' willingness to adopt preventive measures. Clear communication can motivate adherence to guidelines, while misinformation may result in non-compliance or the adoption of ineffective and potentially harmful practices.

Stigmatization and Discrimination: misinformation can contribute to the stigmatization of certain groups or communities. Individuals may alter their behaviors based on unfounded fears, leading to discrimination or social isolation.

Media Framing: the way media outlets frame information can shape public discourse. Sensationalized reporting may amplify certain aspects of the crisis, influencing public perception and attitudes disproportionately.

Social Media Echo Chambers: social media algorithms produce “echo chambers” where people are exposed to content that confirms their preconceived notions. This deliberate dissection of information has the potential to strengthen particular beliefs and actions, creating division among societies.

Risk Perception: the infodemic can impact individuals’ perception of risk. Exaggerated or downplayed information may lead to skewed risk assessments, influencing decisions related to personal safety and health-seeking behaviors.

Credibility of Information Sources: individuals may base their decisions on the perceived credibility of information sources. Trustworthy and transparent communication from authoritative figures is crucial for informed decision-making.

Understanding how the infodemic and infopandemic shape perceptions, attitudes, and behaviors underscores the importance of effective communication strategies. Clear, accurate, and timely information from reliable sources is essential for fostering a well-informed public response during health crises. Additionally, efforts to counter misinformation and enhance media literacy contribute to a more resilient and informed society.

Challenge of Balancing the Need for Transparency with the Potential for Information Overload

Balancing the need for transparency with the potential for information overload poses several challenges, particularly in the context of health crises and the infodemic (Senft and Greenfield 2023). Some of the key challenges associated with this delicate balance are discussed here.

Managing Information Complexity: health crises often involve complex scientific and medical information. Balancing transparency requires presenting detailed information while ensuring it remains comprehensible to the general public.

Timely Communication: the need for transparency may clash with the urgency of timely communication. Providing comprehensive information quickly is essential during a crisis, but it can contribute to information overload if not managed effectively.

Navigating Uncertainty: in the early stages of a crisis information may be incomplete or subject to change as new data emerges. Striking a balance between transparencies and acknowledging uncertainties without causing undue panic is challenging.

Tailoring Information for Diverse Audiences: different audiences have varying levels of familiarity with scientific concepts. Ensuring transparency requires tailoring information for diverse educational backgrounds and literacy levels without oversimplifying or overwhelming.

Addressing Misinformation: transparency efforts may be undermined by the simultaneous spread of misinformation. Therefore, correcting the false information while providing transparent updates may require strategic communication to maintain public trust (Senft and Greenfield 2023).

Maintaining Public Trust: transparency is crucial for building and maintaining public trust. However, the risk of information overload may lead to skepticism or disengagement if individuals feel overwhelmed or perceive inconsistencies in communication.

Media Handling: the media plays a vital role in disseminating information during a crisis. Balancing transparency requires collaboration with media outlets to ensure accurate reporting without sensationalizing or exacerbating information overload.

Avoiding Alarm Fatigue: frequent updates and alerts may lead to “alarm fatigue,” where individuals become desensitized to the information due to its continuous nature. Striking a balance involves providing essential updates without causing unnecessary stress.

Utilizing Effective Communication Channels: choosing the right communication channels is critical. While transparency demands widespread dissemination, relying solely on digital platforms may exclude segments of the population, contributing to disparities in access to information.

Ensuring Two-Way Communication: transparency is a two-way street that involves not only disseminating information but also actively listening to public concerns and feedback. Managing this communication loop effectively amid information overload requires dedicated resources.

Striking a balance between transparency and information overload requires a nuanced approach that considers the nature of the crisis, the diverse needs of the audience, and the evolving dynamics of the information landscape. Clear, concise, and targeted communication strategies are essential to navigate these challenges successfully.

Implications for Vulnerable Populations: introduce the overarching theme of the impact on individuals with disabilities, foreshadowing the subsequent sections that delve into specific challenges faced by this demographic.

Balancing the need for transparency with the potential for information overload poses several challenges, particularly in the context of health crises and the infodemic. There are key challenges associated with this delicate balance which are discussed as follows:

Managing Information Complexity: health crises often involve complex scientific and medical information. Balancing transparency requires presenting detailed information while ensuring it remains comprehensible to the general public.

Timely Communication: the need for transparency may clash with the urgency of timely communication. Providing comprehensive information quickly is essential during a crisis, but it can contribute to information overload if not managed effectively.

Navigating Uncertainty: in the early stages of a crisis, information may be incomplete or subject to change as new data emerges. Striking a balance between transparency and acknowledging uncertainties without causing undue panic is challenging.

Tailoring Information for Diverse Audiences: different audiences have varying levels of familiarity with scientific concepts. Ensuring transparency requires tailoring information for diverse educational backgrounds and literacy levels without oversimplifying or overwhelming.

Addressing Misinformation: transparency efforts may be undermined by the simultaneous spread of misinformation. Correcting false information while providing transparent updates requires strategic communication to maintain public trust.

Maintaining Public Trust: transparency is crucial for building and maintaining public trust. However, the risk of information overload may lead to skepticism or disengagement if individuals feel overwhelmed or perceive inconsistencies in communication.

Media Handling: the media plays a vital role in disseminating information during a crisis. Balancing transparency requires collaboration with media outlets to ensure accurate reporting without sensationalizing or exacerbating information overload.

Avoiding Alarm Fatigue: frequent updates and alerts may lead to "alarm fatigue," where individuals become desensitized to the information due to its continuous nature. Striking a balance involves providing essential updates without causing unnecessary stress.

Utilizing Effective Communication Channels: choosing the right communication channel is critical. While transparency demands widespread dissemination, relying solely on digital platforms may exclude segments of the population, contributing to disparities in access to information.

Ensuring Two-Way Communication: transparency is a two-way street that involves not only disseminating information but also actively listening to public concerns and feedback. Managing this communication loop effectively amid information overload requires dedicated resources.

Striking a balance between transparency and information overload requires a nuanced approach that considers the nature of the crisis, the diverse needs of the audience, and the evolving dynamics of the information landscape. Clear, concise, and targeted communication strategies are essential to navigate these challenges successfully.

Information Accessibility for Persons with Disabilities

Information accessibility includes a wide range of topics related to information availability, affordability, and accessibility, including metadata, interoperability, open-source software, open content, multilingualism, Creative Commons licenses, and meeting the unique needs of individuals with disabilities (UNESCO 2023). United Nation Department of Economics and Social Affairs Disability (2022) added that ensuring information accessibility for people with disabilities is a critical aspect of promoting inclusivity, especially during health crises characterized by infodemics and infopandemics. Several challenges and considerations come into play when addressing the information accessibility needs of individuals with disabilities(Calvert 2021; Martinez and Vemuru 2020).These are discussed in turn as follows:

Accessible Formats: traditional formats of information may not be accessible to individuals with visual or hearing impairments. There should be consideration for the provision of information in alternative formats such as braille, large print, electronic text, and audio format that ensures that individuals with diverse sensory abilities can access essential health information.

Digital Accessibility: online platforms and digital content may pose accessibility barriers for individuals with disabilities, including those using screen readers or assistive technologies. There is therefore the need to consider ensuring digital content adheres to accessibility standards (such as WCAG) and employing accessible design principles which enhances the usability of websites, apps, and online documents for people with disabilities.

Sign Language Interpretation: live broadcasts, press conferences, or video content may lack sign language interpretation, excluding individuals with hearing impairments. There should be consideration of incorporating sign language interpreters in live broadcasts, while providing captioning for pre-recorded content ensures that individuals who are deaf or hard of hearing can access information effectively.

Plain Language and Clarity: complex language and jargon may pose challenges for individuals with cognitive or intellectual disabilities. It is advised that there should be consideration for using plain language, simple graphics, and visual aids which enhances comprehension for individuals with varying cognitive abilities, ensuring that health information is accessible and actionable.

Assistive Technologies: limited access to assistive technologies may hinder individuals with disabilities from utilizing available information. It is important to consider promoting the availability and affordability of assistive technologies, and providing training on their use empowers individuals with disabilities to access and engage with health information independently.

Inclusive Communication Strategies: communication strategies may not be inclusive, leading to the unintentional exclusion of certain disability groups. There is a need to consider implementing inclusive communication strategies which involves consulting with disability advocacy organizations, incorporating feedback from diverse disability communities, and actively involving individuals with disabilities in the planning and dissemination of health information.

Accessible Physical Spaces: physical spaces where information is disseminated, such as health facilities or community centers, may lack accessibility features. Considerations of ensuring physical spaces are wheelchair accessible, have tactile signage, and accommodate individuals with mobility challenges contribute to a more inclusive information dissemination environment.

Tailoring Information for Diversity: individuals with disabilities constitute a diverse group with varying communication preferences and needs. Therefore, it is important to consider tailoring information to address the specific needs of different disability communities to ensure that a broad spectrum of individuals can access and understand health information effectively.

Prioritizing information accessibility for persons with disabilities requires a holistic and inclusive approach that considers the diversity of disabilities and incorporates accessible practices at every stage of information dissemination. This not only upholds the principles of inclusivity but also contributes to a more resilient and equitable response to health crises.

Psychosocial Implications

The psychosocial implications of the infodemic and infopandemic on individuals, communities, and societies during a health crisis are profound and multifaceted. These implications extend beyond the physical aspects of the crisis and encompass the emotional, social, and mental well-being of individuals (WHO 2021; Ibrayeva et al. 2020). The prominent aspects to consider are as follows:

Anxiety and Fear: misinformation, sensationalized content, and uncertainty can contribute to heightened anxiety and fear among the general population. Psychologically, the increased levels of anxiety can lead to stress-related health issues and negatively impact mental well-being.

Confusion and Uncertainty: rapid changes in information, conflicting reports, and evolving guidelines can contribute to confusion and uncertainty. In terms of psychological implication, there may be uncertainty about the situation and this may lead to feelings of helplessness, frustration, and an inability to plan for the future.

Stigmatization and Discrimination: misinformation may contribute to the stigmatization of certain groups or communities. Psychologically, individuals who are unfairly stigmatized may experience social isolation, discrimination, and negative mental health effects.

Trust Issues: inconsistencies in information, the spread of misinformation, or perceived lack of transparency can erode trust in public health institutions and authorities. The psychosocial implication is that distrust can lead to reluctance in following recommended guidelines, reduced cooperation with public health measures, and increased anxiety.

Information Overload and Fatigue: constant exposure to information, especially through digital channels, can lead to information overload and fatigue. This has psychosocial implications which include the fact that information fatigue may result in decreased attention to critical health messages, diminishing the overall effectiveness of public health communication.

Mental Health Challenges: the psychosocial stressors associated with the infodemic may contribute to mental health challenges, including depression, anxiety disorders, and post-traumatic stress symptoms. It is implied psychosocially that individuals may require increased mental health support and resources to cope with the emotional toll of the crisis.

Social Isolation: social distancing measures and fear of contagion may lead to increased social isolation. The social isolation can exacerbate feelings of loneliness, contribute to mental health issues, and impact overall well-being.

Community Resilience and Solidarity: adversity can foster a sense of community resilience and solidarity. It is psychosocially implied that communities that come together to support one another may experience positive psychosocial outcomes, including enhanced social cohesion and collective coping mechanisms.

Impact on Vulnerable Populations: vulnerable populations, including those with pre-existing mental health conditions, may be disproportionately affected. Therefore, special attention and targeted mental health support are crucial for these populations to address unique psychosocial challenges.

Understanding the psychosocial implications of the infodemic and infopandemic is vital for developing comprehensive public health strategies that not only address the physical aspects of a crisis but also prioritize the emotional and mental well-being of individuals and communities.

Health Communication Strategies

According to the Rural Health Information Hub, RHIhub, (2022), health communication encompasses both written and spoken tactics aimed at motivating and influencing people, groups, and communities to adopt healthier decisions. For the purpose of encouraging constructive alterations in attitudes and actions, health communication frequently incorporates elements of several theories and models. The group went on to say that social marketing, which entails creating campaigns and initiatives aimed at altering people's behavior for the better, is connected to health communication. The following elements are part of social marketing and health communication strategies that work well (Rural Health Information Hub, RHIhub 2022);

- Using research-based techniques to design products and materials and choose the best distribution channels to reach the target market.
- Knowledge of common sense, ideas, terminology, and priorities in various contexts and cultures.
- Taking into account the target groups' cultural competency, media exposure, health literacy, and internet availability.
- Creation of content for a variety of media platforms, including workbooks, billboards, newspaper articles, radio and television advertisements, public service announcements, newsletters, pamphlets, films, digital tools, case studies, group discussions, health fairs, field excursions, and health fairs.

Health communication strategies play a pivotal role in effectively disseminating accurate information, fostering public understanding, and promoting positive behavioral responses during a health crisis (Malikhao 2020). These strategies are crucial for managing the infodemic and ensuring that individuals receive clear, actionable, and trustworthy information. Some of the key elements that can be considered in health communication strategies include but are not limited to:

Clear and Accessible Messaging: there should be clear communication in terms of the use of plain language, avoidance of jargon, and presentation of information in a straightforward manner to enhance clarity. There is a need to ensure that information is accessible to individuals with diverse abilities by using alternative formats and providing language options.

Timely and Consistent Updates: provide timely updates as the situation evolves to keep the public informed of the latest developments. There should be maintenance of consistency in messaging across various communication channels to avoid confusion and build trust.

Multichannel Communication, Diverse Platforms: utilize a range of communication channels, including traditional media, social media, websites, and community networks, to reach diverse audiences. Tailoring Messages: customize messages for different platforms and demographics to enhance relevance and resonance.

Engagement with Stakeholders, Collaboration: collaborate with public health authorities, community leaders, and relevant stakeholders to ensure a coordinated and unified communication approach. For inclusivity, it is important to involve representatives from diverse communities, including individuals with disabilities, in the development and dissemination of health messages.

Cultural Sensitivity: consider cultural nuances and sensitivities in communication to ensure that messages are culturally appropriate and respectful. There should be language accessibility in terms of making provision for information in multiple languages to cater to diverse linguistic communities.

Addressing Misinformation: there should be proactive correction by actively identifying and correcting misinformation through official channels to counter the spread of false information.

Community Involvement: engage the community in addressing misinformation by promoting media literacy and critical thinking skills.

Humanizing the Message, Personal Stories: humanize health messages by incorporating personal stories and testimonials to create emotional connections. Emphatically, it is crucial to communicate with empathy and understanding to resonate with the concerns and experiences of the audience.

Risk Communication, Transparent Information: practice transparent communication about risks, uncertainties, and preventive measures. Addressing fears is important in terms of acknowledging and addressing fears and concerns within the community while providing evidence-based information to alleviate anxiety.

Behavioral Guidelines and Empowerment, Actionable Guidance: provide clear, actionable guidelines for preventive measures and health behaviors. Individuals should be empowered to take control of their health by offering practical tips and guidance for self-care and risk reduction.

Feedback Mechanisms, Two-Way Communication: establish channels for two-way communication to receive feedback, address concerns, and adapt communication strategies based on community input. There may be a need to conduct surveys and focus groups to assess the effectiveness of communication strategies and make data-driven adjustments.

Preparedness and Education, Preventive Education: emphasize preventive education to empower individuals with knowledge on risk reduction and early intervention. Promote public health literacy by providing educational resources and materials that enhance understanding of health information.

Effectively navigating health communication during a crisis requires a comprehensive and adaptable approach that prioritizes clarity, inclusivity, and community engagement. By implementing these strategies, public health authorities can foster a more informed, resilient, and empowered society in the face of health challenges.

Socio-Economic Disparities

Socio-economic disparities during a health crisis can exacerbate existing inequalities, influencing individuals' ability to access resources, information, and healthcare services (Ilinca et al. 2018). Ensuring that health services are accessible to all those in need and that those seeking care are not burdened with excessive or prohibitive costs is a critical objective in addressing the ongoing socio-economic disparities in access to care (Ministry of Health, Government of Kenya 2014). Significant spatial and socioeconomic health disparities have remained in Kenya despite significant advancements in recent decades (Achoki et al. 2019; Okiro 2019), which is consistent with a condition that is common in sub-Saharan African nations (Pons-Duran 2019). Not only do people in poverty tend to be less healthy than members of wealthier demographics, but they also have fewer access points to necessary healthcare services. These disparities are often rooted in social and economic structures, and their impact is particularly pronounced during times of crisis. The key aspects to consider when discussing socio-economic disparities are highlighted as follows:

Access to Healthcare Services, Disparities: individuals from lower socio-economic backgrounds may face challenges in accessing healthcare services, including testing, treatment, and preventive measures. Limited access can result in delayed or inadequate medical care, contributing to the spread of the health crisis and exacerbating health outcomes for vulnerable populations.

Digital Divide, Disparities: socio-economic disparities are reflected in the digital divide, with individuals from lower-income groups having limited access to technology and the internet. This digital gap hinders access to online health information, telemedicine services, and remote work or learning opportunities, further marginalizing those with fewer resources.

Employment and Economic Impact, Disparities: vulnerable populations often work in industries with higher exposure to health risks and may have less job security. Impact: job losses, income instability, and economic downturns disproportionately affect individuals with lower socio-economic status, leading to financial stress, housing instability, and food insecurity.

Educational Disparities, Disparities: access to quality education is often unequal, with socio-economic factors influencing educational opportunities. School closures and the shift to remote learning may widen educational gaps, affecting children and students from lower-income families who may lack the necessary resources for online education.

Housing Conditions, Disparities: socio-economic status influences living conditions, with lower-income individuals often residing in crowded or substandard housing. Overcrowded living spaces make physical distancing challenging, increasing the risk of virus transmission within vulnerable communities.

Access to Information, Disparities: individuals with lower socio-economic status may have limited access to reliable health information, relying on informal networks or sources that may not provide accurate guidance. Inadequate information may lead to misconceptions, delayed responses to health advisories, and increased vulnerability to health risks.

Health Literacy, Disparities: socio-economic disparities often correlate with variations in health literacy levels. Lower health literacy can hinder individuals' understanding of health information, impeding their ability to make informed decisions about preventive measures and healthcare utilization.

Vulnerability of Essential Workers, Disparities: essential workers, often from lower socio-economic backgrounds, may face increased exposure to health risks due to the nature of their jobs. These individuals may bear a disproportionate burden of the crisis, facing higher health risks without adequate access to protective measures or resources.

Addressing socio-economic disparities during a health crisis requires targeted interventions and policies that prioritize the needs of vulnerable populations. Efforts to ensure equitable access to healthcare, education, and economic support are essential for fostering resilience and reducing the disproportionate impact of crises on marginalized communities.

Intersectionality and Diverse Experiences

The term "intersectionality" describes how social categories—like race, gender, class, and other identities—are connected to one another and to the manner in which they overlap and intersect. Intersectional approaches, according to Chaplin, Twigg, and Lovell (2021), acknowledge that individuals will have varying identities, needs, priorities, and capacities. These elements are dynamic and will change over time, affecting an individual's capacity to anticipate, manage, and respond to natural hazards and climate variability. Understanding the diverse experiences of individuals during a health crisis requires acknowledging and addressing the intersections of various factors that shape their vulnerabilities and resilience (Chaplin, Twigg, and Lovell 2021).The key considerations when discussing intersectionality and diverse experiences are identified and discussed as follows.

Race and Ethnicity, Intersectional Impact: individuals from different racial and ethnic backgrounds may experience unique challenges during a health crisis due to systemic inequalities. There is a need for diverse experiences where discrimination, access to healthcare resources, and cultural factors can influence how different racial and ethnic groups experience and respond to a crisis.

Gender Identity and Expression, Intersectional Impact: gender identity intersects with various factors, impacting individuals differently during a health crisis. There is a need for diverse experiences where women, transgender individuals, and non-binary individuals may face distinct challenges, including access to reproductive healthcare, increased caregiving responsibilities, and potential gender-based violence.

Socio-Economic Status, Intersectional Impact: socio-economic status intersects with other identities, influencing individuals' access to resources and opportunities. There is a need for diverse experiences where those at the intersection of lower socio-economic status and other marginalized identities may face compounded challenges, such as limited access to healthcare, housing instability, and economic precarity.

Age, Intersectional Impact: age intersects with other factors, affecting the vulnerability of different age groups during a health crisis. Diverse Experiences: older adults and children may face distinct challenges, including increased health risks for older populations and disruptions to education and social development for children.

Disability, Intersectional Impact: individuals with disabilities may experience unique challenges that intersect with other identity factors. There is a need for the diverse experiences in which disabled individuals from different racial, gender, or socio-economic backgrounds may encounter specific barriers in accessing healthcare, information, and support services.

Sexual Orientation, Intersectional Impact: sexual orientation intersects with various identities, influencing experiences during a health crisis. It is important to enable diverse experiences in which LGBTQ+ individuals may face unique challenges, such as discrimination in healthcare settings or disparities in mental health outcomes, based on the intersection of their sexual orientation with other factors (SoundEagle 2023; Krishnan 2024).

Geographic Location, Intersectional Impact: geographic location intersects with other factors, influencing access to resources and the impact of the crisis. Urban and rural communities may experience different challenges, including variations in healthcare infrastructure, information access, and economic support.

Immigration Status, Intersectional Impact: immigration status intersects with various identities, impacting individuals' access to healthcare and support services. The diverse experiences are necessary where immigrant communities may face unique challenges, such as language barriers, fear of deportation, and limited access to government assistance programs.

Addressing the diverse experiences shaped by intersectionality during a health crisis requires tailored and inclusive approaches. Policies and interventions should consider the unique needs and vulnerabilities of individuals at the intersections of various identities to ensure equity and inclusivity in the response to crises (Ma, Qirui, and Lv2023).

Policy and Intervention Frameworks

Policy and intervention frameworks are essential for guiding organized and effective responses to health crises, especially in the context of the infodemic and infopandemic (Tangcharoensathien et al. 2020). These frameworks provide a structured approach to addressing challenges, promoting public health, and ensuring equitable outcomes. The key considerations when discussing policy and intervention frameworks are discussed as follows.

Evidence-Based Decision-Making, Policy Emphasis: prioritize evidence-based decision-making in the formulation of policies and interventions. Implement interventions based on scientific evidence and data-driven insights to ensure their efficacy and relevance.

Inclusive Policy Development, Policy Emphasis: promote inclusive policy development that considers the needs of diverse populations, including vulnerable and marginalized groups. Ensure that interventions are designed to address the specific challenges faced by different communities, taking into account socio-economic, cultural, and intersectional factors.

Community Engagement, Policy Emphasis: emphasize community engagement in policy development to foster trust and collaboration. Intervention Implementation: implement interventions with active involvement from community leaders, organizations, and representatives to ensure culturally sensitive and community-specific approaches.

Health Communication Strategies, Policy Emphasis: develop clear guidelines for health communication strategies that prioritize transparency, accessibility, and accuracy. Implement communication strategies that leverage diverse channels, languages, and formats to reach broad audiences and combat misinformation.

Resource Allocation and Equity, Policy Emphasis: prioritize equitable resource allocation to address socio-economic disparities and ensure access to essential services for all. Implement interventions that specifically target vulnerable populations, allocating resources based on the need to reduce disparities in health outcomes.

Capacity Building, Policy Emphasis: prioritize capacity building for healthcare systems, public health agencies, and community organizations. Intervention Implementation: implement interventions that strengthen the capacity of healthcare providers, enhance infrastructure, and build community resilience to respond effectively to health crises.

Surveillance and Data Collection, Policy Emphasis: establish robust surveillance and data collection systems to monitor the spread of the crisis and assess the impact of interventions. Use real-time data to inform intervention strategies, identify emerging trends, and adjust policies based on the evolving nature of the crisis.

Legal and Ethical Considerations, Policy Emphasis: address legal and ethical considerations in the formulation of policies, ensuring the protection of individual rights and privacy. Implement interventions with a strong ethical foundation, respecting human rights and promoting inclusivity in access to healthcare and information.

Collaboration and Coordination, Policy Emphasis: promote collaboration and coordination among various stakeholders, including government agencies, non-profit organizations, and international bodies. Intervention Implementation: implement interventions through coordinated efforts, leveraging the strengths of different entities to enhance the overall effectiveness of the response.

Adaptability and Flexibility, Policy Emphasis: design policies that are adaptable and flexible to accommodate evolving situations and emerging challenges. Implement interventions that can be adjusted based on real-time feedback and new information, ensuring agility in response strategies.

Public Health Education and Training, Policy Emphasis: prioritize public health education and training initiatives to enhance community resilience and preparedness. Implement interventions that focus on educating the public about preventive measures, health literacy, and the importance of collective action during a crisis.

Policy and intervention frameworks provide the foundation for a comprehensive and coordinated response to health crises, ensuring that efforts are evidence-based, inclusive, and responsive to the diverse needs of populations. Regular evaluation and adaptation of these frameworks are crucial to address emerging challenges and optimize the effectiveness of interventions over time.

Building Resilience in Public Health

Building resilience in public health is a crucial aspect of preparing communities, healthcare systems, and individuals to effectively respond to and recover from health crises (Wulff, Donato, and Lurie 2015). Resilience involves strengthening the ability to anticipate, adapt to, and recover from challenges, including pandemics, natural disasters, and other public health emergencies (Khalil et al. 2022). The main factors necessary to be considered for building resilience in public health are discussed here:

Community Engagement and Empowerment, Inclusivity: engage communities in the planning, decision-making, and implementation of public health initiatives. Empower individuals to actively participate in their health and well-being, fostering a sense of ownership and community resilience.

Education and Health Literacy, Informative Campaigns: conduct educational campaigns to enhance public health literacy, ensuring that individuals have the knowledge and skills to make informed decisions during a crisis. Provide health information in multiple languages, formats, and through various channels to reach diverse populations.

Strengthening Healthcare Systems, Capacity Building: invest in capacity building for healthcare systems, including training healthcare workers, enhancing infrastructure, and ensuring the availability of essential medical supplies. Design healthcare systems that can adapt to increased demand during a crisis, with the flexibility to reallocate resources and adjust operations as needed.

Surveillance and Early Warning Systems, Robust Surveillance: establish robust surveillance systems for monitoring and detecting potential health threats early. Develop effective early warning systems that enable prompt responses to emerging health risks, minimizing the impact of crises.

Community-Based Healthcare Delivery, Accessible Services: promote community-based healthcare delivery models to ensure accessible and decentralized healthcare services. Emphasize preventive care, health promotion, and early detection of health issues within local communities.

Social Support and Mental Health Services, Psychosocial Support: provide social support and mental health services to address the psychological impact of crises. Ensure that mental health resources are accessible and available to all, especially vulnerable populations.

Cross-Sectoral Collaboration, Interdisciplinary Approach: foster collaboration between public health, healthcare, government agencies, non-profit organizations, and other sectors. Develop integrated plans that address social, economic, and environmental factors influencing public health, promoting a holistic and collaborative approach.

Infrastructure Resilience, Critical Infrastructure Planning: build and maintain critical infrastructure with a focus on resilience to withstand and recover from disruptions. Integrate redundancy and contingency planning to ensure the continued functioning of essential services during crises.

Community Resilience Programs, Training and Exercises: implement community resilience programs that include training, drills, and exercises to prepare individuals and communities for potential health threats. Facilitate networking and communication within communities to strengthen social bonds and support systems.

Research and Innovation, Continuous Research: support continuous research and innovation in public health to stay ahead of emerging health threats. Embrace technological advancements for improved disease surveillance, data analysis, and communication during crises(Al-Worafi 2023; Tella2024).

Policy Frameworks for Adaptation, Adaptive Policies: develop adaptive policies that can be adjusted based on lessons learned from previous crises and emerging challenges. Regularly review and update policies to incorporate feedback from communities, healthcare professionals, and stakeholders.

Building resilience in public health requires a comprehensive and sustained effort that spans individual, community, institutional, and systemic levels. By fostering a culture of preparedness, collaboration, and adaptability, societies can better withstand and recover from the complex challenges posed by health crises.

The Impact of Infopandemic on Persons with Disabilities

When considering the pandemic from the perspective of people with disabilities, the detrimental societal repercussions it has caused have frequently been even more severe (Maia., Tommasetti, & Macedo, 2021).

1. People with disabilities may suffer physical and mental health harm from misinformation and disinformation, which can also heighten stigma, jeopardize recent health gains, and result in a lack of adherence to public health measures, thereby decreasing their efficacy and jeopardizing nations' capacity to halt the pandemic (Khan, Khan, and Hussain2022).

2. The costs of the infopandemic to the lives of disabled people: in the absence of proper trust and information, diagnostic tests remain unutilized, immunization efforts (or campaigns to promote effective vaccines) fail to reach the target population of individuals with disabilities, and, as a result, the virus persists.

3. This kind of news not only encourages the spread of racist beliefs and actions (Alegría, Alvarez, Ishikawa, DiMarzio, & McPeck, 2016) but also jeopardizes the health of the disabled population and the capacity of governments to carry out preventative initiatives.

4. People with disabilities may have detrimental health effects from misinformation during a pandemic. Many fake or deceptive tales are made up and circulated without any background research or quality control. Conspiracy theories form the basis of a lot of this misinformation, with some even incorporating components of it into what appears to be mainstream conversation. All facets of the illness, including the origin of the virus, its aetiology, its course of treatment, and

its mode of transmission, have been the subject of misleading and inaccurate information. Infopandemics spread and take hold quickly, altering the behavior of people with disabilities and possibly pushing them to take more risks. Despite already being at higher risk, they are now discovered to be taking even higher risks as a result of false information. In the end, this makes the pandemic even worse for them, inflicting more damage.

5. Higher chances of COVID-19 spreading faster as a result of several obstacles. This, according to Shrestha, Kunwar, and Meierhofer (2022) could include the inability to provide public health information in accessible formats; the inaccessibility of water, sanitation, and hygiene facilities; the impracticality of social distancing and self-isolation measures for certain individuals with disabilities who depend on caregivers; and the situation of persons residing in residential institutions or humanitarian contexts, where they must rely on caregivers or authorities to prevent and respond to outbreaks. Certain individuals are unable to walk, see, talk, or move. They depend on information that regular people offer them, which may be erroneous and incomplete. Because of their condition, it is impossible to verify such, which subsequently accelerates the spread of the Covid-19.

6. Increased risks of dying from COVID-19, as a result of the possibility of aggravating pre-existing medical conditions, in addition to obstacles like physically inaccessible healthcare facilities, a lack of ability on the part of healthcare professionals to treat individuals with disabilities, and stigma and discrimination against individuals with disabilities. Some of these individuals may act or take actions based on false information provided to them, which could result in sudden death (Cmeciu 2021).

7. A rise in pre-existing mental health disorders and psychosocial disabilities brought on by worry and anxiety, financial and economic strains, extended periods of social isolation, demands from families, and conflict.

8. Considering increasing harm to psychological makeup, people with disabilities experience uneasiness and disturbance. They are worried about how they will be able to obtain the treatments and cures that have been shared on social media in order to fend off the epidemic (International Disability Alliance 2020). Many people can be tricked into asking for assistance from others, which could worsen their mental health and weak economic or financial situation.

Conclusion

In conclusion, addressing the multifaceted challenges posed by health crises, including the infodemic and infopandemic, requires a comprehensive and integrated approach. This chapter has explored key aspects such as the impact on individuals with disabilities, the psychosocial implications, socio-economic disparities, intersectionality, and the importance of policy and intervention frameworks in building resilience.

Understanding the diverse experiences of individuals and communities is essential for crafting effective health communication strategies, promoting inclusivity, and mitigating the spread of misinformation. This chapter has looked into the psychosocial implications, emphasizing the importance of empathy, clear communication, and community support to address the emotional and mental well-being of individuals.

Socio-economic disparities and their intersection with various identities underscore the need for equitable policies and interventions. Recognizing the unique challenges faced by vulnerable populations is crucial for fostering resilience and ensuring that resources and support are distributed equitably.

The discussion on policy and intervention frameworks highlighted the importance of evidence-based decision-making, community engagement, and adaptability. These frameworks serve as the foundation for coordinated responses, emphasizing the need for inclusivity, transparency, and ethical considerations.

Building resilience in public health involves a collective effort that spans education, healthcare systems, community engagement, and cross-sectoral collaboration. By addressing the root causes of disparities, promoting accessibility, and integrating lessons learned from previous crises, societies can enhance their capacity to navigate and overcome health challenges.

In essence, fostering resilience requires a holistic understanding of the interconnected factors influencing health outcomes and a commitment to building a robust, adaptable, and inclusive public health infrastructure.

Recommendations

Based on our discussion, here are some recommendations to enhance resilience in the face of health crises, with a focus on the infodemic, psychosocial implications, socio-economic disparities, and inclusive policies:

Strengthen Health Communication: invest in clear, accessible, and culturally sensitive health communication strategies. Prioritize communication channels that reach diverse populations, including those with disabilities and limited digital access.

Address Infodemic Challenges: implement robust strategies to counter misinformation and promote media literacy. Engage with diverse communities to understand information needs and tailor communication accordingly.

Prioritize Mental Health Support: allocate resources for mental health services, especially for vulnerable populations. Integrate psychosocial support into public health initiatives, emphasizing emotional well-being.

Tackle Socio-Economic Disparities: implement policies that address socio-economic disparities, ensuring equitable access to healthcare, education, and economic support. Focus on vulnerable populations, including those in lower-income brackets, to reduce disparities in health outcomes.

Promote Intersectionality and Inclusivity: incorporate intersectionality in policy development and intervention planning. Ensure that interventions are inclusive, considering diverse identities and experiences.

Enhance Community Engagement: foster community involvement in decision-making processes related to public health. Develop initiatives that empower communities to actively participate in their health and well-being.

Invest in Healthcare System Resilience: strengthen healthcare systems through continuous capacity building, training, and infrastructure development. Foster adaptability within healthcare systems to respond effectively to changing circumstances.

Support Research and Innovation: promote ongoing research in public health to stay ahead of emerging challenges. Embrace technological innovations for improved surveillance, data analysis, and communication.

Implement Inclusive Policies: formulate and implement policies that are inclusive, transparent, and adaptable. Address legal and ethical considerations to protect individual rights and privacy.

Build Cross-Sectoral Collaboration: foster collaboration between public health, healthcare, government agencies, NGOs, and other sectors. Develop integrated plans that address the broader social determinants of health.

Educate and empower: invest in public health education to enhance health literacy and preparedness. Empower individuals with the knowledge and skills to actively participate in their health.

Promote Community Resilience Programs: establish and support community resilience programs that include training, drills, and exercises. Facilitate networking and communication within communities to strengthen social bonds.

These recommendations emphasize the importance of a holistic, collaborative, and inclusive approach to public health. Addressing the root causes of disparities, will foster community engagement, and promoting resilience at multiple levels, societies can better navigate and overcome the challenges posed by health crises.

References

Achoki, T, Miller-Petrie., M.K, Glenn, S.D., Kalra, N., Lesego, A., & Gathecha, G.K. (2019). Health disparities across the counties of Kenya and implications for policy makers, 1990–2016: a systematic analysis for the global burden of disease study 2016. *Lancet Global Health*, 7(1), e81–95.

Alam, M. S., & Sultana, R. (2021). Influences of climatic and non-climatic factors on COVID-19 outbreak: A review of existing literature. *Environmental Challenges*, *5*, 100255. https://doi.org/10.1016/j.envc.2021.100255.

Alegría, M., Alvarez, K., Ishikawa, R. Z., DiMarzio, K., & McPeck, S. (2016). Removing obstacles to eliminating racial and ethnic disparities in behavioral health care. *Health Affairs*, 35(6), 991–999. doi:10.1377/hlthaff.2016.0029

Al-Worafi, Y.M. (2023). Importance of Technology for Drug Safety. In: Technology for Drug Safety. Health Informatics. Springer, Cham. https://doi.org/10.1007/978-3-031-34268-4_1.

America Disability Act (2021). What is the definition of disability under the ADA?. Retrieved from https://adata.org/faq/what-definition-disability-under-ada (accessed 12 December 2023).

Ametican Psychological Association (2023). Using Psychological Science to Understand and Fight Health Misinformation An APA consensus statement. Retrieved from: https://www.apa.org/pubs/reports/misinformation-consensus-statement.pdf (accessed 12 January 2024).

Calvert, S. (2021). Challenges for people with disabilities. Retrieved from: https://ballardbrief.byu.edu/issue-briefs/challenges-for-people-with-disabilities (accessed 15 January 2024).

Canestrino, R, Magliocca, P., & Li Y. (2022). The impact of language diversity on knowledge sharing within International University Research Teams: Evidence From TED Project. *Frontier in Psycholo*gy, 21(13), 879154. doi: 10.3389/fpsyg.2022.879154.

Centre for Disease Control and Prevention (2020). Common Barriers to Participation Experienced by People with Disabilities. Retrieved from: https://www.cdc.gov/ncbddd/disabilityandhealth/disability-barriers.html (accessed 12 January 2023).

Chaplin, D., Twigg, J. & Lovell, E. (2021). Intersectional approaches to vulnerability reduction and resilience-building. *Resilience Intel*, 12; https://wrd.unwomen.org/sites/default/files/2021-11/1INTER~1.PDF(access 12 January 2024).

Cmeciu, C. (2021). Keep Calm and Get Informed: Risk Communication and Engagement During the COVID-19 Pandemic in Romania. DOI: 10.4018/978-1-7998-6705-0.ch011.

College of Staten Island Library (2023). Misinformation and Disinformation: Thinking Critically about Information Sources. Retrieved from: https://library.csi.cuny.edu/misinformation (accessed 12 January 2024).

Damme, W. V., Dahake, R., Delamou, A., Ingelbeen, B., Wouters, E., Vanham, G., Dossou, P., Ir, P., Abimbola, S., Narayanan, D., Bloom, G., Engelgem, I. V., Ag Ahmed, M. A.,

Kiendrébéogo, J. A., Verdonck, K., Brouwere, V. D., Bello, K., Kloos, H., Aaby, P., Assefa, Y. (2020). The COVID-19 pandemic: Diverse contexts; different epidemics—How and why? *BMJ Global Health*, *5*(7). https://doi.org/10.1136/bmjgh-2020-003098.

Dutta, B., Peng. M.H., Chen, C.C., Sun, S.L.(2022). Role of infodemics on social media in the development of people's readiness to follow COVID-19 preventive measures. *International Journal of Environmental Resources Public Health*, 19(3):1347. doi: 10.3390/ijerph19031347.

Ferreira Caceres, M. M., Sosa, J. P., Lawrence, J. A., Sestacovschi, C., Tidd-Johnson, A., Ul Rasool, M. H., Gadamidi, V. K., Ozair, S., Pandav, K., Cuevas-Lou, C., Parrish, M., Rodriguez, I., & Fernandez, J. P. (2022). The impact of misinformation on the COVID-19 pandemic. *AIMS Public Health*, *9*(2), 262–277. https://doi.org/10.3934/publichealth.2022018.

Fina, V.D. & Cera, R. & Palmisano, G.(2017). The United Nations convention on the rights of persons with disabilities: A commentary. 10.1007/978-3-319-43790-3.

Grey Group International (2024). Globalization: Dissecting the Global Interconnectedness Maze. Retrieved from https://www.graygroupintl.com/blog/globalization (accessed 27 December 2023).

Goyal, D., Hunt, X., Kuper, H., Shakespeare, T., & Banks, L. M. (2023). Impact of the COVID-19 pandemic on people with disabilities and implications for health services research. *Journal of Health Services Research & Policy*, *28*(2), 77–79. https://doi.org/10.1177/13558196231160047.

Gréaux, M., Moro, M. F., Kamenov, K., Russell, A. M., Barrett, D., & Cieza, A. (2023). Health equity for persons with disabilities: A global scoping review on barriers and interventions in healthcare services. *International Journal for Equity in Health, 22*. https://doi.org/10.1186/s12939-023-02035-w.

Gupta, M., Dennehy, D., Parra, C.M., Mäntymäki, M., & Dwivedi, Y.K. (2023). Fake news believability: The effects of political beliefs and espoused cultural values. *Information & Management*, 60(2), 103745. https://doi.org/10.1016/j.im.2022.103745.

Ibrayeva, L., Sparks, J., Lim, N., Clementi, A., Almukhambetova, A., Nurtayev, Y., & Muratkyzy, A. (2020). Mental Health and Well-Being of University Students: A Bibliometric Mapping of the Literature. *Frontiers in Psychology*, *11*, 540000. https://doi.org/10.3389/fpsyg.2020.01226

Ilinca, S., Di Giorgio, L., Salari, P. (2019). Socio-economic inequality and inequity in use of health care services in Kenya: evidence from the fourth Kenya household health expenditure and utilization survey. *International Journal of Equity Health* 18, 196; https://doi.org/10.1186/s12939-019-1106-z.

Ilinca, A., Samuelsson, S., Piccinelli, P., Soller, M., Kristoffersson, U., & Lindgren, A. (2018). A stroke gene panel for whole-exome sequencing. *European Journal of Human Genetics*. 27(2), DO1. 10.1038/s41431-018-0274-4.

International Disability Alliance (2020). Disability rights during the pandemic A global report on findings of the COVID-19 Disability Rights Monitor. Retrieved from: https://www.internationaldisabilityalliance.org/sites/default/files/disability_rights_during_the_pandemic_report_web_pdf_1.pdf (accessed 12 January 2024).

Ireton C., & Posetti J. (2018). Journalism, 'fake mews' & disinformation: Handbook for Journalism Education and Training. United Nations Educational, Scientific and Cultural Organization; 2018. Retrieved from: https://en.unesco.org/fightfakenews (Accessed 27 December 2023).

Karnay, S., Furqan, R., & Saeni, R. (2023). Inclusive Crisis Communication During COVID-19: Lessons Learned From the Experiences of Persons With Disabilities in Makassar, Indonesia. *Asian Journal for Public Opinion Research, 11*(3), 201–229. https://doi.org/10.15206/ajpor.2023.11.3.201

Kaushal, S., Nyoni, A.M., Sharma, A., & J. Saidi, A.K. 2023). Drawbacks Of Technology As An Antecedent For Knowledge Sharing: Focus On Curbing Misinformation," *2023 9th International Conference on Advanced Computing and Communication Systems (ICACCS)*, Coimbatore, India, 2023, pp. 2483–2487, doi: 10.1109/ICACCS57279.2023.10112895.

Khalil, M., Ravaghi, H., Samhouri, D., Abo, J., Ali, A., Sakr, H., & Camacho, A. (2022). What is "hospital resilience"? A scoping review on conceptualization, operationalization, and evaluation. *Frontiers in Public Health, 10*, 1009400. https://doi.org/10.3389/fpubh.2022.1009400.

Khan,A., Khan, M.K. & Hussain, A. (2022) Mis/disinformation about covid-19 and the position of information professionals in infodemic management, *Journal of Hospital Librarianship*, 22:4, 299–310, DOI: 10.1080/15323269.2022.2124800.

Krishnan, S. (2024). Reframing Teacher Candidates' Views on Disability Through the Writings of Disabled Authors. *Research and Practice for Persons With Severe Disabilities*. https://doi.org/10.1177/15407969231223931.

Ma, C., Qirui, C. & Lv, Y. (2023). "One community at a time": promoting community resilience in the face of natural hazards and public health challenges. *BMC Public Health* **23**, 2510. https://doi.org/10.1186/s12889-023-17458-x

Maia, V.M., Tommasetti, R. and Macedo, M.A.d.S. (2021). Australian market response to COVID-19 as moderated by social media. *Pacific Accounting Review*, 33(5), 625–635. https://doi.org/10.1108/PAR-09-2020-0138.

Mailman School of Public Health (2021). Epidemic, endemic, pandemic: What are the differences? Retrieved from: https://www.publichealth.columbia.edu/news/epidemic-endemic-pandemic-what-are-differences (accessed 12 January 2024).

Malikhao, P. (2020). Health Communication: Approaches, Strategies, and Ways to Sustainability on Health or Health for All. *Handbook of Communication for Development and Social Change*, 1015–1037. https://doi.org/10.1007/978-981-15-2014-3_137.

Martinez, R.M. & Vemuru, V. (2020). Social inclusion of persons with disabilities in Nigeria: Challenges and opportunities. Retrieved from World Bank Blog: https://blogs.worldbank.org/nasikiliza/social-inclusion-persons-disabilities-nigeria-challenges-and-opportunities (accessed 12 January 2024).

McKee, M., & Krentel, A. (2022). *Issues in public health: challenges for the 21st century*. McGraw-Hill Education.

Media Defence (2020). *Module 8: 'False News', Misinformation and Propaganda*: Misinformation, Disinformation and Mal-Information. Retrieved from: https://www.mediadefence.org/ereader/publications/introductory-modules-on-digital-rights-and-freedom-of-expression-online/module-8-false-news-misinformation-and-propaganda/misinformation-disinformation-and-mal-information/

Ministry of Health Government of Kenya. Kenya Health Policy 2014–2030: Towards attaining the highest standard of health. Nairobi: Ministry of Health; 2014. https://www.afidep.org/?wpfb_dl=80.

OECD (2020). The territorial impact of COVID-19: Managing the crisis across levels of government. Retrieved from: https://www.oecd.org/coronavirus/policy-responses/the-territorial-impact-of-covid-19-managing-the-crisis-across-levels-of-government-d3e314e1/ (accessed 12 January 2023).

Okiro EA. (2019). Estimates of subnational health trends in Kenya. *Lancet Global Health*, 7(1), e8–9.

PAHO (2020). Annual Report 2020. Universal Health and the Pandemic. Resilient Health Systems. The Bahamas and Turks and Caicos Islands. Retrieved from https://iris.paho.org/handle/10665.2/54604 (accessed 15 August 2024).
Pan American Health Organization, (2020). Some rights reserved.
Pian, W., Chi, J., & Ma, F. (2021). The causes, impacts and countermeasures of COVID-19 "Infodemic": A systematic review using narrative synthesis. *Information Processing & Management, 58*(6), 102713. https://doi.org/10.1016/j.ipm.2021.102713.
Pons-Duran, C, Lucas A, Narayan A, Dabalen A, Menéndez C. (2019). Inequalities in sub-Saharan African women's and girls' health opportunities and outcomes: Evidence from the Demographic Health Surveys. *Journal of Global Health*, 9(1):010410.
Posetti J., & Bontcheva, K. (2020). Deciphering COVID-19 disinformation. Paris: United Nations Educational, Scientific and Cultural Organization. Retrieved from: https://en.unesco.org/covid19/disinfodemic (accessed 12 December 2023).
Pulido, C.M., Villarejo-Carballido., Redondo-Sama, G., & Aitor, G. (2020). COVID-19 infodemic: More retweets for science-based information on coronavirus than for false information. *International Sociology*, 35(4) 377–392.
RHIhub (2022). Your first stop for rural health information. Retrieved from https://www.ruralhealthinfo.org/ (accessed 15 August 2024).
Rocha, Y.M., de Moura., G.A., Desidério, G.A, de Oliveira, C.H., Lourenço, F.D., de Figueiredo-Nicolete, L.D.(2021). The impact of fake news on social media and its influence on health during the COVID-19 pandemic: a systematic review. Z Gesundh Wiss. 2021 Oct 9:1–10. doi: 10.1007/s10389-021-01658-z. Epub ahead of print.
Rockville, M.D. (2018). Research Protocol: Communication and dissemination Strategies to facilitate the use of health-related evidence: Effective Health Care Program. Agency for Healthcare Research and Quality. Retrieved from https://effectivehealthcare.ahrq.gov/products/medical-evidence-communication/research-protocol
Rohwerder, B., Wong, S., Pokharel, S., Khadka, D., Poudyal, N., Prasai, S., Shrestha, N., Wickenden, M., & Morrison, J. (2022). Describing adolescents with disabilities' experiences of COVID-19 and other humanitarian emergencies in low- and middle-income countries: A scoping review. *Global Health Action, 15*(1). https://doi.org/10.1080/16549716.2022.2107350.
Senft, T. & Greenfield, S. (2023). People's experience of information overload and its impact on infodemic harms; In book: Managing Infodemics in the 21st Century (pp. 27–40). 10.1007/978-3-031-27789-4_3.
Shrestha, A., Kunwar, B. M., & Meierhofer, R. (2022). Water, sanitation, hygiene practices, health and nutritional status among children before and during the COVID-19 pandemic: Longitudinal evidence from remote areas of Dailekh and Achham districts in Nepal. *BMC Public Health, 22*. https://doi.org/10.1186/s12889-022-14346-8.
SoundEagle (2023).Misquotation pandemic and disinformation polemic:mind pollution by viral falsity.Retrieved from https://soundeagle.wordpress.com/2020/12/19/misquotation-pandemic-and-disinformation-polemic-mind-pollution-by-viral-falsity/ (accessed 15 January 2024).
Tangcharoensathien, V, Calleja, N., Nguyen, T., Purnat, T., D'Agostino, M, Garcia-Saiso, S., Landry, M., Rashidian A, Hamilton C, AbdAllah A, Ghiga I, Hill A, Hougendobler D, van Andel J, Nunn M, Brooks I, Sacco PL, De Domenico M, Mai P, Gruzd A, Alaphilippe A, Briand S. (2020). Framework for Managing the COVID-19 Infodemic: Methods and Results of an Online, Crowdsourced WHO Technical Consultation. *Journal of Medical Internet Resources*, 26;22(6):e19659. doi: 10.2196/19659.
Tella, A. (2024). Future information dynamics. Reference Module in Social Sciences, Elsevier, https://doi.org/10.1016/B978-0-323-95689-5.00060-2.

UNESCO (2023). Information accessibility. Retrieved from https://www.unesco.org/en/ifap/information-accessibility#:~:text=Information%20Accessibility%20encompasses%20the%20many,needs%20of%20people%20with%20disabilities (accessed 26 December 2023).

United Nation Department of Economics and Social Affairs Disability (2022). **Article 9 – Accessibility. Retrieved from** https://www.un.org/development/desa/disabilities/convention-on-the-rights-of-persons-with-disabilities/article-9-accessibility.html (accessed 26 December 2023).

UN Office of the High Commissioner (2006). Convention on the rights of the persons with disabilities. Retrieved from https://www.ohchr.org/en/instruments-mechanisms/instruments/convention-rights-persons-disabilities (accessed 12 December 2023).

Wardle C, & Derakhshan H. (2017). Information disorder: Toward an interdisciplinary framework for research and policy making. Council of Europe 2017. Report DGI(2017)09 https://rm.coe.int/information-disorder-report-november-2017/1680764666

WHO (2021). International Classification of Functioning, Disability and Health (ICF). Retrieved from https://www.who.int/standards/classifications/international-classification-of-functioning-disability-and-health(accessed 12 December 2023).

World Health Organization. (2020). Immunizing the public against misinformation, Press Release 25 August 2020. Internet]. Retrieved from: https://www.who.int/news-room/feature-stories/detail/immu nizing-the-public-against-misinformation (Access on 27 December 2023).

WHO (2023). Environment, climate change and health: **Disease outbreaks**. https://www.who.int/teams/environment-climate-change-and-health/emergencies/disease-outbreaks

WHO (2023). Infodemic. https://www.who.int/health-topics/infodemic#tab=tab_1

World Bank (2021). Disability inclusion. https://www.worldbank.org/en/topic/disability

Wulff, K., Donato, D., & Lurie, N. (2015). What is health resilience and how can we build it? *Annual Review of Public Health*, 36; 361–374. https://doi.org/10.1146/annurev-publhealth-031914-122829.

Zielinski, C. (2021). Infodemics and infodemiology: A short history, a long future. *Revista Panamericana de Salud PúBlica*, *45*. https://doi.org/10.26633/RPSP.2021.40

Aderinola Ololade Dunmade and Samuel C. Avemaria Utulu

Rethinking Theoretical Foundations of Scientific Inquiry into Fake News Phenomena in the Informing Sciences

Abstract: This study delves into the burgeoning significance of fake news within informing sciences, encompassing various disciplines. The integration of information technology, particularly social media, has propelled the dissemination of unverified information due to factors like confirmation bias and filter bubbles. Scholarly attention across disciplines such as computer science, information systems, and information science is accentuating the impact of fake news on professional practices and societal welfare. While computer science is devising machine learning tools to combat fake news, the library and information science discipline assumes a distinct role, pivoting towards educating users rather than discerning the authenticity of news. The foundational principles, such as Ranganathan's laws, attribute the judgment of news credibility to users, underscoring the discipline's emphasis on information literacy and user education programs. This study elucidates the divergence in theoretical frameworks within informing sciences, underscoring the chapter's aim to bridge theoretical gaps across disciplines grappling with fake news phenomena.

Keywords: Fake news, Informing sciences, Social media, Information technology, Information literacy, Theoretical frameworks

Introduction

Fake news has become a popular subject among researchers in some of the disciplines domiciled in the informing sciences (Campbell 2023). The informing sciences comprise many academic disciplines that study the processes of creating, organizing, and disseminating, as well as using information. The disciplines include computer science, information systems, information science, and library and in-

***Corresponding author: Samuel C. Avemaria Utulu,** Department of Information Systems, School of Information Technology and Computing, American University of Nigeria, Lamido Zubairu Way, Yola-Bypass, Yola, PMB 2250, Adamawa State, e-mail: samuel.utulu@aun.edu.ng
Aderinola Ololade Dunmade, Centre for Open and Distance Learning, University of Ilorin, PMB 1515, Ilorin, Nigeria, e-mail: derin_d@unilorin.edu.ng, Orcid: 0000-0002-7745-0494

https://doi.org/10.1515/9783111176345-007

formation science. Studying fake news became popular in the informing science disciplines due to the relationship they share with information technology (IT), particularly social media (Kondamundi et al. 2023), with this popularity also reinforced by the emergent transformation in the act of informing as a social phenomenon. The absence of censorship also makes social media platforms ideal for the transmission of rumors because of their widespread use (Tariq et al. 2023). Furthermore, the consequences of confirmation bias and filter bubbles both contribute to an enhanced spread of information that has not been independently verified. In response to the outcry from the public, independent fact-checking organizations have emerged in various parts of the world, and many platforms are making efforts to counteract the dissemination of fake news.

For scholars in the computer science field, accommodating fake news stands against the reasons why computers were invented, namely, to help replenish the processes for producing and using accurate data and information. Therefore, tools including those for sentiment analysis tools which are based on machine learning techniques are being invented to trace, expose, and control the occurrence and consumption of fake news. In recent years, there has been a growing amount of interest in computational methods for automating fact-checking, particularly in the fields of data mining, natural language processing, and artificial intelligence (Das et al. 2023), which has been spurred on by the rise of the internet as a primary source of information. The currently available methods place a significant amount of reliance on training classifiers. During this phase of the process, historical events or claims are assembled, classified as authentic or fake, and key characteristics are retrieved to construct an adequate data representation.

While what one may consider worthwhile efforts have been made by disciplines whose theoretical foundations provide their scholars with grounds for inquiring into fake news phenomena, the same cannot be said about the library and information science discipline. Foundational theoretical notions of the library and information science discipline, a good example being that of the Ranganathan's laws (books are for use, every reader his/her book, every book its reader, save the time of the reader, and a library is a growing organism), seem to suggest that library and information professionals' responsibilities do not include deciding whether the news produced and disseminated to users are fake or not (Chhetri 2023). The underlying claim seems to make sense going by the core responsibilities of library and information service professionals: acquisition, organization, dissemination, preservation, and conservation of information resources. By professional demands, library and information professionals are obliged to acquire information resources that are recommended by users, the implication being that the moral responsibility to determine the credibility of the information they consume, including news, lies on the shoulders of library and information service users. However, library and information professionals also offer information literacy and other user education programs (Sharun 2023) which ensure that library and information service users gain the analytical skills they need to adjudge the credibility of the news they consume. It, therefore, follows that much of what library and information service professionals do to help solve problems related to fake news is linked to information literacy and user education programs. The underlying theoretical foundations of the library and information science discipline are therefore different from those of the other disciplines in the informing sciences in two ways. First, is that the library and information science discipline puts the responsibility of deciding whether news is fake or real in the hands of users (Scheufele and Krause 2019). Second, the theoretical foundations of the discipline place the responsibility of training users to make the right judgment in the hands of library and information service practitioners. This means that the discipline acknowledges that, given some conditions, the credibility of the information it acquires and disseminates may be questionable (Chukwusa and Onowhakphor 2023).

Theoretical Foundations of Information Science in the Context of Fake News

Theoretical foundations of information science are also highly relevant to the study of fake news phenomena in the information sciences. Information science is an interdisciplinary field that draws on theories and methods from computer science, library science, communication studies, cognitive psychology, and other fields to study information in all its forms and contexts (De Paor and Heravi 2020; Di Domenico et al. 2021; Mehra 2022). The proliferation of fake news in the digital age poses a significant challenge to society, and understanding its dynamics and impacts has become a crucial area of research for informing sciences (Howell and Brossard 2021). Information science, as a discipline that studies the creation, organization, dissemination, and utilization of information, plays a critical role in understanding fake news phenomena (Ansar and Goswami 2021). To effectively address the issue of fake news, rethinking the theoretical foundations of information science becomes essential (Olan et al. 2022). This section aims to discuss the key theoretical frameworks in information science and their relevance to the study of fake news.

Information Theory

Information theory, developed by Claude Shannon, provides a mathematical foundation for understanding the limits of signal processing and the quantification of information (Guizzo 2003). It focuses on the transmission, storage, and retrieval of information in the presence of noise and other disturbances (Renes 2022). Although information theory does not directly address the issue of fake news, understanding the limits of signal processing and the efficiency of information channels can help identify potential vulnerabilities exploited by fake news dissemination.

Information Behavior

Information behavior refers to how people seek, use, and share information (Savolainen 2007). Understanding information behavior is crucial to understanding how fake news spreads and is consumed by different groups of people (Majerczak and Strzelecki 2022). Researchers can use theories of information behavior to identify the factors that influence people's decision-making around the consumption and sharing of news and to develop more effective strategies for promoting critical thinking and media literacy.

Information Retrieval

Information retrieval is concerned with the development of techniques for searching and retrieving information from large datasets (Singhal 2001). This is particularly relevant to the study of fake news, as researchers need to be able to identify and analyze large volumes of news articles and social media posts to detect patterns and trends in the dissemination of fake news. Information retrieval techniques can be used to develop algorithms for identifying fake news stories and tracking their spread across different platforms (Jain and Kasbe 2018).

Human-Computer Interaction

Human-computer interaction (HCI) is concerned with the study of how people interact with technology (Issa and Isaias 2022). HCI theories can be used to develop interfaces and tools that help people identify and avoid fake news, such as browser extensions that highlight unreliable sources or news literacy games that teach critical thinking skills (Musi et al. 2023).

Information Ethics

Information ethics is concerned with the ethical and moral implications of information technologies and practices (Dunmade 2022; Yilmaz and Celebi 2022). Understanding information ethics is crucial to developing strategies for combating fake news that are grounded in principles of social responsibility and democratic values (Lee and Wei 2022). Information ethics theories can be used to identify the ethical considerations involved in the production, dissemination, and consumption of fake news, and to develop guidelines and best practices for responsible news reporting and consumption.

Social Epistemology

Social epistemology investigates the social dimensions of knowledge production and dissemination, emphasizing the role of social factors in shaping the reliability and credibility of information sources (Oliphant 2021). In the context of fake news, social epistemology helps us understand how social interactions and networks influence the spread of misinformation and disinformation (Lorusso 2022). By examining the credibility of information sources and the dynamics of information sharing, social epistemology provides valuable insights into the mechanisms behind the propagation of fake news.

Actor-Network Theory (ANT)

Actor-network theory, developed by Bruno Latour and his colleagues, is an approach that emphasizes the role of both human and non-human actors in shaping the dynamics of sociotechnical systems (Henttu-Ato Järvinen, and Lassila 2023; Mihaylova 2022). ANT can be applied to the study of fake news by examining the complex interplay of actors, such as individuals, social networks, algorithms, and media platforms, involved in the production and dissemination of fake news (Hajli et al. 2022). This perspective allows for a more comprehensive understanding of how various factors contribute to the prevalence of fake news in the digital age.

Rethinking the theoretical foundations of information science is crucial for effectively addressing the challenges posed by fake news phenomena. By incorporating interdisciplinary and integrative approaches, researchers can develop a more comprehensive understanding of the complex dynamics involved in the production, dissemination, and impact of fake news. This can inform the individual of the development of effective interventions and strategies to mitigate the harmful effects of fake news on individuals and society. Ultimately, a robust theoretical foundation in information science is essential for advancing knowledge and informing individuals of evidence-based policies to combat fake news.

Theoretical Foundations of Computer Science in the Context of Fake News Phenomena

Theoretical foundations of computer science refer to the fundamental principles, concepts, and techniques that underlie the design, analysis, and implementation of algorithms, data structures, programming languages, operating systems, and other software systems. These foundations are based on mathematical and logical theories, and they provide the framework for understanding the capabilities and limitations of computers, as well as how they can be programmed and controlled.

Algorithms and Computational Complexity

The theory of computation is concerned with the study of algorithms and computational processes (McCarthy 1959). It provides a mathematical framework for understanding the limits and capabilities of computing systems and helps researchers identify the computational complexity of different types of problems (Lewis and Papadimitriou 1998). In the context of fake news, the theory of computation can help researchers to identify the computational challenges involved in detecting and combating the spread of fake news, such as the need to process and analyze large volumes of data in real time.

The most famous computational model is the Turing machine, which is a theoretical machine that can perform any computation that can be carried out by a physical computer (Blum and Blum 2022). The theory of computation also includes the study of formal languages and automata, which are mathematical models of the syntax and semantics of programming languages (Burgin and Dodig-Crnkovic 2020). The study of algorithms, their efficiency, and computational complexity is integral to understanding fake news phenomena (Jarrahi and Safari 2023). The increasing scale and speed of information dissemination through social media platforms and other digital channels are powered by complex algorithms that determine what content is displayed, recommended, and shared (Nukuna and Bernstein 2022). These algorithms often prioritize engagement and virality, inadvertently amplifying fake news stories that provoke strong reactions (Curmi 2022; Duan et al. 2022).

By examining the mathematical properties and limitations of these algorithms, researchers can identify potential biases and vulnerabilities that enable the spread of fake news. Moreover, understanding computational complexity can inform the development of more sophisticated detection algorithms that can efficiently identify and flag fake news in real-time.

Machine Learning and Artificial Intelligence

Machine learning is a subfield of computer science that is concerned with the development of algorithms that can learn from data (Koçak et al. 2023). It has become an important tool for analyzing and understanding complex datasets, including those related to fake news. Machine learning algorithms can be used to identify patterns and trends in the dissemination of fake news, and to develop predictive models that can help researchers to anticipate and mitigate the impact of fake news (Mishra et al. 2022; Palani et al. 2022). Machine learning (ML) and artificial intelligence (AI) have become indispensable tools in the fight against fake news. ML models can be trained to automatically detect and classify fake news based on textual, visual, and metadata features (Mallik and Kumar 2023). In addition, AI-powered tools can help automate the fact-checking process, enabling faster and more accurate identification of misinformation (Tide et al. 2022).

However, the same technologies can also be exploited by malicious actors to create and disseminate fake news more effectively. For instance, generative models like GPT and StyleGAN can produce realistic-looking synthetic text and images that are difficult to distinguish from authentic content (Bauwelinckx et al. 2023; Rossi et al. 2022). As a result, the arms race between fake news generators and detectors continues to escalate, demanding a deeper understanding of the theoretical foundations of ML and AI (Illia et al. 2022i).

Network Theory and Graph Analysis

Network theory and graph analysis offer valuable insights into the structure and dynamics of social networks that facilitate the spread of fake news (Papadopoulou et al. 2022). By studying the properties of these networks, such as their topology, clustering, and community structure, researchers can pinpoint influential nodes and pathways through which misinformation propagates (Chen, Wei, and Yang 2022). Furthermore, graph-based algorithms can be employed to detect and mitigate the spread of fake news (Inan 2022). For example, algorithms like PageRank and HITS can be used to assess the importance and credibility of information sources, while community detection algorithms can help identify and isolate echo chambers that amplify misinformation (Pastor-Galindo, Mármol, and Pérez 2022).

Information Theory and Data Science

Information theory and data science provide the necessary tools for quantifying and analyzing the vast amounts of data generated by fake news phenomena (Luo, Chen, and Guo 2022; Xing et al. 2022). Concepts like entropy, mutual information, and Bayesian inference can be applied to understand the uncertainty and dependencies in the data, as well as to develop probabilistic models for fake news detection (Saha et al. 2022; Varley 2023. Moreover, data science techniques like clustering, dimensionality reduction, and anomaly detection can help identify patterns and trends in the propagation of fake news, revealing insights into the strategies employed by malicious actors and the vulnerabilities of information ecosystems (Mishra et al. 2022).

Natural Language Processing

Natural language processing is a subfield of computer science that is concerned with the development of algorithms that can analyze and generate natural language (Hamilton and Lahne 2023). It has become an important tool for analyzing the linguistic features of fake news, such as the use of emotive language and sensationalist headlines (Nanath et al. 2022). Natural language processing techniques can also be used to identify the sources of fake news, by analyzing the linguistic patterns and characteristics of the authors (Sakketou et al. 2022; Sharifani et al. 2022).

Implications for Theory and Practice

In re-examining the foundations that underpin our understanding of the informing sciences, a crucial need arises to reevaluate the theoretical frameworks, especially within the realm of library and information science. This reassessment is essential to effectively address the modern challenges presented by the proliferation of fake news.

A pivotal aspect of this reevaluation involves acknowledging the imperative role of users in assessing the credibility of information. Central to this concept is the recognition and emphasis placed on user-centric responsibility. It is vital to align this approach with the fundamental ethos of library and information science, thereby fostering a collective responsibility shared between users and information professionals. By doing so, a collaborative effort emerges, empowering users to critically evaluate the information they encounter, while also reinforcing the guidance and expertise provided by information professionals.

Furthermore, a commitment to adaptability is paramount in the face of rapidly evolving technologies. As the landscape of fake news dissemination continues to transform, it is imperative to continuously evolve computational methods and technologies. This adaptive approach enables the proactive mitigation of the detrimental effects of fake news, ensuring that information professionals remain at the forefront of combatting misinformation.

In essence, the implications for theory and practice in addressing fake news within the informing sciences necessitate a comprehensive and dynamic approach. By revisiting theoretical foundations, embracing user-centric responsibility, and continually adapting to evolving technologies, the field of library and information science can effectively navigate the challenges posed by fake news, ultimately fostering a more informed and discerning society.

Recommendations and Conclusions

To effectively combat the spread of fake news, it is imperative to foster interdisciplinary collaboration among a variety of fields, such as computer science, information systems, and library sciences. By promoting teamwork and knowledge-sharing across these disciplines, we can develop comprehensive strategies to address the complex challenges posed by misinformation.

Furthermore, the development of enhanced fact-checking tools represents a crucial step in the fight against fake news. Leveraging further advancements in machine learning and AI-driven technologies can significantly improve the accuracy and efficiency of identifying and mitigating false information. It is essential to prioritize the integration of deep learning methods to enable more nuanced analysis, ultimately fortifying our defenses against misleading content.

In addition to technological advancements, a user-centric approach is vital in empowering individuals to navigate the modern information landscape with confidence. By implementing robust education programs, we can equip users with enhanced information literacy skills, emphasizing critical thinking and analytical abilities. These initiatives are pivotal in enabling individuals to discern credible information from misinformation, ultimately fostering a more informed and discerning society.

In conclusion, by promoting interdisciplinary collaboration, advancing fact-checking technologies, and prioritizing user-centric educational initiatives, we can take significant strides toward addressing the pervasive issue of fake news. These recommendations lay the groundwork for a multifaceted approach that acknowledges the complexities of misinformation while empowering individuals and communities to make informed decisions in an increasingly digital world.

References

Ansar, W., & Goswami, S. (2021). Combating the menace: A survey on characterization and detection of fake news from a data science perspective. *International Journal of Information Management Data Insights*, *1*(2), 100052.

Bauwelinckx, Y. C., Dhaene, J., Verdonck, T., & Heuvel, M. V. D. (2023). On the causality-preservation capabilities of generative modeling. *arXiv preprint arXiv:2301.01109*.

Blum, L., & Blum, M. (2022). A theory of consciousness from a theoretical computer science perspective: Insights from the Conscious Turing Machine. *Proceedings of the National Academy of Sciences*, *119*(21), e2115934119.

Burgin, M., & Dodig-Crnkovic, G. (2020). Prolegomena to an Operator Theory of Computation. *Information*, *11*(7), 349.

Campbell, D. G. Fake News as an Emergent Subject Domain: conceptual and ethical perspectives for the development of a critical knowledge organisation. *The H m Position in an Artifical World: Creativity, Ethics and AI in Knowledge Organization*, 208.

Chen, J., Wei, N., & Yang, H. (2022). Immune Algorithm to Suppress Rumor Propagation Based on Influence Maximization. *Security and Communication Networks, 2022*.

Chhetri, P. (2023). Rethinking Ranganathan's Five Laws of Library Science in the Artificial Intelligence Era. *File. Lislinks*, *9*(1), 10–16.

Chukwusa, J., &Onowhakpor, E. J. (2023). Scholarly Communication Channels Easily Reached by Library and Information Science (LIS) Students. *UNIZIK Journal of Educational Research and Policy Studies*, *15*(4), 56–69.

Curmi, F. (2022). Social Contagion and Virality in Online Networks. *The SAGE Handbook of Social Media Marketing*, 146.

Das, A., Liu, H., Kovatchev, V., & Lease, M. (2023). The state of human-centered NLP technology for fact-checking. *Information processing & management*, *60*(2), 103219.

Di Domenico, G., Sit, J., Ishizaka, A., & Nunan, D. (2021). Fake news, social media and marketing: A systematic review. *Journal of Business Research*, *124*, 329–341.

De Paor, S., & Heravi, B. (2020). Information literacy and fake news: How the field of librarianship can help combat the epidemic of fake news. *The Journal of Academic Librarianship*, *46*(5), 102218.

Duan, Z., Li, J., Lukito, J., Yang, K. C., Chen, F., Shah, D. V., & Yang, S. (2022). Algorithmic agents in the hybrid media system: Social bots, selective amplification, and partisan news about COVID-19. *Human Communication Research*, *48*(3), 516–542.

Dunmade, A. O. (2022). Perception, Awareness, and Attitude of female postgraduate students towards cyberethical behaviour in North-Central Nigeria universities. *PhD dissertation, Adeleke University.* https://www.adelekeuniversity.edu.ng.

Guizzo, E. M. (2003). *The essential message: Claude Shannon and the making of information theory* (Doctoral dissertation, Massachusetts Institute of Technology).

Hajli, N., Saeed, U., Tajvidi, M., & Shirazi, F. (2022). Social bots and the spread of disinformation in social media: the challenges of artificial intelligence. *British Journal of Management*, *33*(3), 1238–1253.

Hamilton, L. M., & Lahne, J. (2023). Natural language processing. In *Rapid sensory profiling techniques* (pp. 371–410). Woodhead Publishing.

Henttu-Aho, T., Järvinen, J. T., & Lassila, E. M. (2023). Constructing the accurate forecast: an actor-network theory approach. *Meditari Accountancy Research*, *31*(7), 116–132.

Howell, E. L., & Brossard, D. (2021). (Mis) informed about what? What it means to be a science-literate citizen in a digital world. *Proceedings of the National Academy of Sciences*, *118*(15), e1912436117.

Illia, L., Colleoni, E., &Zyglidopoulos, S. (2023). Ethical implications of text generation in the age of artificial intelligence. *Business Ethics, the Environment & Responsibility*, *32*(1), 201–210.

Inan, E. (2022). ZoKa: a fake news detection method using edge-weighted graph attention network with transfer models. *Neural Computing and Applications*, *34*(14), 11669–11677.

Issa, T., & Isaias, P. (2022). Usability and Human-Computer Interaction (HCI). In *Sustainable Design: HCI, Usability and Environmental Concerns* (pp. 23–40). London: Springer London.

Jain, A., &Kasbe, A. (2018, February). Fake news detection. In *2018 IEEE International Students' Conference on Electrical, Electronics and Computer Science (SCEECS)* (pp. 1–5). IEEE.

Jarrahi, A., & Safari, L. (2023). Evaluating the effectiveness of publishers' features in fake news detection on social media. *Multimedia Tools and Applications*, *82*(2), 2913–2939.

Koçak, B., Cuocolo, R., Pinto dos Santos, D., Stanzione, A., &Ugga, L. (2023). Must-have Qualities of Clinical Research on Artificial Intelligence and Machine Learning. *Balkan Medical Journal*, *40*(1).

Kondamudi, M. R., Sahoo, S. R., Chouhan, L., & Yadav, N. (2023). A comprehensive survey of fake news in social networks: Attributes, features, and detection approaches. *Journal of King Saud University-Computer and Information Sciences*, *35*(6), 101571.

Lee, K. S., & Wei, H. (2022). Design Factors of Ethics and Responsibility in Social Media: A Systematic Review of Literature and Expert Review of Guiding Principles. *Journal of Media Ethics*, *37*(3), 156–178.

Lewis, H. R., & Papadimitriou, C. H. (1998). Elements of the Theory of Computation. *ACM SIGACT News*, *29*(3), 62–78.

Lorusso, A. M. (2022). Fake News as Discursive Genre: Between Hermetic Semiosis and Gossip. *Social Epistemology*, 1–13.

Luo, X., Chen, H. H., & Guo, Q. (2022). Semantic communications: Overview, open issues, and future research directions. *IEEE Wireless Communications*, *29*(1), 210–219.

Majerczak, P., & Strzelecki, A. (2022). Trust, Media Credibility, Social Ties, and the Intention to Share towards Information Verification in an Age of Fake News. *Behavioral Sciences*, *12*(2), 51.

Mallik, A., & Kumar, S. (2023). Word2Vec and LSTM-based deep learning technique for context-free fake news detection. *Multimedia Tools and Applications*, 1–22.

McCarthy, J. (1959). A basis for a mathematical theory of computation. In *Studies in Logic and the Foundations of Mathematics* (Vol. 26, pp. 33–70). Elsevier.

Mehra, B. (2022). Toward an impact-driven framework to operationalize social justice and implement ICT4D in the field of information. *Journal of the Association for Information Science and Technology*.

Mihaylova, R. (2022). Researching Education Through Actor-Network Theory. *Критика и хуманизъм*, *2*(57), 135–146.

Mishra, S., Shukla, P., & Agarwal, R. (2022). Analyzing machine learning enabled fake news detection techniques for diversified datasets. *Wireless Communications and Mobile Computing*, *2022*, 1–18.

Musi, E., Carmi, E., Reed, C., Yates, S., & O'Halloran, K. (2023). Developing Misinformation Immunity: How to Reason-Check Fallacious News in a Human-Computer Interaction Environment. *Social Media+ Society*, *9*(1), 20563051221150407.

Nanath, K., Kaitheri, S., Malik, S., & Mustafa, S. (2022). Examination of fake news from a viral perspective: an interplay of emotions, resonance, and sentiments. *Journal of Systems and Information Technology*, *24*(2), 131–155.

Nukuna, N., & Bernstein, B. (2022). The Role of Social Media in Spreading Information and Influencing Elections. *Brave New World*, 36.

Olan, F., Jayawickrama, U., Arakpogun, E. O., Suklan, J., & Liu, S. (2022). Fake news on social media: The impact on Society. *Information Systems Frontiers*, 1–16.

Oliphant, T. (2021). Emerging (information) realities and epistemic injustice. *Journal of the Association for Information Science and Technology*, *72*(8), 951–962.

Palani, B., Elango, S., & Viswanathan K, V. (2022). CB-Fake: A multimodal deep learning framework for automatic fake news detection using capsule neural network and BERT. *Multimedia Tools and Applications*, *81*(4), 5587–5620.

Papadopoulou, O., Makedas, T., Apostolidis, L., Poldi, F., Papadopoulos, S., &Kompatsiaris, I. (2022). Mevernetworkx: Network analysis and visualization for tracing disinformation. *Future Internet*, *14*(5), 147.

Pastor-Galindo, J., Mármol, F. G., & Pérez, G. M. (2022). Profiling users and bots in Twitter through social media analysis. *Information Sciences*, *613*, 161–183.

Renes, J. (2022). *Quantum Information Theory: Concepts and Methods*. Walter de Gruyter GmbH & Co KG.

Rossi, S., Kwon, Y., Auglend, O. H., Mukkamala, R. R., Rossi, M., & Thatcher, J. (2022). Are Deep Learning-Generated Social Media Profiles Indistinguishable from Real Profiles? *arXiv preprint arXiv:2209.07214*.

Saha, A., Tawhid, K., Miah, M. B., Somudro, A. A., & Nandi, D. (2022, March). A Comparative Analysis on Fake News Detection Methods. In *Proceedings of the 2nd International Conference on Computing Advancements* (pp. 392–399).

Sahu, J. K. The Role of Social Media in Media Trials: Exploring the Impact and Challenges in the Indian Criminal Justice System.https://jurynesia.com/wp-content/uploads/2023/08/The-Role-of-Social-Media-in-Media-Trials-Exploring-the-Impact-and-Challenges-in-the-Indian-Criminal-Justice-System.pdf

Sakketou, F., Plepi, J., Geiss, H. J., & Flek, L. (2022, October). Temporal Graph Analysis of Misinformation Spreaders in Social Media. In *Proceedings of TextGraphs-16: Graph-based Methods for Natural Language Processing* (pp. 89–104).
Savolainen, R. (2007). Information behavior and information practice: reviewing the “umbrella concepts” of information-seeking studies. *The Library Quarterly*, *77*(2), 109–132.
Scheufele, D. A., & Krause, N. M. (2019). Science audiences, misinformation, and fake news. *Proceedings of the National Academy of Sciences*, *116*(16), 7662–7669.
Sharifani, K., Amini, M., Akbari, Y., &Aghajanzadeh-Godarzi, J. (2022). Operating Machine Learning across Natural Language Processing Techniques for Improvement of Fabricated News Model. *International Journal of Science and Information System Research*, *12*(9), 20–44.
Sharun, S. (2023). The critical information literacy of social workers: information literacy as interpersonal practice. *Journal of Information Literacy*, *17*(1), 186–203.
Singhal, A. (2001). Modern information retrieval: A brief overview. *IEEE Data Eng. Bull.*, *24*(4), 35–43.
Tariq, M., Shakoor, A., Waheed, A., & Khan, S. 2023. The Proliferation of Inaccurate and Misleading Information through the Use of Social Media: The Impact on Pakistani Society.https://www.qlantic.com/qjss/papers/3AJ0eHmDcd.pdf
Tida, V. S., Hsu, D. S., & Hei, D. X. (2022). A Unified Training Process for Fake News Detection based on Fine-Tuned BERT Model. *arXiv preprint arXiv:2202.01907.*
Xing, Y., Wang, X., Qiu, C., Li, Y., & He, W. (2022). Research on opinion polarization by big data analytics capabilities in online social networks. *Technology in Society*, *68*, 101902.
Yilmaz, Y., & Çelebi, C. (2022). Views of Information Technologies Teachers on the Effects of Values Education on Informatics Ethics. *International Online Journal of Education and Teaching*, *9*(1), 506–530.

Vusi Tsabedze

Library and Information Science Students' Perception of Fake News on Social Media: Sanitizing the Sanitizable Information Landscape to Enhance Information Hygienic Experiences

Abstract: The ease of accessing information through the internet has revolutionized the way students gather knowledge, with a few keystrokes providing access to vast amounts of information. However, this convenience comes with a downside: the dissemination of fabricated information, commonly known as fake news. The impact of fake news can be detrimental to both individuals and society as a whole. This chapter aims to assess the perception of Library and Information Science (LIS) students regarding the spread of news through social media platforms. The findings of this study will facilitate the development of appropriate recommendations to counter fake news and promote informed news consumption. A quantitative research approach was employed to collect and analyze data for this study, with a total of 75 students surveyed using questionnaires. The analysis of the responses revealed that students tend to trust social media platforms as credible sources of news. Additionally, the results indicated that students are inclined to recommend social media as news sources to their peers. Furthermore, the study found that students generally have trust in the news they acquire from social media. While some respondents expressed a willingness to share news on social media if they were convinced of its credibility, others admitted to sharing interesting news despite doubts about its authenticity.

Furthermore, this chapter presents the perception of students regarding the security of social media applications. Finally, the researcher provides recommendations for countering fake news and enhancing individuals' ability to critically consume news.

Keywords: fake news, social media, news consumption, information literacy, Library and Information Science students, credibility, perception, Eswatini

Vusi Tsabedze, Department of Computer Science, University of Eswatini, Department Information Science, University of South Africa

https://doi.org/10.1515/9783111176345-008

Introduction

The concept of fake news has undergone significant transformations since the advent of print media and continues to evolve in the current era of social media. The rapid advancement of technology has facilitated the seamless and inconspicuous dissemination of false information and thus the evolution of fake news will likely persist, accompanied by the emergence of novel technologies aimed at enhancing its propagation. However, it is equally conceivable that newer technologies will be developed to combat the spread of fake news. Amongst these technological advancements, social media has emerged as the primary platform for students to access information. Notably, platforms such as Facebook are consistently ranked as preferred sources for news consumption, particularly among students (Newman et al. 2019).

The proliferation of fake news in the digital realm, facilitated by platforms such as Facebook and Twitter, is a growing concern. These platforms have contributed to the spread of false information by employing strategies like creating echo chambers through segregated groups or utilizing recommendation algorithms (Zimmer, Scheibe, and Stock 2019). To address this issue, numerous studies have been conducted to develop effective solutions specifically targeting fake news on social media. One such study by Burkhardt (2017) explores the implementation of technological programs capable of detecting fake news on social media platforms. Alternatively, other initiatives focus on raising awareness among social media users and curbing the dissemination of falsehoods by employing automated fact-checking mechanisms (Graves 2018).

The prevalence of fake news on social media presents a multifaceted challenge, as detecting such misinformation can be arduous (Staffers and Hackett 2017). As highlighted by Borel (2018), social media platforms offer numerous avenues through which users can encounter fake news. However, it remains uncertain how frequently individuals unknowingly succumb to false information disseminated on these platforms.

According to Agosto (2018) and De Paor and Heravi (2020), libraries have important roles to play in educating students on how to find and evaluate online information sources. Libraries can also teach how to detect fake news. Every student, more or less, uses smartphones and computers to access the internet or social media to be up to date with the continuous generation of information. It is important to assess their potential to identify fake news (El Rayess et al. 2018).

Furthermore, it is not known how library and information science (LIS) students in Eswatini perceive the concept of news on social media platforms. Therefore, this study intended to identify the students' level of trust in the information that they read on social media platforms and the security measures that are integrated into their social media platforms. This chapter also aimed to identify the perceptions that LIS students have on the concepts of fake news dissemination on social media platforms.

Problem Statement

The widespread usage of social media platforms has facilitated easy access to information for students, however, it has also led to the proliferation of fabricated information, commonly known as fake news, which can have disastrous effects on individuals and society as a whole. This phenomenon raises concerns about the perception and behaviour of LIS students regarding the propagation of news on social media and their ability to discern reliable sources of information. It is essential to assess the perception of LIS students on the credibility of news from social media platforms, their sharing behaviors, and their understanding of the security implications associated with these applications. Understanding these factors can allow appropriate strategies to be developed to counter fake news and enhance the students' abilities to critically evaluate and consume news from social media platforms.

Aim and Research Objectives

The study aimed to assess the perception of LIS students towards fake news on social media at the University of Eswatini and to make appropriate recommendations that will assist the university in developing strategies that will help students avoid fake news and become better news consumers. Specifically, this study sought to:

- ascertain the perception of LIS students towards fake news on social media
- establish the level to which LIS students consider social media platforms as reliable sources for news acquisition
- determine the level of trust that LIS students put in the news that they acquire from social media platforms.
- ascertain LIS students' behavior towards the news that they acquire from social media
- determine the challenges associated with fake news on the social media platform
- suggest recommendations to overcome the challenges associated with fake news on social media.

Literature Review

To meet the objectives of the study stated in the preceding section, the researcher reviewed several studies carried out on fake news on social media platforms.

Social Media and Fake News

Over the last decade, the popularity of social media platforms such as Facebook, WhatsApp, YouTube, Instagram, and TikTok has rapidly increased. With Internet penetration and affordability of personal devices such as smartphones, people have increasingly become dependent on social media for information and communication (Ummah and Fajri 2020).

Although these platforms were initially created to connect people worldwide, their applications have expanded, with more people using them to access news and other crucial information.

Presently, according to Foresti, Varvakis and Viera (2018), interactions are primarily mediated by technology in contemporary society, which is surrounded by information overload on the internet and other media, leading to a shift away from the physical world and a strengthening of the virtual realm (Pinto et al. 2020). People are glued to smartphones and tablets, and the willingness to acquire and share information quickly, decontextualized and without careful attention to the sources and how information is disseminated, makes them vulnerable and conducive to the acquisition and dissemination of misinformation, disinformation, and fake news.

Students and Fake News on Social Media

The Internet has created a wide-open door for spreading fake news, with social media specifically seen as one of the most powerful drivers for this (Burkhardt 2017). Nowadays, students' lives are dominated by the internet, as they spend more of their time learning about the world on social media platforms than through traditional media (McGrew et al. 2017). It is becoming increasingly difficult for students to know how to detect fake news on social media. Fake news, in some cases, originates from a fake website that looks similar to a genuine news outlet but was created by an individual or organization to promote a specific view or political philosophy. Therefore, students must know how to differentiate between fake and true information.

In a study conducted by Fernandez (2017) on the technology behind fake news, the findings showed how technology encourages fake news through web-based publishing platforms. The researcher discussed the role of social media and technology companies in the information society, and libraries' role in creating digitally literate users. Some librarians designed modules for skill development in the context of information literacy and fake news detection in a learning management system, aiming for users to improve their capacity to discover accurate information (Auberry 2018).

Bahige (2019) indicated that most students from the University of Kwazulu Natal receive their news from social media. With students spending more time online nowadays, these platforms have become critical information dissemination channels, seen as students explore these developments by leveraging social media to complete school assignments and conduct research activities. The threat of fake news has also become a challenge affecting the credibility and quality of online content, with more people becoming sceptical of online information due to the increased dissemination of fake information. Students use social media platforms, particularly Facebook and Twitter, in the acquisition and circulation of online news. Social media simplifies and facilitates the sharing of such news, considering its convenience and tools that are easy to use when posting content for individuals or even for media organizations.

Staffers and Hackett (2017) studied how bad data and misleading graphs fuel fake news. Their findings showed that students from middle school, high school, and tertiary education from 12 states in the United States of America found it challenging to differentiate between factual news and fabricated news, which was evidenced when the participants were presented with information from articles, tweets, and comments.

Domonoske (2016) stated that a shocking observation was made by different scholars about the failure of students to distinguish factual news from fake news and how consistent their responses were in this regard. The majority of high school students (80%) believed that sponsored contents were factual news; while many people assume that because students spend most of their time on social media, they might have more knowledge of what they find on these platforms, this was contradicted by the scholars' findings. Another observation made by the scholars was that an "about" section of a webpage, if well-presented and polished, was able to persuade the students that the site was genuine, and they would tend to naïvely believe the content without any supporting evidence (Domonoske 2016).

In Indonesia, Syam and Nurrahmi (2020) conducted a study on social media literacy among students of two renowned universities. The study revealed that university students were unaware of fake news; while they were skilled at finding news on social media, they lacked the skill to differentiate between fake news and genuine news. Syam and Nurrahmi (2020) recommended that universities need to include social media literacy as a mandatory subject for students.

In conclusion, the literature reviewed shows that students mostly use Facebook, YouTube, and WhatsApp to acquire information from the Internet and social networking sites. The literature review also showed that information literacy and media information literacy are recognized as the most effective techniques to fight against fake news on social media. The need for social media literacy and political news literacy has also been fascinating in some studies.

The current study attempts to build on these studies to assess the perception of LIS students on fake news on social media at the University of Eswatini, a topic that has not been covered in any of the studies conducted before.

Detecting Fake News

Studies have been conducted on how to detect fake news on social media, and it has been revealed that if a social media user gets delivery of fake news by bots, they have a slightly better than 50% chance of detecting that the news is fake (Burkhardt 2017). To increase the probability of detecting fake news, information and communication professionals have been trying to develop a set of methods to improve the automatic computerized recognition of fake news (Burkhardt 2017). Eva and Shea (2018) and Crate (2017), in their studies on how to detect fake news on social media, recommended the following:

- It is important to check the source, namely domain names that look strange or webpages that end with strange syllables.
- Readers should click on the supporting sources provided to see if they are talking about the same thing. If not, then the page might contain fake news.
- When reading online news, it is necessary to learn more about who the author of the news is and who the sponsors might be. Sponsors may influence the news that is being presented.
- Fake news pages usually use web addresses that make them look like real sites, but they sometimes end in strange ways, for example, ".com. co". It is very important to check the URL.
- There are visual clues that can help to spot fake news. A sloppy design or overuse of all caps might be used for fake news sites.
- If a story seems to be too alarming that it makes the readers angry, there is a need to consult other news sites or simply perform fact checks using various fact-checking sites available on the internet.

- People need to read beyond headlines before they can think of sharing stories, as headlines can be outrageous in the quest for clicks. Readers need to get the whole story. According to Burkhardt (2017), one of the reasons why fake news on social media goes viral is because people share news without reading beyond the headlines, and without thinking about the content of the message. Headlines are designed to capture the attention of the readers, and they are usually written to incite strong reactions.
- People also need to make use of their browsers. An installation of a browser plugin to help flag fake news might be of great help in the task of spotting fake news.
- Lastly, it is important to consider when and where the news was published.

The above literature review reveals the pressing need for increased digital literacy, critical thinking skills, and educational interventions to address the challenges posed by fake news and misinformation on social media, particularly among student populations. These implications can guide policymakers, educators, and researchers in developing effective strategies to combat the spread of fake news in the digital era.

Methodology

The study aimed to assess the perception of LIS students at the University of Eswatini regarding fake news on social media. The target population for this study consisted of 75 LIS students enrolled at the University of Eswatini. To collect data, a survey questionnaire was employed as the primary research instrument.

The construction of the questionnaire was guided by a thorough analysis of relevant literature, ensuring that the questions were designed to elicit authentic and transparent responses from the participants. Recognizing that perceptions towards fake news on social media may vary among respondents, a Likert five-point scale was utilized to capture a range of perspectives. The scaling options employed in the questionnaire were as follows: (1) strongly disagree, (2) disagree, (3) neither agree nor disagree, (4) agree, and (5) strongly agree.

To establish the validity and reliability of the data collection tool, a pilot test was conducted among expert researchers specializing in social media. Their feedback and insights were invaluable in refining the questionnaire, ensuring its clarity and effectiveness in capturing the intended information. Adjustments were made based on the pre-test results, enhancing the questionnaire's quality and suitability for administration to the target respondents.

Adhering to these methodological procedures, the study aimed to gather robust and reliable data that would provide valuable insights into the perception of LIS students regarding fake news on social media platforms.

Findings

The study findings that are reported and discussed in this section focused on the following aspects: acquisition of news from social media; trust in social media as a news source; behavior towards news from social media; social media bots; and social media applications security. In the subsequent section, possible solutions relating to mitigating fake news on social media platforms are suggested.

Acquisition of News From Social Media

The study sought to establish the types of social media platforms used by students to access news. The findings revealed that all of the respondents 75 (100%), use Facebook, YouTube, and WhatsApp to access news. Instagram and TikTok also have a significant user base, with 63 (84%) and 58 (77%) of the respondents using them, respectively. Snapchat, Pinterest, Twitter, and LinkedIn have slightly lower usage rates, ranging from 51 (68%) to 30 (40%). Reddit has the lowest usage rate, with only 10 (13%) of the respondents indicating that they use it. These findings are presented in Table 1.

Table 1: Social media platforms.

Social Media Platforms	Frequency	Percentage
Facebook	75	100
YouTube	75	100
WhatsApp	75	100
Instagram	63	84
TikTok	58	77
Snapchat	51	68
Pinterest	41	55
LinkedIn	30	40
Twitter	41	55
Reddit	10	13

*Multiple responses were possible.
Source: field data (2023).

Overall, the above results imply that students heavily rely on social media platforms, particularly Facebook, YouTube, and WhatsApp, as their primary sources of news. Instagram and TikTok also play a significant role in news consumption, possibly indicating the increasing influence of visually oriented platforms in delivering news content to younger audiences.

Trust in Social Media as a News Source

The study sought to establish if the respondents believe that social media is a reliable source for the acquisition of news. The results revealed that 46 (61%) of the respondents indicated that they believe that social media is a reliable source for the acquisition of news, while 29 (49%) of the respondents indicated that they do not believe this to be the case (see Figure 1).

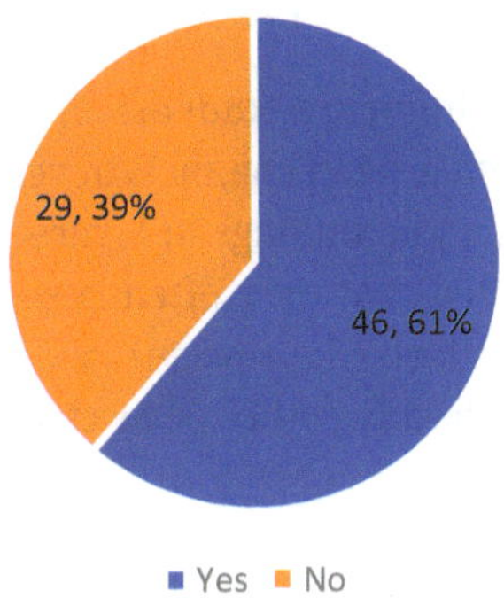

Figure 1: Trust in social media as a news source.
Source: field data (2023).

The respondents were then asked if they would recommend their friends to use social media as a source of information. Forty-two (56%) indicated that they would recommend their friend to use social media as a source of information, while 33 (44%) of the respondents indicated that they would not do so (see Figure 2).

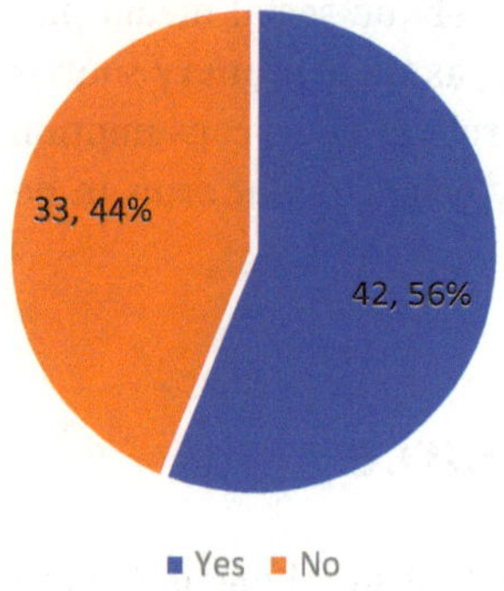

Figure 2: Recommend social media as a source of information.
Source: field data (2023).

Fact-checking

Fact-checking is important because misinformation can influence students' opinions. The study sought to establish if the students check other news channels to verify the news that they read from their social media pages. The findings in Figure 3 below showed that 72 (96%) of the respondents indicated that they often check other news channels to verify the news that they read from their social media pages, while three (4%) of the respondents indicated not checking other news channels to verify what they read on their social media pages.

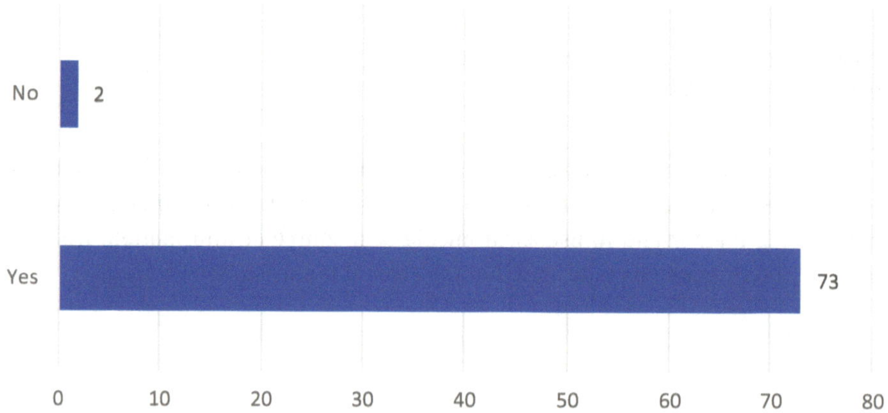

Figure 3: Fact-checking.
Source: field data (2023).

Behaviour Towards News From Social Media

The respondents were asked if they shared the news that they read on social media. The results in Table 2 below showed that 42 (56%) of the respondents share the news that they read on social media if they are sure that the news is accurate, authentic, and reliable while 29 (38.7%) of the respondents indicated that they do share interesting news on social media without checking its veracity. However, 46 (61.3%) of the respondents indicated that they do not share interesting news on their social media platforms without checking its veracity, whereas 63 (84%) indicated that they would not share news, no matter how interesting it is, if they thought that it was fake. Twelve (16%) of the respondents indicated that they would share interesting news even if they think that it is fake. Furthermore, 15 (20%) of the respondents indicated that they do not share news on social media even if they are convinced that it is true. A total of 60 (80%) of the respondents indicated that if they are convinced that some news is true, they share it on social media.

Table 2: Behavior towards news from social media.

Statement	Strongly Agree	Agree	Disagree	Strongly Disagree
If I am sure that the news, I read on social media is true, I share it so that my social media followers can also read it.	20 (26.7%)	22 (29.3%)	19(25.3%)	14 (18.7%)
If I find some news interesting, I share it on my social media without checking its veracity.	9 (12%)	20 (26.7%)	30 (40%)	16 (21.3%)
If I think that the news is fake, I do not share it, no matter how interesting it is.	38 (51%)	25(33%)	7 (9.3%)	5 (6.7%)
I do not share news on social media, even if I am convinced it is true.	8 (10.7%)	7(9.3%)	35 (46.7)	25 (33.3%)

Source: field data (2023).

Social Media Bots

Less than half of the respondents (23, 30.6%) indicated that social media bots are implemented to ensure the security of social media applications, while 52 (69.3%) of the respondents indicated that social media bots are not implemented to ensure the security of social media applications.

In Table 3 below, Forty-seven (62.6%) of the respondents expressed the belief that social media bots are implemented to prevent people from reading fake news, while 28 (37.3%) disbelieve that social media bots are used for this. Forty-five (60%) of the respondents indicated the belief that social media bots are used to spy on social media users, while 30 (40%) of the respondents indicated that they do not this is true. In addition, 51 (68%) of the respondents indicated the belief that social media bots are put in place to learn people's interests from pages they access through social media, as opposed to 24 (32%) who indicated that they do not think this is the case. Furthermore, 56 (74.7%) of the respondents indicated that social media bots provide users with news that is in line with their interests, as opposed to 19 (25.3%) who do not believe this is provided.

Table 3: Social media bots.

Statement	Strongly Agree	Agree	Disagree	Strongly Disagree
Social media bots are implemented to ensure the security of social media applications.	8 (10.6%)	15 (20%)	12 (16%)	40 (53.3%)
Social media bots are implemented to prevent people from reading fake news.	27 (36%)	20 (26.6 %)	12 (16%)	16 (21.3%)
Social media bots are used to spy on social media users.	20 (26.7%)	25 (33.3%)	20 (26.7%)	10 (13.3%)
Social media bots are put in place to learn people's interests from pages they access through social media.	16 (21.3%)	35 (46.7%)	14 (18.7)	10 (13.3%)
Social media bots provide users with news that is in line with their interests.	25 (33.4%)	31 (41.3%)	12 (16%)	7 (9.3%)

Source: field data (2023).

Social Media Applications for Security

The study sought to establish if the respondents were aware of the security settings that are present on social media applications. As shown in Figure 4, the findings revealed that the majority of the respondents (65, 86.7%), indicated that they are aware of the security settings that are present on social media applications, while 10 (13.3%) of the respondents indicated that they were not aware of the security settings available on their social media applications.

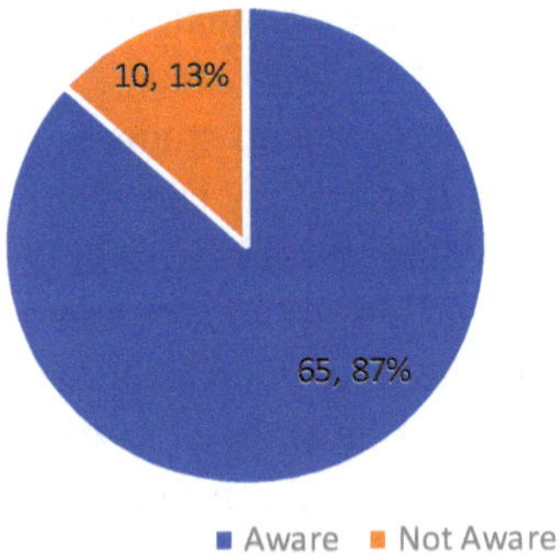

Figure 4: Social media applications security.
Source: field data (2023).

Furthermore, the respondents were asked if they have security settings implemented on most of their social media applications. The findings revealed that 60 (80%) of the respondents have security settings implemented on their social media applications, while 15 (20%) of the respondents indicated not having security settings implemented on most of their social media applications.

Conclusion and Recommendations

With regards to perceptions of LIS students on the confirmation of the news acquired from social media platforms, the majority of students indicated that they often check other news channels to confirm news received from their social media pages. A considerable number of the students indicated that they only share news on social media if they are sure that the news is true, while some students admitted to sharing news on social media even if they are not sure that the news is accurate and authentic. The findings further revealed that the majority of the respondents would not share news on social media even when convinced that such news was truthful. Furthermore, the study revealed that students are confused about and do not understand the concept of social bots, which exposes them further to fake news and makes them more active in its spread through social media platforms. However, the majority of the respondents are aware of security measures available on social media applications and indicated that they have security measures implemented on their social media applications.

Students are likely to learn about the world through their social media platforms, where there is a large spreading of fake news. Students must know how to differentiate between false and true information. It has become pertinent for the University of Eswatini to know the perception of its students on the dissemination of news on social media platforms since the students' lives are dominated by the internet, and they are online most of the time.

Therefore, the University of Eswatini has to provide lessons to students on how to verify news that they have received from social media. They can also be taught about the concept of social media bots and how they are involved in the dissemination of fake news on social media. Students can also be taught how to evaluate online information sources, which can help them to avoid fake news and become better news consumers. In addition to this, the University of Eswatini should also organize seminars for academic staff so that they can also learn about information literacy. Furthermore, the challenges of fake news diffusion on social media do not affect only students but everyone in the country with access to social media. Therefore, the study recommends that the government of Eswatini, through the Ministry of Information Communication and Technology (ICT), organize documentary series and campaigns to inform society about fake news and provide them with information that can help them be more cautious about the news they are getting from social media. This can help reduce the spread of fake news on social media platforms.

Implications of the study

The findings of this study have several implications for both research and practice in the context of fake news and social media.

Firstly, the study highlights the need for increased awareness and critical thinking among social media users regarding the authenticity and credibility of news sources. The fact that students trust social media platforms as news sources underscores the importance of promoting media literacy and information literacy skills among students and the wider population. Educating users about the potential pitfalls of fake news and providing them with tools to identify and evaluate reliable news sources can empower individuals to make informed decisions and minimize the impact of false information.

Secondly, the study emphasizes the role of technology in both the dissemination and detection of fake news. The recommendation of strategies to counter fake news, such as technological programs for detection and automated fact-checking, suggests the importance of leveraging technological advancements to combat the spread of misinformation. Further research and development in this area can contribute to the creation of more effective and efficient tools for identifying and flagging fake news on social media platforms.

Additionally, the study underscores the significance of understanding users' behaviors and motivations when it comes to sharing news on social media. The finding that some respondents were willing to share news even if unsure of its credibility highlights the need for interventions that encourage responsible sharing practices. Strategies that promote critical evaluation of news content and encourage users to verify information before sharing can help mitigate the inadvertent spread of fake news.

Lastly, the study's implications extend to policymakers and social media platforms themselves. The insights gained from this research can inform the development of policies and guidelines aimed at curbing the dissemination of fake news. Social media platforms can utilize these findings to enhance their algorithms and features to prioritize credible sources and flag potentially misleading content, ultimately creating a safer and more trustworthy information environment.

Overall, the implications of this study contribute to the broader efforts to combat fake news, promote media literacy, and foster a more informed and responsible use of social media platforms.

References

Agosto, D. E. (2018). Information Literacy and Libraries in the Age of Fake News. ABC-CLIO.

Auberry, Kendra (2018). Increasing students' ability to identify fake news through information literacy education and content management systems. The Reference Librarian, 59: 179–187.

Burkhardt, Joanna M. (2017) History of Fake News. Library Technology Reports 53: 5–9.

Bahige, N (2019). The perception of university students on the diffusion of fake news on social media: A case of the University of KwaZulu-Natal. MA Thesis. Kwazulu Natal: University of Kwazulu Natal

Borel, B. (2018). Last year there were 8,164 fake news stories. Popular Science, 290(2), 64–124.

Crate, L. (2017). Fake News vs. Real News. Education Digest, 83(1), 4.

De Paor, Saoirse & Heravi, Bahareh (2020). Information literacy and fake news: How the field librarianship can help combat the epidemic of fake news. The Journal of Academic Librarianship, 46: 102218.

Domonoske, C. (2016). Students have a "dismaying" inability to tell fake news from real, a study finds. National Public Radio, 23.

El Rayess, M., Chebl, C., Mhanna, J., & Hage, R.-M. (2018). Fake news judgement: The case of Undergraduate students at Notre Dame University-Louaize, Lebanon. Reference Services Review, 46: 146–149.

Eva, N & Shea, E (2018). Marketing Libraries in an Era of "Fake News". Reference & User Services Quarterly, 57(3), 168–171.

Fernandez, P. (2017). The technology behind fake news. Library Hi Tech News, 34: 1–5.

Foresti, F., Varvakis, G., & Viera, A. (2018). A importância do contexto na Ciência da Informação. *Biblios*, 72. doi: 10.5195/biblios.2018.383

Graves, D. (2018). Understanding the Promise and Limits of Automated Fact-Checking. Reuters Institute for the Study of Journalism Factsheets. Reuters Institute for the Study of Journalism.
McGrew, S., Ortega, T., Breakstone, J., & Wineburg, S. (2017). The Challenge That's Bigger than Fake News: Civic Reasoning in a Social Media Environment. American Educator, 41(3), 4–9.
Newman, Nic, Richard Fletcher, Antonis Kalogeropoulos, and Rasmus Nielsen. (2019). Reuters Institute Digital News Report 2019. Oxford: Reuters Institute for the Study of Journalism, vol. 2019.
Pinto, C., Carvalho, M., Martins, S., & Braga, I. (2020). How misinformation and disinformation affect the organization and its employees' performance in the digital era. https://www.researchgate.net/profile/Claudia-Pinto-11/publication/338868113_
Staffers, C., & Hackett, J. (2017). Fake News, Fake Data: How bad data and misleading graphs are fueling fake news. Scholastic Math, 38(2), 8.
Syam, H. M., & Nurrahmi, F (2020). "I Don't Know If It Is Fake or Real News" How Little Indonesian University Students Understand Social Media Literacy. Jurnal Komunikasi: Malaysian Journal of Communication, 36: Article 2.
Ummah, N. H., & Fajri, M. S. A. (2020). Communication strategies used in teaching media information literacy for combating hoaxes in Indonesia: A case study of Indonesian national movements. Informacijos Mokslai, 2020, 90, 26–41. doi:10.15388/Im.2020.90.48
Wineburg, S., & McGrew, S. (2016). Why Students Can't Google Their Way to the Truth. Education Week, 36(11), 22–28.
Zimmer, F, Katrin Scheibe, M, & Stock, W. (2019) Fake News in Social Media: Bad Algorithms or Biased Users? Journal of Information Science Theory and Practice 7: 40–53

Bolaji David Oladokun and Millie Nne Horsfall

Fake News and Misinformation on Social Media: The Roles of Librarians

Abstract: This research seeks to investigate the roles of librarians in addressing the challenges posed by fake news and misinformation on social media. As the prevalence of fake news and misinformation grows in the digital age, librarians' roles as information professionals become even more critical. The paper systematically analyzed the literature to determine the roles of librarians in combating fake news and misinformation in the social media era. By equipping individuals with the skills to critically assess information and distinguish between credible and unreliable sources, librarians can empower users to navigate the complexities of social media and make informed decisions. This research aims to contribute to the ongoing efforts to combat misinformation and foster a more informed and digitally literate society. This chapter futuristically suggests that librarians need to closely monitor and anticipate potential future trends in misinformation on social media, as trends such as the rise of deepfake technology, AI-generated misinformation, manipulation of social media algorithms, and disinformation campaigns during major events may become even more prominent in the future.

Keywords: Fake news, misinformation, libraries, social media, librarians

Introduction

In recent years, the proliferation of social media platforms has revolutionized the way people access and share information. While social media offers numerous benefits, it has also given rise to a concerning phenomenon: the spread of fake news and misinformation. The dissemination of false or misleading information on social media can have far-reaching consequences, including undermining public trust in reliable sources, shaping public opinion, and even influencing political and social outcomes.

Bolaji David Oladokun, Library and Information Technology, Federal University of Technology, Ikot Abasi, Akwa Ibom, Nigeria, e-mail: Bolaji.oladokun@yahoo.com, ORCID: 0000-0002-7826-9187
Millie Nne Horsfall, Donald Ekong Library, University of Port Harcourt, Rivers State, Nigeria, e-mail: millie.horsfall@uniport.edu.ng, ORCID:0000-0002-0115-0735

https://doi.org/10.1515/9783111176345-009

Amidst this digital information landscape, librarians have emerged as essential gatekeepers of credible and accurate information. Traditionally seen as custodians of knowledge and guardians of information integrity, librarians are uniquely positioned to play a crucial role in combating the spread of fake news and misinformation on social media. Social media platforms have become breeding grounds for the rapid dissemination of information, where unverified or intentionally false information can quickly reach millions of users (Inobemhe 2021). The rise of echo chambers and confirmation bias further exacerbates the problem, as individuals tend to be exposed to information that reinforces their existing beliefs, regardless of its accuracy. As purveyors of reliable and credible information, librarians possess a unique set of skills that can be leveraged to address this growing concern.

This research seeks to investigate the roles of librarians in addressing the challenges posed by fake news and misinformation on social media. As the prevalence of fake news and misinformation continues to grow in the digital age, librarians' roles as information professionals become even more critical. By equipping individuals with the skills to critically assess information and distinguish between credible and unreliable sources, librarians can empower users to navigate the complexities of social media and make informed decisions. This research aims to contribute to the ongoing efforts to combat misinformation and foster a more informed and digitally literate society.

Context of Fake News and Misinformation on Social Media

Fake news refers to intentionally fabricated or misleading information presented as genuine news with the purpose of deceiving and misleading the public. It is deliberately designed to misinform, create confusion, or advance specific agendas, often for political, social, or financial gains. Fake news can take various forms, such as misleading headlines, manipulated images or videos, and false claims presented as factual information.

Misinformation, on the other hand, according to Oladokun et al. (2024), refers to the dissemination of false or inaccurate information, often unintentionally, without the intention of misleading or deceiving the audience. Unlike fake news, which is deliberately created with malicious intent, misinformation may result from errors, misunderstandings, misinterpretations, or the lack of verification of facts before sharing information.

In the context of social media, both fake news and misinformation can spread rapidly due to the ease and speed of sharing information on these platforms. The viral nature of social media amplifies the reach of such content, making it essential for users to critically assess the credibility of information before sharing or believing it (Durodolu, Nduka, and Dube 2022). Addressing fake news and misinformation on social media is crucial for maintaining the integrity of information dissemination and promoting informed and responsible digital citizenship.

The Increasing Influence and Challenges Posed by Misinformation on Social Media

Social media platforms have become integral parts of modern life, reshaping how people communicate, share information, and access news. While social media offers numerous benefits, it has also become a breeding ground for the spread of misinformation. Misinformation on social media refers to false or inaccurate information that is shared and disseminated widely, often without proper fact-checking or verification (Oladokun et al. 2023). Several authors have outlined the increasing influence of misinformation on social media:

- Misinformation on social media is increasingly pervasive due to the ease and speed of sharing content. Social media algorithms prioritize engaging and viral content, leading to the rapid dissemination of false information. A study by Vziatysheva (2020) found that falsehoods are more likely to be retweeted than accurate information, indicating the prevalence of misinformation in online spaces.
- Misinformation erodes public trust in traditional media and information sources. As social media becomes a primary source of news for many users, the presence of misinformation compromises the credibility of news outlets. Majerczak and Strzelecki (2022) revealed that 59% of respondents believe that the media they consume is contaminated with untrustworthy information.
- Misinformation on social media can significantly influence public opinion and behavior. False or misleading information about public health, politics, or social issues can sway public sentiment and lead to harmful consequences. A study by Allcott and Gentzkow (2017) highlighted the influence of misinformation on political beliefs and voting behavior.

On the other hand, misinformation poses a significant threat to democratic processes, as it can influence elections and democratic decision-making. The dissemination of false information during electoral campaigns can manipulate public opinion, leading to distorted outcomes. Delwiche (20 illustrates the use of social media for spreading misinformation during elections in various countries. Also, social media platforms face challenges in moderating and curbing the spread of misinformation. The sheer volume of user-generated content and the evolving tactics of those spreading misinformation make it difficult to detect and remove false information effectively, which raises concerns about the platforms' responsibility in maintaining a healthy information ecosystem (Tandoc 2020). Furthermore, social media algorithms often reinforce users' preexisting beliefs, creating echo chambers and filter bubbles. This phenomenon limits users' exposure to diverse perspectives and increases susceptibility to misinformation that aligns with their existing views (Rhodes 2022).

In view of the foregoing, exploring the roles of librarians in addressing the issue of misinformation on social media has become important. Librarians, as information professionals, possess a unique skill set that includes expertise in information literacy, research, and critical thinking (Oyetola, Aderibibge, and Oladokun 2023). By investigating their involvement, strategies, and initiatives, we can gain valuable insights into how their knowledge and resources can be harnessed to promote accurate information, media literacy, and responsible information consumption.

Understanding Fake News and Misinformation on Social Media

The advent of social media has revolutionized information dissemination, enabling users to access news and content instantaneously. However, this rapid flow of information has also led to the spread of fake news and misinformation. Fake news refers to intentionally fabricated or misleading information presented as genuine news, while misinformation refers to false or inaccurate information shared without the intention of deceiving. Both have become significant challenges on social media platforms, impacting individuals, society, and democratic processes.

The prevalence of fake news and misinformation on social media is alarming. A study by Allcott and Gentzkow (2017) found that false information spreads faster and more widely than accurate information on platforms like Twitter. This rapid spread can have profound consequences. For instance, false information related to COVID-19 led to misinformation about treatments and prevention methods, impacting public health (Vziatysheva 2020).

The spread of fake news and misinformation can influence public opinion and behavior. False information about political candidates, policies, or events can sway public sentiment and even influence election outcomes (Wang 2020). The viral nature of social media amplifies the reach of such content, making it challenging to correct misconceptions once they have gained traction.

Furthermore, fake news and misinformation contribute to social and political polarization. Social media algorithms often create echo chambers, where users are exposed only to information that aligns with their preexisting beliefs (Kitchens, Johnson, and Gray 2020). This selective exposure intensifies divisions within society and inhibits open dialogue and understanding. However, efforts to combat fake news and misinformation face several challenges. The decentralized nature of social media allows for the proliferation of false information with limited accountability (Dhall et al. 2021). The sheer volume of user-generated content makes it difficult to detect and fact-check every piece of information, straining the resources of social media platforms and fact-checking organizations.

Difference Between Misinformation, Disinformation, and Fake News

Misinformation, disinformation, and fake news are terms used to describe different types of false or misleading information, each with distinct characteristics and intentions. Misinformation refers to false or inaccurate information that is shared without the intention of deceiving or misleading others (Oladokun et al. 2024). It can be spread inadvertently, often due to a lack of fact-checking or verification before sharing information. Misinformation can result from errors, misunderstandings, or misinterpretations and is not necessarily created to cause harm.

Disinformation, on the other hand, is deliberately false or misleading information that is intentionally spread to deceive, mislead, or manipulate others (Aïmeur, Amri, and Brassard 2023). Unlike misinformation, disinformation is created with a malicious intent to deceive and influence public opinion. It often serves specific political, social, or financial agendas.

Fake news is a subset of disinformation, referring specifically to false or misleading information presented as genuine news with the intention of deceiving readers or viewers (Allcott and Gentzkow 2017). Fake news is often sensationalized and designed to attract attention and engagement. It can be spread through various media channels, including social media platforms.

Additionally, Pelemo, Horsfall, and Osedo (2021) noted that misinformation falls under the category of Information Disorder (ID) which has three arms: misinformation, which is fake information shared not to cause harm; disinformation, which is fake information shared to cause harm; and malinformation, which is correct information deliberately shared to cause harm.

The Psychology Behind the Spread and Acceptance of Fake News on Social Media

The spread and acceptance of fake news on social media can be attributed to various psychological factors that influence individuals' information processing and decision-making. Understanding these psychological mechanisms is crucial in addressing the challenges posed by misinformation and developing effective interventions to promote media literacy and critical thinking. Some key psychological factors behind the spread and acceptance of fake news on social media include:

- Confirmation Bias: confirmation bias is a cognitive bias that leads individuals to seek, interpret, and remember information that confirms their preexisting beliefs and values (Zhou and Shen 2022). When encountering fake news that aligns with their existing beliefs, individuals are more likely to accept and share it, reinforcing their own perspectives and contributing to the spread of misinformation.
- Selective Exposure: selective exposure refers to individuals' tendency to seek out information that aligns with their beliefs while avoiding information that challenges their views (Mukerjee and Yang 2021). On social media, algorithms often present users with content that matches their preferences, creating echo chambers where users are exposed to a limited range of perspectives, including fake news that supports their existing beliefs.
- Emotional Appeals: fake news often employs emotional appeals to grab attention and elicit strong reactions from users (Oladokun et al. 2024). Content that evokes strong emotions, such as fear, anger, or excitement, is more likely to be shared widely on social media, even if its factual accuracy is questionable.
- Social Proof: Social proof is a psychological phenomenon where individuals look to others' actions or behaviors as a guide for their own actions (Naeem 2021). When fake news appears to have gained significant attention and traction on social media, others may be more likely to share it, assuming that the information is trustworthy due to its apparent popularity.
- Cognitive Laziness: the rapid flow of information on social media can lead to cognitive laziness, where users may not critically evaluate the accuracy of information before sharing it (Nyemezu and Nsirim 2022). This lack of effort in verifying information can contribute to the unintentional spread of misinformation.
- Overwhelming Information Environment: the overwhelming amount of information on social media can make it challenging for users to discern credible sources from unreliable ones (Guess, and Nyhan 2020). Users may feel inundated with information and resort to quick judgments, leading to the sharing of fake news without proper verification.

While it is important to state that the spread and acceptance of fake news on social media can be attributed to various psychological factors, some other key factors that contribute to the virality of misinformation are helpful in addressing the challenges posed by fake news and misinformation which include:

- Emotional Appeal: misinformation often employs emotional language and appeals to evoke strong emotions in users (Oladokun et al. 2024). Content that triggers emotional responses, such as fear, anger, or excitement, is more likely to be shared widely on social media, contributing to its virality.
- Novelty and Sensationalism: misinformation that presents shocking or sensational claims is more likely to capture users' attention and drive them to share the content (Vziatysheva 2020). Novel or surprising information tends to stand out in users' feeds, increasing the likelihood of sharing.
- Social Identity and Belongingness: misinformation that reinforces users' existing beliefs or aligns with their social identity is more likely to be shared within their social networks (Guess et al. 2020). Users may share misinformation to strengthen their affiliation with like-minded individuals or groups.
- Algorithmic Amplification: social media algorithms prioritize engaging content and often amplify content that generates high levels of user engagement (Bakshy, Messing, and Adamic 2015). Misinformation that garners early engagement may be further promoted by algorithms, leading to increased visibility and potential virality.
- Lack of Verification: the fast-paced nature of social media encourages users to share information quickly without thoroughly verifying its accuracy (Oladokun et al. 2023). Misinformation may spread rapidly due to users' willingness to share content without critically evaluating its credibility (Nyemezu and Nsirim 2022).
- Influential Social Actors: misinformation shared by influential social actors, such as celebrities, public figures, or prominent individuals, may gain more traction and virality due to their large following and influence (Vziatysheva 2020).

The Evolving Role of Librarians in the Digital Age

The role of librarians has evolved significantly in the digital age, adapting to the changing information landscape and emerging technologies. Librarians are no longer just custodians of physical collections; they have become vital facilitators of information access, digital literacy, and knowledge management.

In the digital age, librarians have embraced digital resources, including e-books, online journals, and databases, to provide users with easy access to a vast array of information (Dahl 2021). Librarians curate and manage digital collections, ensuring that users can access relevant and up-to-date information online. Also, as stated by Aharony et al. (2017), librarians play a crucial role in promoting digital literacy by offering workshops, training sessions, and resources that teach users how to navigate the digital world effectively. They help users develop skills in information searching, evaluating online sources, and using digital tools.

While emphasizing the role of librarians, Chiware (2020) states that librarians assist researchers in managing and preserving research data, ensuring compliance with data sharing and open access policies and as such they offer guidance on data management planning and provide support for data analysis and visualization. Following emerging trends in librarianship, librarians have become technology integrators, exploring and incorporating emerging technologies like artificial intelligence, virtual reality, and data analytics into library services (Ali, Naeem and Bhatti 2020).

Librarians in the digital age focus on fostering information literacy and critical thinking skills to help users navigate the abundance of information online (Vinuta 2020). In so doing, librarians teach users how to assess the credibility and reliability of online sources. Similarly, Sabharwal (2021) adds that librarians engage with the community through outreach initiatives, leveraging social media and digital platforms to connect with users beyond the physical library space.

The digital era has brought about significant changes in the roles of librarians, necessitating adaptation to new technologies and digital resources. Traditionally, their roles centered around managing physical collections, offering reference services, and promoting literacy within the community (Mehta and Wang 2020). However, with the rapid advancement of technology and the digital age, the roles of librarians have undergone significant transformation to adapt to the changing information landscape. This transition has seen librarians embracing digital resources, promoting digital literacy, and leveraging technology to enhance their services.

The Impact of Fake News and Misinformation on Society

The impact of fake news and misinformation on society is far-reaching and multifaceted. Misleading information can lead to public confusion, polarization, and distrust in institutions and media. It can also influence public opinions, attitudes, and behaviors, often with serious consequences. Given the foregoing, several authors have provided empirical evidence:

- The spread of fake news and misinformation erodes trust in traditional media and institutions. When individuals encounter false information that appears credible, they may question the reliability of news sources and become more susceptible to conspiracy theories and alternative narratives (Guess et al. 2020).
- Fake news and misinformation can contribute to societal polarization by disseminating biased or extreme views, reinforcing existing beliefs, and creating echo chambers where individuals are exposed only to information that aligns with their perspectives (Vziatysheva 2020).
- False information is often exploited in disinformation campaigns aimed at manipulating public opinion or influencing political outcomes (Aïmeur, Amri, and Brassard 2023). This can have significant implications for democratic processes and public decision-making.
- Misinformation can have severe consequences for public health and safety. False medical information, for example, can lead to the spread of harmful treatments or discourage individuals from seeking appropriate medical care (MacFarlane, Hurlstone, and Ecker 2020).
- The spread of misinformation can also have economic repercussions. False information about businesses or financial markets can lead to stock market fluctuations and affect investor confidence (Fong 2021).
- Exposure to misinformation can cause cognitive dissonance, anxiety, and feelings of confusion among individuals (Taddicken and Wolff 2020). This can lead to a sense of uncertainty and helplessness in distinguishing fact from fiction.

The impact of fake news and misinformation on society is significant, affecting trust in institutions, fostering polarization, and influencing public opinions and behaviors. As misinformation continues to spread through various media channels, understanding its implications is essential in developing strategies to combat its effects and promote media literacy, critical thinking, and responsible information consumption. Efforts to address fake news must involve collaboration between media platforms, educators, and society at large to foster a more informed and resilient public.

Case Studies of Misinformation Campaigns and Their Consequences

The following case studies illustrated the stories of misinformation and fake news on social media and their grave consequences:

Case Study 1: "Pizzagate" Misinformation Campaign

In 2016, a misinformation campaign known as "Pizzagate" emerged on social media platforms and conspiracy theory websites (Bleakley 2023). The campaign falsely claimed that high-ranking officials, including members of the Democratic Party, were involved in a child sex trafficking ring operating out of a Washington, D.C. pizzeria called Comet Ping Pong. The baseless allegations spread rapidly through social media, forums, and message boards, fueled by fake news articles and manipulated images. The consequences were:

- Violent Incident: the Pizzagate conspiracy theory had real-world consequences when an individual, influenced by the false information, entered Comet Ping Pong with a firearm, believing he was "self-investigating" the alleged child trafficking (Brown 2018). Thankfully, no one was injured, but the incident highlighted the dangers of misinformation.
- Harm to Reputation: the pizzeria and the people implicated in the conspiracy theory faced severe reputational damage. Innocent individuals were falsely accused and targeted with threats and harassment, causing distress to them and their families (Napolitano 2020).

Case Study 2: "2019–20 COVID-19 Misinformation"

During the COVID-19 pandemic, misinformation about the virus spread rapidly on various online platforms. False claims about treatments, origins, and prevention methods circulated widely, leading to confusion and misinformation among the public. The consequences were:

- Health Risks: misinformation about COVID-19 treatments and preventive measures led some individuals to use unproven remedies, leading to potential health risks and harm (Chavda et al. 2022).
- Resistance to Public Health Measures: misinformation about the severity of the pandemic and the effectiveness of public health measures contributed to resistance against mask-wearing, social distancing, and vaccination, hindering efforts to control the spread of the virus (Romer and Jamieson 2020).

Case Study 3: "Myanmar Genocide: Disinformation and Incitement to Violence"

In Myanmar, misinformation campaigns propagated through social media contributed to the persecution of the Rohingya Muslim minority. False narratives and incendiary content on Facebook fueled hate speech and violence against the Rohingya community, leading to a humanitarian crisis. The consequences were:

- Humanitarian Crisis: the spread of misinformation on social media contributed to widespread violence and forced displacement of the Rohingya population, leading to a humanitarian crisis with the loss of lives and displacement of hundreds of thousands of people (Kyaw 2021).
- Ethnic Tensions: the use of disinformation to fuel hatred and discrimination exacerbated existing ethnic tensions and deepened divisions within Myanmar (Banaji and Bhat 2022).

Based on these case studies, it is thus believed that false information can lead to real-world harm, violence, reputational damage, and public health risks. It is crucial to remain vigilant, critically evaluate information sources, and combat misinformation through media literacy, fact-checking, and responsible information sharing. Efforts from social media platforms, government, civil society, and individuals are necessary to address the far-reaching consequences of misinformation campaigns and promote accurate information dissemination.

Role of Social Media Platforms in Facilitating the Spread of Fake News

Social media platforms play a significant role in facilitating the spread of fake news due to their reach, speed, and algorithmic mechanisms. These platforms have become fertile ground for the rapid dissemination of information, including misinformation and fake news. Social media platforms have millions of active users, allowing fake news to reach a vast audience quickly (Vziatysheva 2020). Information, regardless of its veracity, can be easily shared, liked, and retweeted, increasing its visibility and potential to go viral. Another issue is that social media platforms use algorithms to tailor content to individual users' preferences. This personalized content delivery can inadvertently amplify misinformation, as users are more likely to be shown content that aligns with their existing beliefs and interests (Swart 2021).

Interestingly, social media platforms offer easy and seamless sharing functionalities, making it effortless for users to repost information without verifying its accuracy, which amplifies the spread of fake news and misinformation (Taddicken and Wolff 2020). Furthermore, social media algorithms can create echo chambers and filter bubbles, reinforcing users' existing beliefs and limiting exposure to diverse viewpoints (Bakshy, Messing, and Adamic 2015). This phenomenon can lead to the reinforcement of misinformation among like-minded individuals.

Libraries and Librarians as Gatekeepers of Credible Information

Libraries and librarians have long been recognized as gatekeepers of credible information, ensuring that users have access to reliable and authoritative sources. Their role in evaluating, selecting, and curating information helps users navigate the vast sea of information and find trustworthy resources. Libraries play a critical role in curating collections that include credible and reputable information sources. They carefully select materials from reputable publishers and academic sources to ensure that users have access to high-quality information (Wong and Chiu 2024).

Also, librarians actively engage in information literacy instruction, teaching users how to evaluate sources critically and determine their credibility. Through workshops and tutorials, librarians empower users to become discerning consumers of information (Panda and Kaur, 2024). Furthermore, librarians provide reference services, offering personalized assistance to users in finding credible and relevant information. They guide users to authoritative sources and help them verify the accuracy of information (Guo et al. 2021).

One of these gatekeeping functions is that libraries advocate for open access to scholarly research and promote the use of reputable scholarly publishing platforms and as such libraries support researchers in sharing their work openly and ensure access to peer-reviewed and reliable information (Boufarss, and Harviainen 2021).

Importantly, libraries play a key role in preserving digital information for future generations. By archiving and maintaining digital collections, librarians ensure that credible information remains accessible and preserved for posterity (Kiszi and Fodor 2021). De Paor and Heravi (2020) also stated that libraries organize information in a logical and structured manner, making it easier for users to find credible sources on specific topics. This led Horsfall (2023) to affirm that libraries throughout the world have the mandate to provide citizens with access to networked information to alleviate poverty and deprivation due to the paradoxical situation of scarcity in an era of information overflow.

It is believed that libraries and librarians will continue to serve as gatekeepers of credible information sources. Through their expertise in information evaluation, curation, and information literacy instruction, librarians play a vital role in promoting media literacy, critical thinking, and responsible information consumption among library users. As guardians of knowledge, libraries continue to be trusted sources of credible information in an increasingly digital and information-rich world.

Information Literacy Initiatives by Librarians

Librarians play a vital role in promoting information literacy among library users. Nzomo and Fehrmann (2020) state that librarians design and implement various information literacy initiatives to empower individuals with the skills to locate, evaluate, and use information effectively and responsibly. It is believed that information literacy skills facilitate the ability to detect fake news and misinformation. The authors further outline examples of information literacy initiatives by librarians. They include:

- Librarians conduct information literacy workshops to teach users how to navigate library resources, search databases, and critically evaluate information sources. These workshops often focus on developing skills for fact-checking, source verification, and understanding different types of information.
- Librarians collaborate with faculty to integrate information literacy instruction into specific courses. Through embedded librarianship, librarians work closely with students in assignments and research projects, guiding them in finding and evaluating relevant and credible sources.
- Librarians create online tutorials and guides to support users in developing information literacy skills. These resources cover topics such as information-searching strategies, source evaluation, and citation practices.

Supporting this, Aharony, Julien, and Nadel-Kritz (2020) observed that librarians offer one-on-one consultations to help users with specific research needs and information literacy challenges. These personalized sessions allow users to receive tailored guidance in navigating complex information landscapes. Also, librarians design and conduct information literacy assessment activities to measure users' proficiency in information literacy skills. These assessments help librarians identify areas for improvement and tailor future information literacy initiatives accordingly (Dunagan 2023).

Challenges Faced by Librarians in Tackling Misinformation

Librarians play a crucial role in tackling misinformation and promoting information literacy among their users. However, they encounter various challenges in their efforts to combat the spread of false information. The rapid dissemination of misinformation through digital channels makes it challenging for librarians to keep up with the sheer volume of false information (Panezi 2021). False narratives can spread quickly on social media platforms and other online spaces, leading to the potential for widespread misinformation.

Misinformation often thrives within disinformation ecosystems, where false information is supported by echo chambers, filter bubbles, and coordinated efforts to spread disinformation (Aïmeur, Amri, and Brassard 2023). Sadly, many libraries operate with limited budgets and may not have dedicated fact-checking resources. This may hinder their ability to verify the accuracy of information effectively (Wong and Chiu 2024).

Another challenge is that the algorithmic nature of social media platforms can contribute to the spread of misinformation. Algorithmic amplification can expose users to false information that aligns with their pre-existing beliefs, reinforcing echo chambers and filter bubbles (Crinnion, Yannopoulou, and Bhattacharya 2024). Adding to the foregoing, misinformation can be perpetuated due to users' cognitive biases, such as confirmation bias and the "backfire effect" (Morris 2021). Overcoming these biases and effectively correcting false beliefs can be challenging for librarians.

Summarily, librarians face significant challenges in tackling misinformation, including the rapid proliferation of false information, the complexity of disinformation ecosystems, limited resources for fact-checking, algorithmic amplification, addressing cognitive biases, and respecting intellectual freedom. Despite these challenges, librarians continue to play a vital role in promoting information literacy, critical thinking, and responsible information consumption to empower users with the skills needed to discern reliable information from misinformation.

Ethical Considerations in Addressing Fake News

Ethical considerations in addressing fake news are crucial for librarians and information professionals as they navigate the complex landscape of information dissemination. Balancing freedom of expression with the need to combat misinformation, ensuring user privacy and data protection in information literacy efforts, and addressing biases while promoting diverse perspectives in information curation are essential aspects of ethical decision-making.

- **Balancing Freedom of Expression and Combating Misinformation:** librarians uphold the principle of intellectual freedom, which safeguards individuals' right to access information without censorship. However, they also have a responsibility to combat misinformation that can harm society and undermine democratic processes. Striking a balance between these two principles involves promoting media literacy and critical thinking, empowering users to discern reliable sources and debunk misinformation without infringing upon their right to access diverse viewpoints (Crinnion, Yannopoulou, and Bhattacharya 2024).
- **Ensuring User Privacy and Data Protection:** as librarians engage with users through information literacy efforts, they collect and handle user data. Ethical considerations demand that librarians prioritize user privacy and data protection. Obtaining informed consent from users for data collection, using secure and encrypted platforms, and practicing data anonymization when possible are essential steps to safeguard user privacy and prevent unauthorized access or breaches.
- **Addressing Biases and Promoting Diverse Perspectives**: librarians act as information gatekeepers, curators, and educators. Ethical information curation requires them to be mindful of biases and ensure the presentation of a wide range of perspectives (Panezi 2021). By actively seeking and promoting diverse voices, librarians counteract the perpetuation of misinformation through echo chambers and support a more inclusive information ecosystem.

Conclusion

The research on librarians' role in the fight against fake news and misinformation highlights their critical importance in promoting a more informed society. Librarians serve as information experts, educators, and advocates for media literacy, playing a pivotal role in equipping users with the skills to discern reliable sources from false information. Through sustainable and scalable information literacy strategies, librarians can empower individuals to become critical thinkers and responsible consumers of information in the digital age.

This chapter upholds that librarians will need to continuously update their skills and knowledge to navigate the rapidly changing information landscape. Embracing emerging technologies, such as AI-powered fact-checking tools and mobile information literacy apps, can enhance librarians' effectiveness in combatting misinformation. Additionally, librarians should prioritize community engagement and outreach, extending their reach to diverse populations and addressing the unique information needs of different communities. By forming alliances with media literacy advocates, educators, and policymakers, librarians can amplify their impact and effect meaningful change.

In the face of misinformation challenges, librarians remain at the forefront of promoting critical thinking, media literacy, and responsible information consumption. Their dedication to the principles of intellectual freedom, transparency, and user empowerment positions them as key drivers in creating a more informed, resilient, and trustworthy information ecosystem. Through their continued efforts, librarians will continue to be invaluable partners in the fight against fake news and misinformation, paving the way for a better-informed society.

Future Directions

- Librarians need to closely monitor and anticipate potential future trends in misinformation on social media. Trends such as the rise of deepfake technology, AI-generated misinformation, manipulation of social media algorithms, and disinformation campaigns during major events may emerge. By staying informed about these trends, librarians can tailor their information literacy efforts to address evolving challenges.
- Librarians should develop sustainable information literacy strategies that can adapt to changing technologies and user behaviors. Online resources, virtual workshops, and mobile apps are scalable approaches that can reach a wider audience. Collaborating with educational institutions, fact-checking organizations, and community partners strengthens the impact of information literacy initiatives.
- Librarians must advocate for policy changes that promote media literacy education in schools, open access to information, and transparent algorithms on social media platforms. Collaborating with social media platforms to develop user-friendly fact-checking tools and flagging systems can help curb the spread of misinformation. By engaging in collaborative efforts, librarians can influence the design and implementation of initiatives to combat misinformation.

References

Aharony, N., Julien, H., & Nadel-Kritz, N. (2020). Survey of information literacy instructional practices in academic libraries. *Journal of Librarianship and Information Science*, *52*(4), 964–971.

Aharony, N., Badke, W., Bradley, C., Bury, S., Coonan, E., Dorner, D. G., . . . & Wang, L. (2017). *Global perspectives on information literacy: Fostering a dialogue for international understanding.* Association of College and Research Libraries.

Ali, M. Y., Naeem, S. B., & Bhatti, R. (2020). Artificial intelligence tools and perspectives of university librarians: An overview. *Business Information Review*, *37*(3), 116–124.

Allcott, H., & Gentzkow, M. (2017). Social media and fake news in the 2016 election. *Journal of Economic Perspectives*, *31*(2), 211–236.

Bakshy, E., Messing, S., & Adamic, L. A. (2015). Exposure to ideologically diverse news and opinion on Facebook. Science, 348(6239), 1130–1132.

Banaji, S., & Bhat, R. (2022). *Social media and hate* (p. 140). Taylor & Francis.

Bleakley, P. (2023). Panic, pizza and mainstreaming the alt-right: A social media analysis of Pizzagate and the rise of the QAnon conspiracy. *Current sociology*, *71*(3), 509–525.

Boufarss, M., & Harviainen, J. T. (2021). Librarians as gate-openers in open access publishing: A case study in the United Arab Emirates. *The Journal of Academic Librarianship*, *47*(5), 102425.

Brown, J. (January, 2018). Is social media bad for you? The evidence and the unknowns. BBC News. Retrieved from https://www.bbc.com/future/article/20180104-is-social-media-bad-for-you-the-evidence-and-the-unknowns

Chavda, V. P., Sonak, S. S., Munshi, N. K., & Dhamade, P. N. (2022). Pseudoscience and fraudulent products for COVID-19 management. *Environmental Science and Pollution Research*, *29*(42), 62887–62912.

Chiware, E. R. (2020). Open research data in African academic and research libraries: a literature analysis. *Library Management*, *41*(6/7), 383–399.

Crinnion, F., Yannopoulou, N., & Bhattacharya, S. (2024). Fake news inside ideological social media echo chambers. In *Handbook of Social Media in Education Consumer Behavior and Politics* (pp. 139–187). Academic Press.

Dahl, M. (2021). The evolving role of library collections in the broader information ecosystem. In *Future Directions in Digital Information* (pp. 161–174). Chandos Publishing.

Delwiche, A. (2020). Computational Propaganda and the Rise of the Fake Audience. *P. Baines, N. O'Shaughnessy, & N. Snow, The SAGE Handbook of Propaganda*, 105–125.

De Paor, S., & Heravi, B. (2020). Information literacy and fake news: How the field of librarianship can help combat the epidemic of fake news. *The Journal of Academic Librarianship*, *46*(5), 102218.

Dhall, S., Dwivedi, A. D., Pal, S. K., & Srivastava, G. (2021). Blockchain-based framework for reducing fake or vicious news spread on social media/messaging platforms. *Transactions on Asian and Low-Resource Language Information Processing*, *21*(1), 1–33.

Dunagan, J. A. (2023). *Information literacy at the crossroads: Fake news influence and student behavior* (Doctoral dissertation, University of Illinois at Urbana-Champaign).

Durodolu, O. O., Nduka, S. C., & Dube, T. V. (2022). Media Literacy and Fake News: Evaluating the Roles and Responsibilities of Radio Stations in Combating Fake News in the COVID-19 Era. *International Journal of Information Management Sciences*, *6*(1), 33–49.

Fong, B. (2021). Analysing the behavioural finance impact of'fake news' phenomena on financial markets: a representative agent model and empirical validation. *Financial Innovation*, *7*(1), 53.

Guo, Y., Yang, Z., Yang, Z., Liu, Y. Q., Bielefield, A., & Tharp, G. (2021). The provision of patron services in Chinese academic libraries responding to the COVID-19 pandemic. *Library Hi Tech*, *39*(2), 533–548.

Guess, A. M., & Lyons, B. A. (2020). Misinformation, disinformation, and online propaganda. *Social media and democracy: The state of the field, prospects for reform*, *10*.

Guess, A. M., Nyhan, B., O'Keeffe, Z., & Reifler, J. (2020). The sources and correlates of exposure to vaccine-related (mis) information online. *Vaccine*, *38*(49), 7799–7805.

Horsfall, M. N. (2023). Librarians as change agents for sustainable library and information service delivery in a digital society. *Information Impact: Journal of Information and Knowledge Management. 14* (1), 113–125.

Inobemhe, K. (2021). Social Media Fake News, Myths and Conspiracies on COVID-19: Threat to Public Health in Nigeria. *Media & Communication Currents*, *5*(1), 38–55.

Kitchens, B., Johnson, S. L., & Gray, P. (2020). Understanding echo chambers and filter bubbles: The impact of social media on diversification and partisan shifts in news consumption. *MIS quarterly*, *44*(4).

Kiszl, P., & Fodor, J. (2021). Remaining futureproof: lasting librarian roles in managing digital collections. *The Reference Librarian*, *62*(3–4), 165–192.

Kyaw, N. N. (2021). Social media, hate speech and fake news during Myanmar's political transition. *From Grassroots Activism to Disinformation: Social Media in Southeast Asia. Edited by A. Sinpeng and R. Tapsell. Singapore*: *Yusof Ishak Institute (ISEAS)*, 86–104.

MacFarlane, D., Hurlstone, M. J., & Ecker, U. K. (2020). Protecting consumers from fraudulent health claims: A taxonomy of psychological drivers, interventions, barriers, and treatments. *Social Science & Medicine*, *259*, 112790.

Majerczak, P., & Strzelecki, A. (2022). Trust, media credibility, social ties, and the intention to share towards information verification in an age of fake news. *Behavioral Sciences*, *12*(2), 51.

Mehta, D., & Wang, X. (2020). COVID-19 and digital library services–a case study of a university library. *Digital library perspectives*, *36*(4), 351–363.

Morris, S. (2021). The Museum of Misinformation: An Interactive Exhibit Illuminating the Effects of Social Media Algorithms on Cognitive Bias and the Spread of Misinformation. *Masters Theses*. 822. https://digitalcommons.liberty.edu/masters/822

Mukerjee, S., & Yang, T. (2021). Choosing to avoid? A conjoint experimental study to understand selective exposure and avoidance on social media. *Political Communication*, *38*(3), 222–240.

Naeem, M. (2021). The role of social media to generate social proof as engaged society for stockpiling behaviour of customers during Covid-19 pandemic. *Qualitative Market Research*: *An International Journal*, *24*(3), 281–301.

Napolitano, M. G. (2022). *Conspiracy theories and resistance to evidence* (Doctoral dissertation, UC Irvine).

Nyemezu, C.O., & Nsirim, O. (2022). Managing big data in the era of misinformation: The role university library. *International Journal of Advanced Research and Learning. 1*(1).

Nzomo, P., & Fehrmann, P. (2020). Advocacy engagement: The role of information literacy skills. *Journal of Information Literacy*, *14*(1), 41–65.

Oladokun, B. D., Aruwa, J. E., Ottah, G. A., & Ajani, Y. A. (2024). Misinformation and Disinformation in the Era of Social Media: The Need for Fact-Checking Skills. *Journal of Information and Knowledge*, 1–7.

Oladokun, B. D., Nsirim, O., Otebe, F., Agih, A., Mbazu, E. C., & Mohammed, J. D. (2023). Influence of Social Media on Political News Credibility in Kogi State, Nigeria. *Covenant Journal of Communication*, *9*(2). Retrieved from https://journals.covenantuniversity.edu.ng/index.php/cjoc/article/view/3211

Oyetola, S. O., Aderibibge, N. A., & Oladokun, B. D. (2023). Implications of social media technologies (SMTs) to library services. *Library Hi Tech News*. DOI: https://doi.org/10.1108/LHTN-03-2023-0037

Panda, S., & Kaur, N. (2024). Empowered Minds: Navigating Digital Seas With Emerging Information Literacy Framework. In *Examining Information Literacy in Academic Libraries* (pp. 48–82). IGI Global.

Panezi, A. (2021). A Public Service Role for Digital Libraries: A Case of Emergency Electronic Access to Library Material and the Unequal Battle Against Misinformation Through Copyright Law Reform. *Cornell JL & Pub. Pol'y*, *31*, 65.

Pelemo, G. D., Horsfall, M. N., & Osedo, O. A. (2021). Roles of Librarians in combating misinformation on Corona Virus Disease (COVID-19). *Library Philosophy and Practice (e-journal).* 5539. https://digitalcommons.unl.edu/libphilprac/5539

Rhodes, S. C. (2022). Filter bubbles, echo chambers, and fake news: How social media conditions individuals to be less critical of political misinformation. *Political Communication*, *39*(1), 1–22.

Romer, D., & Jamieson, K. H. (2020). Conspiracy theories as barriers to controlling the spread of COVID-19 in the US. *Social science & medicine*, *263*, 113356.

Sabharwal, A. (2021). Functional frameworks for socialized digital curation: Curatorial interventions and curation spaces in archives and libraries. *Library Trends*, *69*(3), 672–695.

Swart, J. (2021). Experiencing algorithms: How young people understand, feel about, and engage with algorithmic news selection on social media. *Social media+ society*, *7*(2), 20563051211008828.

Taddicken, M., & Wolff, L. (2020). 'Fake News' in science communication: emotions and strategies of coping with dissonance online. *Media and Communication*, *8*(1), 206–217.

Tandoc Jr, E. C. (2020). Tools of disinformation: How fake news gets to deceive. In *Disinformation and fake news* (pp. 35–46). Singapore: Springer Singapore.

Vinuta, J. S. (2020). Revitalizing Librarianship In The Age Of Information Abundance: A Critical Examination Of The Profession's Evolution In The Digital Era. *Webology (ISSN: 1735–188X)*, *17*(1).

Vziatysheva, V. (2020). How fake news spreads online?. *International Journal of Media & Information Literacy*, *5*(2).

Wang, T. L. (2020). Does Fake News Matter to Election Outcomes?: The Case Study of Taiwan's 2018 Local Elections. *Asian Journal for Public Opinion Research*, *8*(2), 67–104.

Wong, A. K. K., & Chiu, D. K. (2024). Digital curation practices on web and social media archiving in libraries and archives. *Journal of Librarianship and Information Science*, 09610006241252661.

Zhou, Y., & Shen, L. (2022). Confirmation bias and the persistence of misinformation on climate change. *Communication Research*, *49*(4), 500–523.

Helen O. Jaiyeola Akinade and Adeyemi Adewale Akinola

The Role of Information Technology in Disseminating and Combating Misinformation

Abstract: Misinformation is erroneous, incomplete, or deceptive information disseminated into society through various communication channels and information technology. This chapter examined the role information technology plays in combating dissemination of misinformation. The chapter carried out a thorough review on the growing prevalence of misinformation and why "misinformation" is a hard problem to solve. It also examined strengthening the fight against misinformation, technological challenges in addressing misinformation, solutions to addressing the challenge of misinformation, the impact of social media platforms on misinformation spread, the role of search engines in disseminating and combating misinformation, leveraging artificial intelligence to detect and combat misinformation, and challenges and ethical considerations in combating misinformation online. The study concluded that it is necessary to build increasingly complex but effective models and approaches in order to stop misinformation. Measures such as media literacy, accountability and transparency, promoting responsibility in the use of social media, machine learning, and transparency reports will prevent misinformation.

Keywords: ICT, misinformation, combating, dissemination, Fake News, Social Media

Introduction

The information technology (IT) industry has made significant strides in the last two decades. These advancements have become more noticeable due to the emergence of web-based platforms and social media applications, which have completely changed how we receive, interpret, and respond to information from different sour-

Helen O. Jaiyeola Akinade, University Library, Mountain Top University, Ogun State, Nigeria, e-mail: hojakinade@mtu.edu.ng, hellenakinade88@gmail.com
Adeyemi Adewale Akinola, University Library, Mountain Top University, Ogun State, Nigeria, e-mail: akinolaadeyemi@yahoo.com, aaakinola@mtu.edu.ng

https://doi.org/10.1515/9783111176345-010

ces. The proliferation of misleading information on the internet has witnessed a significant increase, and the rapid dissemination of such information through social media networks has exacerbated this issue at an alarming rate. False information can now travel more quickly than ever before, with far-reaching effects on many facets of society (Aderonmu, Oke, and Olufade 2021).

Misinformation has always been present throughout history. In the past, governments and other agencies frequently used propaganda to manipulate public opinion and maintain power. Falsehoods and rumors were disseminated by pamphlets and word of mouth during the Middle Ages (Olufemi 2020). Later, the development of mass media and new media platforms like radio and television in the twentieth century made it possible for false information to spread more widely. More recently, the internet and social media platforms have made it simpler than ever for false information to spread quickly and extensively in the digital era (Durodolu, Nduka and Tinyiko, 2022). According to Roozenbeek and Linden (2018), once misinformation has gained its stronghold, it becomes more difficult to rectify because it can be manufactured and spread quickly, frequently by anonymous sources.

Further attesting to this fact, Tucker et al. (2018) opined that misinformation can be disseminated for a variety of reasons, including self-interest, harming someone else's reputation, dividing or confusing a community, or undermining public confidence in a particular institution or branch of government. Misinformation primarily focuses on the intent of the individual or entity disseminating it, often disregarding its original source or purpose (Asemah and Oluwatosin 2019). The fast spread of misinformation in the information age has threatened institutions. People can now find, absorb, and exchange knowledge more easily than ever before. However, these same elements have also played a significant role in the broad distribution of false information, misinformation, and fake news, which has had a negative impact on public opinion, the legitimacy of reliable sources, and confidence in institutions (Cosentino 2020)

The term "misinformation" according to Wu et al. (2019) refers to any incorrect or inaccurate information (fake news, urban legends, rumor, and misinformation that is purposely or accidentally communicated). False or misleading claims, gossip, conspiracy theories, photoshopped or edited movies, and biased or selective reporting are just a few examples of the various ways that misinformation can materialize. Using shaky sources, misinterpreting data, or extracting facts out of context are some possible consequences. Though the term "fake news" is frequently used informally in everyday speech to refer to all of the aforementioned features, there are definitions and theoretical frameworks that identify many subtleties of the misinformation dilemma, with misinformation frequently distinguished in many different ways (Guess and Lyons 2020). When people propagate inaccurate or misleading information, especially on social media, they are typically acting ignorantly. It is in view of this that the study seeks to investigate and establish the role of information and communication technology in disseminating and combating misinformation.

Understanding the Pervasive Growth of Misinformation

Misinformation is erroneous, incomplete, or deceptive information that circulates through a variety of media, including but not restricted to written works, news sources, social media, and informal word-of-mouth. Misinformation has the amazing power to affect and impact people's perceptions, beliefs, and behaviors as well as reinforce their own confirmation biases. It differs from disinformation in that it does not intentionally deceive (Faix and Fyn 2020).

Misinformation has existed for centuries. The proliferation of mass communication technologies, including print, radio, television, and the internet, has made misinformation much harder to stop and much more pervasive (Ogbuehi 2019). Especially in recent times, social media platforms have played a major role in the spread of misinformation by enabling the quick and widespread distribution of such inaccurate and misleading content.

The spread of false information can have extremely negative effects, such as escalating social and political unrest, confusing the public, encouraging dangerous behaviour, and eroding confidence in democratic institutions. Extreme examples of misinformation's ability to physically damage individuals and subvert democratic processes include situations where people have refused life-saving medical care due to misleading health information and conspiracy theories surrounding vaccines (Hotez 2020).

On the other hand, content generation and quick information exchange between sources has been made possible by information technology (Fasanya and Ogundare 2018). Newly developed information technologies expand human potential and boost ability, particularly in terms of message sharing and communication. During online conversations and exchanges, individuals are compelled to convey information to other members of their network that they believe is important and pertinent.

Figure 1 differentiates between respondents who share fake news in Germany and the United Kingdom, through intentional (columns 1 and 2) and unintentional sharing (columns 3 and 4). The data does not reveal a distinct trend for intentional sharing, except for a statistically significant inclination among younger individuals in the United Kingdom. It is important to emphasize that the pro-

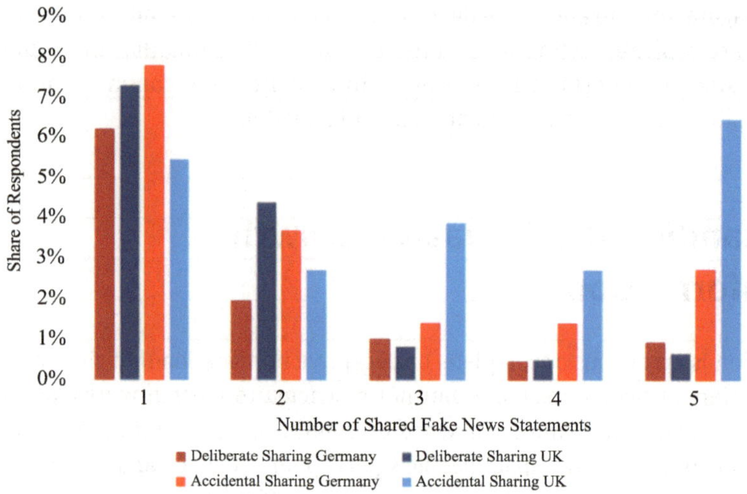

Notes. The diagram shows the share of respondents who wanted to share 1...5 fake news statements either deliberately or accidentally.

Figure 1: The diagram below shows the difference between respondents who shares fake news deliberately or accidentally in Germany and United Kingdom.
Source: Arin, Mazrekaj & Thum (2023). Ability of detecting and willingness to share fake news. F.1; P. 6

portion of individuals who can identify fake news but still choose to share it is relatively low. On the other hand, there is a higher incidence of unintentional sharing. Across both countries, it is consistent that right-leaning participants share more frequently, and accidental sharing decreases with age. In the UK, women are less likely to share unintentionally compared to men, while centrist participants show a higher inclination for accidental dissemination of fake news compared to the baseline group of left-leaning respondents.

Why is "Misinformation" a Hard Problem to Solve?

Misinformation is a very challenging problem to handle since it is hard to recognize, monitor, and regulate content that is not trustworthy. Advanced technological tools and standards for content assessment are necessary for a suitable solution. Additionally, because this will take a long time and many resources, it is very difficult at present to raise awareness of the issue among social media users (Shu et al. 2020).

Managing potentially false narratives can also be challenging; even in cases when there is the preliminary indication that misleading information is being shared online, little can be done until the matter is proven false. If not, actions such as taking down an article or stopping others from spreading it because of early warning signs can be interpreted as an attempt at censorship and intervention (Pinnaparaju, Indurthi, and Varma 2020). Conversely, fake content is intended to go viral, with its producers wanting it to spread quickly and naturally. Misinformed news is intentionally designed to grab readers' attention and elicit strong feelings, making them want to share the article with other individuals in their social networks who share their views. A fake tale can become viral in a matter of hours with the correct strategies and timing. More precisely, the following gaps in our online world are exploited by the false news industry.

The state of the evidence, expert opinions, and the information environment in which misinformation occurs all play a role in whether or not an item of information can be classified as misinformation, according to Vraga and Bode (2020), who also discuss the difficulties in doing so. As a result, different contexts may assign different meanings to the same piece of information.

Information that is purposefully inaccurate or misleading and presented as factual or newsworthy is referred to as misinformation or fake news. With the growth of social networking sites and internet news sources in particular, this issue has become more widespread in recent years. Misinformation and fake news can have detrimental effects on society and individuals, among other things, with the capacity for misinformation to proliferate swiftly and extensively on social media platforms among the most alarming features. It might take minutes for false information to spread millions of times, making it challenging for reliable sources to stop its spread (Igwebuike, Ebuka, and Chimuanya 2021). This may result in a situation where people accept lies as truth even in the face of contradicting information.

Misinformation can also be employed for ideological or political ends. During political campaigns, for example, some politicians and organizations purposefully disseminate misleading information to advance their objectives. This may result in a situation where the general public is uninformed about crucial topics and where democracy and societal order are threatened (Mian and Khan 2020). Misinformation and fake news can have a serious negative influence on public health and safety. For instance, people may make decisions that endanger their health as a result of receiving false information on health and medical matters. False information concerning public safety has the ability to exacerbate a harmful situation by causing panic and uncertainty.

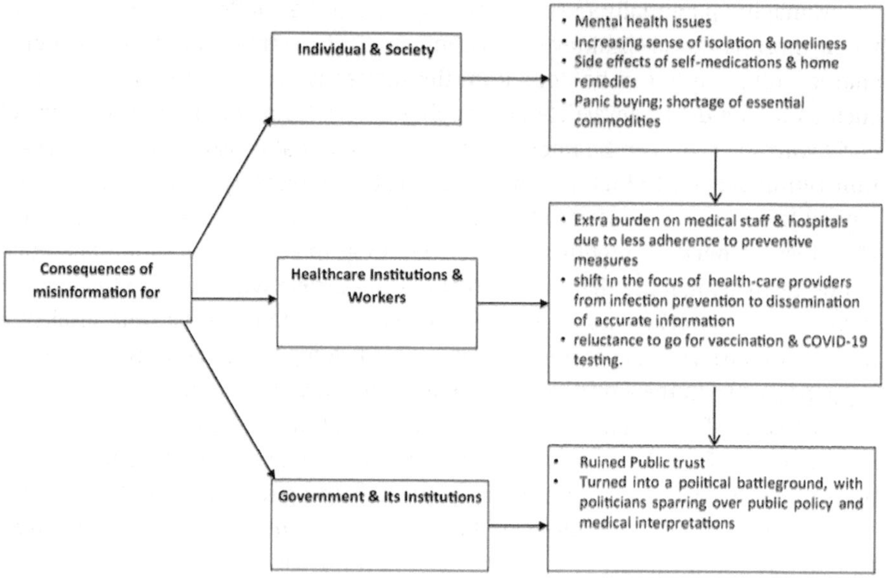

Figure 2: Consequences of misinformation.
Source: Malik, A., Bashir, F., & Mahmood, K. (2023). Antecedents and Consequences of Misinformation Sharing Behavior among Adults on Social Media during COVID-19. F.7; P. 9.

Information Technology and Misinformation

Information and communication technology is a global village where information can spread quickly amongst users. It makes it simple for anyone to participate and voice their ideas on a range of topics. It is renowned for its few restrictions on the exchange of information, affordability, and ease of use. However, because users can post their own, often incomplete, content, social media platforms could be considered the birthplace of hate speech, malicious intent, and fake news (Humprecht, Esser, and Van Aelst 2020). The deliberate dissemination of incorrect information to damage a person's reputation is misinformation.

People hardly ever consult traditional media to verify facts in the modern era because ICT platforms have emerged as important information providers. It is true that ICT is the source of some information that appears in traditional media. The impulse to use ICT to look up news is nearly compulsive. By posting on ICT and publishing fictitious news articles, people are willing to fuel this addiction, primarily in an effort to increase web traffic (Chou, Gaysynsky, and Cappella 2020). Many people using ICT believe things without giving them any thought.

Also, as ICT users are unlikely to verify the information they receive, it has become quite convenient for those who take advantage of this deficiency to distribute misleading information on platforms. People have made money for themselves by taking advantage of this heavy traffic volume. The audience continues to spread the word about the supposed facts on various social media platforms since they are so persuasive, but real news does not always spread as quickly as misleading information (Langin 2018). Misleading information employs a wide range of strategies and instruments to draw viewers in, such as attention-grabbing headlines and captions to promote their phony news.

False or misleading tales can be quickly written and shared across the world's web networks with just a few clicks. Misinformation finds its way into the endless, tailored feeds that individuals see every day. This issue is exacerbated by the fact that a sizable portion of social media users distribute content automatically and readily without exercising the necessary critical thought.

However, even with critical thinking skills, it is not always simple to identify a false narrative. It can be very challenging to determine whether or not what you are seeing is real. Up-to-date technology makes it possible to hack real videos or create fake ones that show people acting very realistically in situations they have never been in. Furthermore, it is possible to assert phrases or statements that were never made by using speech synthesis to seem like a well-known person said these things.

Analysis of Information Technological platforms in the Dissemination of Misinformation

The diagram represents a hierarchical structure or flowchart, showing relationships between different elements or steps in a process. Without color or additional details, the black-and-white structure suggests a decision tree, organizational chart, or a structured process flow.

To provide a more precise explanation, please describe the specific labels or context of the diagram, such as whether it pertains to management, research methodology, technology, or another field.

The internet emerged as a significant news source, but it is not the sole source of news, as indicated in the accompanying Figure, with TV the primary source for over half of the respondents. The internet served as the top source, while social media was the primary source for 14% of respondents.

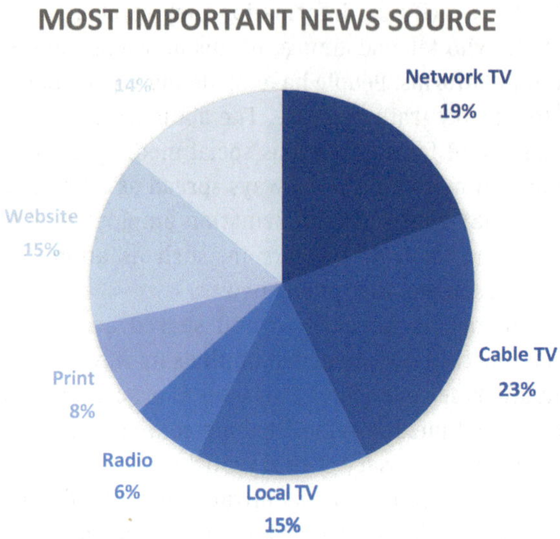

Figure 3: Analysis of Information Technological platforms in the dissemination of misinformation. Source: Allcott & Gentzkow (2017). Social Media and Fake News in the 2016 Election. F. 4; P. 224.

The Impact of Social Media Platforms on Misinformation Spread

Misinformation spread on social media platforms has become a pressing issue in today's digitally connected world. The rapid dissemination of false information through platforms such as Facebook, Twitter, and Instagram has the potential to influence public opinion, undermine trust in institutions, and even incite violence. Understanding the impact of social media on misinformation spread is thus crucial for developing strategies to mitigate its harmful effects. Several studies have examined the role of social media platforms in the spread of misinformation. Vosoughi, Roy, and Aral (2018) analyzed the spread of true and false news stories on Twitter and found that false stories spread significantly faster, further, and broader than true stories. The researchers attributed this phenomenon to the novelty and emotional appeal of false information, which makes it more likely to be shared by users.

Researchers have also identified various factors that contribute to the spread of misinformation on social media platforms. For instance, Pennycook and Rand (2019) found that cognitive factors such as cognitive laziness and confirmation bias make individuals more susceptible to believing and sharing false information. Moreover, the algorithms used by social media platforms to prioritize content based on user engagement can amplify the dissemination of misinformation.

The spread of misinformation on social media can have significant implications for public opinion and behavior. Attesting to this view, Allcott and Gentzkow (2017) examined the impact of fake news stories on the 2016 U.S. presidential election and found that exposure to misinformation on social media may have influenced voter behavior. Similarly, false information about health-related topics on social media can lead to misconceptions and harmful behavior among the public.

Various strategies have been proposed to combat the spread of misinformation on social media platforms. For example, fact-checking initiatives and the promotion of media literacy skills can help users discern between true and false information. Additionally, social media platforms can implement algorithms that prioritize accurate and reliable sources of information to reduce the visibility of misinformation. However, the impact of social media platforms on misinformation spread is a complex and multifaceted issue that requires a comprehensive approach. By understanding the factors contributing to misinformation spread and its impact on public opinion and behavior and implementing effective strategies to combat it, we can work towards creating a more informed and trustworthy digital landscape.

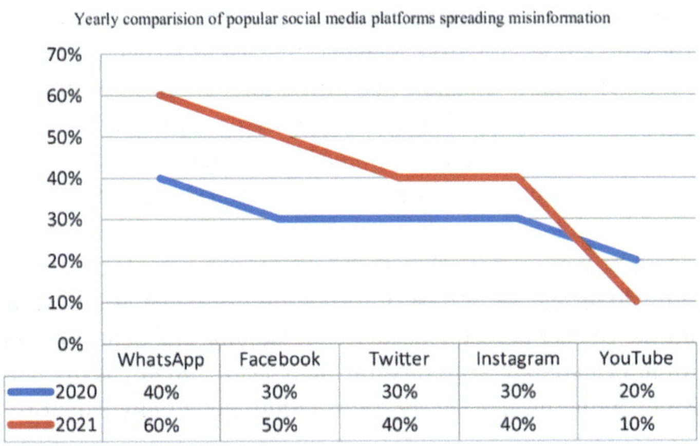

Figure 4: Annual comparison of popular social media platforms spreading misinformation.
Source: Malik, Bashir & Mahmood (2023). Antecedents and Consequences of Misinformation Sharing Behavior among Adults on Social Media during COVID-19. F. 3; P. 5.

The study indicates that WhatsApp and Facebook are the social media applications that spread misinformation the most in 2020 and 2021. Twitter and Instagram are relatively low but they do also contribute to the spread of misinformation and fake news.

Role of Search Engines in Disseminating and Combating Misinformation

The role of search engines in disseminating and combating misinformation is a critical area of study in the realm of information dissemination and credibility. Search engines play a central role in shaping the information landscape by directing users to sources of information based on search queries. However, search engines also face challenges in distinguishing between accurate and inaccurate information, thereby contributing to the spread of misinformation. Research has highlighted the susceptibility of search engines to the spread of misinformation. Bright et al. (2022) examined the presence of misinformation in search engine results and found that search engines sometimes prioritize sensational or inaccurate information over reliable sources. This phenomenon can be attributed to the algorithms used by search engines, which may prioritize content based on popularity or engagement metrics rather than accuracy.

Several factors contribute to the dissemination of misinformation in search engine results. A study by Liu et al. (2019) revealed that factors such as search engine optimization (SEO) tactics, clickbait headlines, and the proliferation of fake websites and content farms can influence the prevalence of misinformation in search results. Additionally, the personalization of search engine results based on user data can create filter bubbles that reinforce users' existing beliefs and expose them to misinformation.

The spread of misinformation through search engines can have far-reaching consequences for individuals and society. Misinformation presented at the top of search results can influence user perceptions, decision-making, and behavior. For example, false health information in search results may lead individuals to make health-related decisions based on inaccurate information, potentially jeopardizing their well-being.

Efforts to combat misinformation in search engine results have focused on implementing algorithmic changes, enhancing fact-checking mechanisms, and promoting media literacy among users. For example, search engines can prioritize authoritative sources of information, highlight fact-checked content, and provide users with transparency about the credibility of search results. Collaborations between search engines, fact-checkers, and researchers have also been proposed to address the challenge of misinformation. The role of search engines in disseminating and combating misinformation is a complex and evolving area of research. By understanding the factors influencing misinformation in search results and the impact of misinformation from search engines, implementing strategies to mitigate its spread, stakeholders can work towards creating a more reliable and trustworthy information ecosystem online.

Leveraging Artificial Intelligence to Detect and Combat Misinformation

The proliferation of misinformation and fake news in the digital age has raised significant concerns about the credibility and reliability of online information. In response to this challenge, researchers and practitioners have turned to artificial intelligence (AI) as a powerful tool to detect and combat misinformation. Artificial intelligence technologies, such as natural language processing (NLP) and machine learning, have been widely employed to detect misinformation in online content. Shu et al. (2017) demonstrated the effectiveness of using NLP algorithms to analyze the linguistic patterns and features of fake news articles, enabling the automatic identification of deceptive content. Similarly, Wang, Yin, and Zeng (2019) utilized machine learning models to differentiate between true and false information based on textual and contextual cues.

AI-powered fact-checking systems have emerged as promising tools to combat misinformation by verifying the accuracy of online claims and debunking false information. Guo, Schlichtkrull, and Vlachos (2022) developed a fact-checking framework that leverages machine learning techniques to analyze the credibility of claims and source reliability. By cross-referencing information with trusted sources, AI fact-checking systems can provide users with accurate and up-to-date information.

Recent advancements in deep learning approaches, such as neural networks and deep neural language models, have enabled more sophisticated methods for detecting misinformation. Asghar et al. (2021) explored the application of deep learning architectures for identifying fake news and rumor propagation on social media platforms. By analyzing large-scale datasets and leveraging deep learning algorithms, researchers have achieved higher accuracy in detecting and classifying misinformation. While AI presents significant potential in combating misinformation, researchers have also highlighted challenges and ethical considerations associated with using AI in this context. Issues such as bias in AI algorithms, explainability of decision-making processes, and the potential for AI systems to inadvertently contribute to censorship have been raised in the literature. Ad-

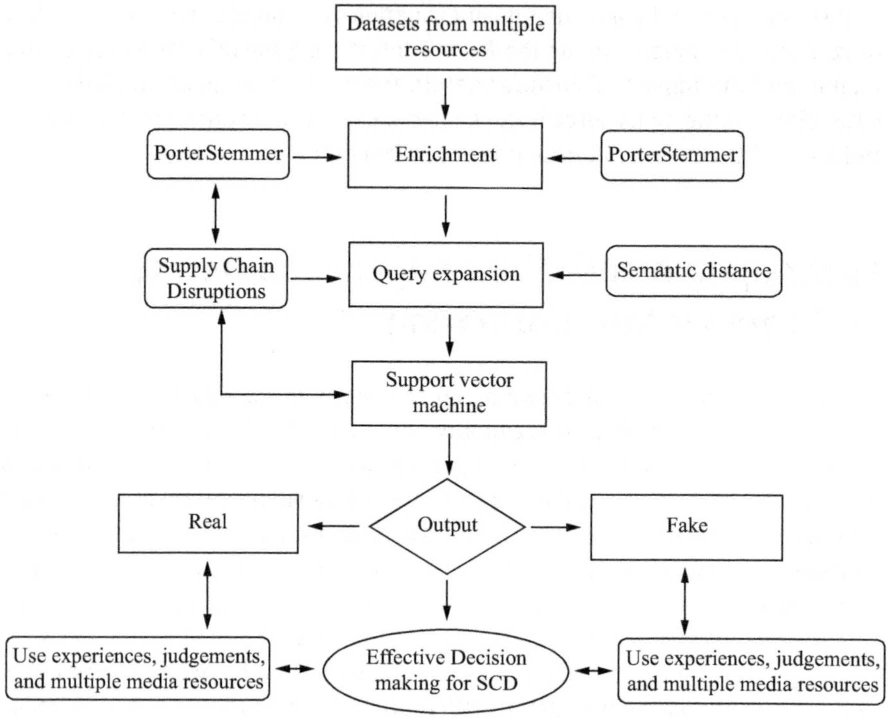

Figure 5: A fake news and misinformation detection model that uses AI and ML.
Source: Akhtar, Ghouri, Khan, Amin ul Haq, Awan, Zahoor, & Ashraf, A. (2023). Detecting fake news and disinformation using artificial intelligence and machine learning to avoid supply chain disruptions. F. 1; P. 648.

dressing these challenges is crucial to ensure the responsible and effective use of AI in detecting and combating misinformation.

The integration of artificial intelligence technologies holds great promise in the fight against misinformation by enabling scalable and efficient detection methods. By leveraging AI for fact-checking, linguistic analysis, and deep learning approaches, researchers are advancing the field of misinformation detection and mitigation. Moving forward, continued research and collaboration between AI experts, data scientists, and information professionals will be essential to further enhance the effectiveness of AI-driven solutions in combating misinformation in the digital age.

Strengthening the Role ICT Fight Against Misinformation

In the current digital era, fighting misinformation is essential, and information and ICT may greatly support these efforts. Here are strategies ICT can use to combat misinformation:

Implementing Measures for Fact Checking

A number of fact-checking programs have been launched by various organizations in response to misinformation. Technological applications have been implemented by organizations in an effort to counteract the spread of misinformation.

Nurturing Critical Thinking and Digital Literacy

Enacting policies that place high priority on encouraging critical thinking and improving the public's digital literacy is essential. The ability to locate, assess, and effectively communicate information via digital media is referred to as digital literacy, which is essential in the fight against disinformation because it helps people evaluate information critically, separate reliable sources from fraudulent ones, and spot misinformation. People who acquire these abilities are better equipped to make educated decisions and defend themselves against the influence of false information.

Putting in Place User Reporting Systems

ICT platforms ought to put in place reliable reporting mechanisms that let users report information that they feel is inaccurate or deceptive. Clear criteria for content control and prompt action against verified disinformation should go hand in hand with these tools.

Investing in AI-Powered Content Moderation

ICT firms can purchase content moderation systems that use artificial intelligence to detect and eliminate false information instantly. These technologies can assist human moderators by swiftly identifying and resolving objectionable information on a large scale.

Encouraging Cross-Platform Collaboration

It might be difficult to successfully counteract misinformation because it frequently circulates across several platforms. ICT businesses ought to collaborate to exchange knowledge on best practices, statistics, and technological solutions for identifying and removing false content from various platforms.

Stakeholder Perspective Strategy

An essential instrument in the ongoing fight against misinformation is a multi-stakeholder strategy. Stakeholders here include people, teachers, media outlets, technology platforms, and governments. In order to counteract misinformation and foster an informed public, each stakeholder has a crucial role to play.

People: the people concerned need to sharpen their critical thinking abilities, critically examine information sources, and properly distribute information within their networks of contacts.

Teachers: teachers must incorporate digital literacy into their lesson plans.

Media outlets: media outlets must uphold strict journalistic standards by fact-checking and reporting in an open and transparent manner.

Technology: technology and social media platforms ought to promote openness, incorporate fact-checking resources, and take action to prevent misleading material from being widely visible. Similarly, they may use machine learning (ML) and artificial intelligence (AI) techniques to create methods that can dispel false information and advance knowledge in society.

Government: governments have an obligation to create laws and regulations that assist independent fact-checking organizations and warn the public about the dangers of spreading false information in order to promote accountability, ethics, and transparency in the use of technology.

Challenges and Ethical Considerations in Combating Misinformation Online

The widespread dissemination of misinformation and fake news on online platforms poses a significant threat to the integrity of information and public discourse. As efforts to combat misinformation intensify, researchers and practitioners are increasingly recognizing the complex challenges and ethical considerations inherent in developing strategies to address this issue. One of the primary challenges in combating misinformation lies in the complexity of the online information ecosystem. Pennycook, Cannon, and Rand (2018) emphasized the interconnected nature of misinformation, highlighting that false information often spreads rapidly through social networks and online communities. This interconnectedness makes it challenging to track and debunk misinformation effectively, as it can quickly adapt and evolve in response to detection efforts.

The use of algorithms and recommendation systems in online platforms can contribute to the spread of misinformation by amplifying biases and creating filter bubbles. Studies by Amazeen, Vargo, and Hopp (2019) have shown how algorithmic personalization can reinforce users' pre-existing beliefs and preferences, leading to the polarization of information consumption. This algorithmic bias can inadvertently promote the dissemination of false or misleading information to a targeted audience.

The absence of robust regulatory frameworks and accountability mechanisms for online content poses a significant challenge in combating misinformation. Diakopoulos and Koliska (2017) highlighted the limited oversight and transparency in content moderation practices on social media platforms, where misinformation can proliferate unchecked. The lack of clear guidelines and enforcement mechanisms creates a fertile ground for the dissemination of false information without consequences.

Ethical dilemmas arise in the context of content moderation and misinformation detection, particularly concerning issues of censorship, free speech, and the balance between combating misinformation and preserving platform openness. A study by Milano, Taddeo, and Floridi (2020) underscored the ethical complexities of content moderation practices, emphasizing the need to uphold principles of transparency, accountability, and respect for diverse viewpoints while combating misinformation.

Addressing the challenges and ethical considerations in combating misinformation online requires a multi-faceted approach that encompasses technological solutions, regulatory interventions, and ethical frameworks. By acknowledging the complexity of the misinformation ecosystem, addressing algorithmic biases, promoting regulatory oversight, and upholding ethical principles in content moderation, researchers and practitioners can work towards building more resilient and ethical strategies to combat misinformation. Continued research and collaboration across disciplines will be essential to navigate these challenges and promote a more informed and trustworthy online information environment.

Technological Challenges in Combating Misinformation

There are substantial technological obstacles in combating misinformation because of the dynamic nature of online platforms and the speed at which information is created and distributed. A comprehensive strategy including technological, governmental, and educational measures is needed to combat the growing ubiquity of false information (Adesanya, Sotubo, and Ighalo 2018). Though there are still difficulties in striking a balance between accuracy and freedom of speech, technological solutions like fact-checking software and algorithmic modifications can assist in spotting and flagging false material. Regulations that target the spread of misleading information may be implemented, as well as norms of responsibility and openness for internet platforms (Aderonmu, Oke, and Olufade 2021). Moreover, encouraging media literacy and critical thinking abilities might enable people to recognize reliable sources and assess information more skillfully (Asemah and Oluwatosin 2019).

Quick Spread: more than one petabyte of material is created every day on the internet and social media platforms, marking an unprecedented increase in content production. As a result, the quick spread of false information as well as true information presents a serious problem that requires researchers' attention and efforts.

Algorithmic Constraints: even though IT applications and online platforms use algorithms to recognize and flag misleading content, it can still be difficult to properly identify erroneous information. Language and cultural hurdles, massive amounts of continuously created data, and a lack of contextual knowledge are some of the reasons behind the limits of current algorithms.

Intricate Architecture on the Internet: the internet is becoming more and more complicated. As a result, the potential for spreading misinformation internationally is growing at a rate never seen before. However, the swift dissemination of rules and norms across many jurisdictions presents noteworthy obstacles. The issue for internet platforms is further complicated by the fact that different cultural and societal contexts have varying perspectives on what constitutes being misleading. These platforms have to negotiate a variety of viewpoints and misinformation-related sensitivities.

Constantly Adapting Tactics: the constantly changing strategies and procedures used to disseminate misinformation present another difficulty. To be effective in the face of misinformation, online platforms and social media networks need to continuously improve their detection and prevention systems. This means that continuous research and development expenditure are required to enhance the capabilities of algorithms.

Solutions to Addressing the Challenge of Misinformation

Encourage media literacy and critical thinking abilities: one of the most important ways to counteract misinformation is to teach people how to assess the reliability of sources, verify the accuracy of information, and recognize biases. This involves providing users with fact checking tools and limiting the dissemination of incorrect information by lowering the visibility of unverified or untrustworthy sources. Technology companies can create algorithms that can identify and flag misleading content.

Boost accountability and transparency: organizations and media outlets should be held responsible for the veracity and accuracy of their reporting. They ought to follow moral guidelines and make an effort to offer objective, accurate information.

Promote responsible use of social media: by giving users the means to report fake news, limiting the dissemination of incorrect information, and encouraging fact checking, social media companies may encourage responsible use of their services.

Improved Content Labeling: this gives content a clearer label that identifies its source and level of reliability. This can include labels like "opinion piece," "satire," or "verified by fact-checkers," which can assist users in determining the credibility of the content they come across.

Crowd sourced Verification: promote initiatives that allow users to submit false information and take part in fact-checking procedures. This can make it easier for platforms to recognize and remove false information.

Machine Learning for Detection: to automatically identify and highlight possibly erroneous or misleading material, machine learning techniques could be used for detection. These algorithms are more effective at identifying false information by examining patterns in user behavior and content.

Transparency reports: publish periodic transparency reports outlining the platform's efforts to counter false information, including the quantity of content that has been detected, the steps taken to address it, and the enhancements made to its policies and algorithms.

Regulatory measures: support and abide by regulations designed to stop the dissemination of false information. Examples of these regulations include rules mandating political advertising on the internet to be transparent and sanctions for platforms that don't adequately stop the spread of false information.

Conclusion

Misinformation is anticipated to spread more widely in our society as many people perceive it as a money-making scheme used to gain an advantage over citizens. One method or model cannot identify all forms of misinformation because it comes in diverse forms and with different motivations. If only one viewpoint is considered, it will be ineffective to detect misinformation. When identifying false information, several viewpoints must be taken into consideration. This study reveals that identifying false information will continue to be a difficult task and a complicated problem in the near future. Therefore, it is necessary to build increasingly complex models and approaches in order to stop misinformation. It is important to never undervalue the potential harm that false information might inflict.

References

Adesanya, Y. O., Sotubo, A. O., & Ighalo, J. (2018). Misinformation and Public Health: A Study of Nigerian Social Media Users. *Global Media Journal*, *16*(30), 1–18.

Aderonmu, F. A., Oke, J. O., & Olufade, F. O. (2021). Understanding the Dynamics and Implications of Social Media Misinformation during the COVID-19 Pandemic in Nigeria. *International Journal of Digital Literacy and Digital Competence (IJDLDC)*, *12*(1), 34–52.

Allcott, H., & Gentzkow, M. (2017). Social media and fake news in the 2016 election. *Journal of economic perspectives*, *31*(2), 211–236.

Amazeen, M. A., Vargo, C. J., & Hopp, T. (2019). Reinforcing attitudes in a gatewatching news era: Individual-level antecedents to sharing fact-checks on social media. *Communication Monographs*, *86*(1), 112–132.

Arin, K. P., Mazrekaj, D., & Thum, M. (2023). Ability of detecting and willingness to share fake news. *Scientific Reports*, *13*(1), 7298.

Asemah, E. S., & Oluwatosin, T. (2019). Fake News and Disinformation in the Nigerian Social Media: A Study of Facebook Users in Edo State. *Global Journal of Arts, Humanities and Social Sciences*, *7*(3), 34–52.

Asghar, M. Z., Habib, A., Habib, A., Khan, A., Ali, R., & Khattak, A. (2021). Exploring deep neural networks for rumor detection. *Journal of Ambient Intelligence and Humanized Computing*, *12*, 4315–4333.

Bright, L. F., Sussman, K. L., Hendricks, A., Archer, L., & Wilcox, G. B. (2022). A deeper look at the 2020 Facebook boycott and related themes of misinformation: A text mining analysis of topics, emotion and sentiment. *Journal of Brand Strategy*, *11*(1), 65–79.

Chou, W. Y. S., Gaysynsky, A., & Cappella, J. N. (2020). Where we go from here: Health misinformation on social media. *American Journal of Public Health*, *110*(53), S273–S275. https://doi.org/10.2105/AJPH.2020.305905

Cosentino, G. (2020). *Social media and the post-truth world order*. London; Cham: Palgrave Pivot.

Diakopoulos, N., & Koliska, M. (2017). Algorithmic transparency in the news media. *Digital journalism*, *5*(7), 809–828.

Durodolu, O. O., Nduka, S. C. and Tinyiko, T.V (2022). Media Literacy and Fake News: Evaluating the Roles and Responsibilities of Radio Stations in Combating Fake News in the COVID-19 Era. International Journal of Information Sciences Management (IJIMS). Vol. 6 (1), pp. 32 –49

Guo, Z., Schlichtkrull, M., & Vlachos, A. (2022). A survey on automated fact- checking. *Transactions of the Association for Computational Linguistics*, *10*, 178–206.

Fasanya, I. O., & Ogundare, A. E. (2018). Trivariate analysis of oil revenue, government spending and economic growth in Nigeria. *OPEC Energy Review*, *42*(2), 107–122.

Faix, A., & Fyn, A. (2020). Framing fake news: Misinformation and the ACRL framework. *portal: Libraries and the Academy*, *20*(3), 495–508.

Guess, A. M., & Lyons, B. A. (2020). Misinformation, disinformation, and online propaganda. *Social media and democracy: The state of the field, prospects for reform*, *10*.

Hotez, P. J. (2020). Combating antiscience: Are we preparing for the 2020s?

PLOS Biology, 18(3), e3000683. https://doi.org/10.1371/journal.pbio.3000683

Humprecht, E., Esser, F., & Van Aelst, P. (2020). Resilience to online disinformation: A framework for cross-national comparative research. *The International Journal of Press/ Politics*, *25*(3), 493–516.

Malik, A., Bashir, F., & Mahmood, K. (2023). Antecedents and consequences of misinformation sharing behavior among adults on social media during COVID- 19. *SAGE open*, *13*(1), 21582440221147022.

Mian, A., & Khan, S. (2020). Coronavirus: The spread of misinformation. *BMC Medicine, 18*(1). https://doi.org/10.1186/s12916-020-01556-3

Milano, S., Taddeo, M., & Floridi, L. (2020). Recommender systems and their ethical challenges. *Ai & Society, 35*, 957–967.

Langin, K. (2018). Fake news spread faster than true news on Twitter- thanks to people not bots", [Online], www.sciencemag.org/news/2018/03/fake-news-spreads-faster-true-news-twitter-thanks-people-not-bots

Liu, Y., Yu, K., Wu, X., Qing, L., & Peng, Y. (2019). Analysis and detection of health-related misinformation on Chinese social media. *IEEE Access, 7*, 154480–154489.

Ogbuehi, I. H. (2019). Media literacy, social media and the challenge of misinformation in Nigeria. *Journal of Media Literacy Education, 11*(1), 18–31.

Olufemi, O. (2020). Disinformation in Nigeria: The Danger of Fake News. *Journal of Communication Research, 12*(1), 38–53.

Pennycook, G., Cannon, T. D., & Rand, D. G. (2018). Prior exposure increases perceived accuracy of fake news. *Journal of experimental psychology: general, 147*(12), 1865.

Pennycook, G., & Rand, D. G. (2019). Lazy, not biased: Susceptibility to partisan fake news is better explained by lack of reasoning than by motivated reasoning. *Cognition, 188*, 39–50.

Pinnaparaju, N., Indurthi, V., & Varma, V. (2020). Identifying Fake News Spreaders in Social Media. In CLEF.

Roozenbeek, J. and Linden, S. (2018) The Fake news game: Actively Inoculating against the risk of misinformation. *Journal of Risk Research, 22* (5) 1–28.

Shu, K., Bhattacharjee, A., Alatawi, F., Nazer, T. H., Ding, K., Karami, M., & Liu, H. (2020). Combating disinformation in a social media age. Wiley Interdisciplinary Reviews: Data Mining and Knowledge Discovery, 10(6), e1385.

Shu, K., Sliva, A., Wang, S., Tang, J., & Liu, H. (2017). Fake news detection on social media: A data mining perspective. *ACM SIGKDD explorations newsletter, 19*(1), 22–36.

Tucker, J. A., Guess, A., Barberá, P., Vaccari, C., Siegel, A., Sanovich, S., & Nyhan, B. (2018). Social media, political polarization, and political disinformation: A review of the scientific literature. *Political polarization, and political disinformation: a review of the scientific literature (March 19, 2018).*

Vosoughi, S., Roy, D., & Aral, S. (2018). The spread of true and false news online. *science, 359*(6380), 1146–1151.

Vraga, E. K., & Bode, L. (2020). Defining misinformation and understanding its bounded nature: Using expertise and evidence for describing misinformation. *Political Communication, 37*(1), 136–144.

Wang, Y., Yin, W., & Zeng, J. (2019). Global convergence of ADMM in nonconvex nonsmooth optimization. *Journal of Scientific Computing, 78*, 29–63.

Wu, L., Morstatter, F., Carley, K., & Liu, H. (2019). Misinformation in social media: Definition, manipulation and detection. *ACM SIKDD Explorations Newsletter, 21*(2), 80–90

Mathew Adesanmi Farukuoye and Kolawole Akinjide Aramide

Combating Fake News in the Information Age: The Roles of Library and Information Professionals in Diverse Societies

Abstract: The present age has been rightly called an information age because information has become the most important ingredient of national growth and development. This age is also marked by a rapid and dynamic evolution of access to and use of digital devices to manipulate and use information, which has empowered professionals and non-professionals alike to engage in the business of information sharing and dissemination which has oftentimes resulted in the spread of fake news. Using current empirical and non-empirical literature, this chapter examined ways of combating fake news in this era of information explosion with an emphasis on the roles of library and information professionals in ensuring social stability in multi-cultural and diverse societies. Data for this chapter was collected through a literature review and it was done using various online tools with relevant information including LibHub and Google Scholar. The findings show that fake news is information disorder that undermines the credibility of objective thought and decision, as well as that fake news is fueling electoral violence, ethno-religious conflicts, jungle justice and leadership mistrust, among other things. librarians and libraries thus have to play roles as media information literacy teachers and fact-checkers, and support governments in making policies to regulate the information environment without undermining civil liberties and freedom of expression. The chapter recommends among other things, an interdisciplinary approach that will involve information professionals and all relevant stakeholders in addressing the menace of fake news.

Keywords: Fake News, Information Age, Library, Librarians, Information Professionals

---

*Corresponding author: **Mathew Adesanmi Farukuoye**, College Library, College of Health Sciences and Technology Ijero-Ekiti, Ekiti State, Nigeria, e-mail: farukson2000@gmail.com
Kolawole Akinjide Aramide, Department of School Media and Technology, University of Ibadan, Oyo State, e-mail: kolaakinjide@gmail.com

https://doi.org/10.1515/9783111176345-011

1 Introduction

Fake news which is misinformation and disinformation has been a disturbing phenomenon in recent times. Its negative impacts cut across every fiber of communal life, from politics to religion, from business to social life. Perhaps the most pronounced area of concern over fake news is national security, which unarguably is, the greatest challenge of nations worldwide at present. This state of acknowledged insecurity has continued to impact negatively on every sector of national life especially the economy. Several factors such as the absence of institutional capacity, disconnection between the people and government, the failure of security agencies in term of sharing information, and ethno-religious/tribal conflicts have been identified as responsible for the state of insecurity in some nations of the world. However, one seemingly silent factor that has emerged as the greatest threat to security is the phenomenon of fake news.

Fake news is a fabricated media content deliberately created and circulated with the intent to deceive. It is a medium through which social vices like hate speech are circulated to fuel ethno-religious crises, political tension, tribal sentiment, character assassination and defamation. The primary aim of the promoter of fake news is to instigate acts capable of causing disunity among people. It was hate speech and fake news that contributed largely to the genocide that ravaged Rwanda in 1994. The damaging implication of fake news to national security is unquantifiable, however, more than ever before, the media for spreading fake news is created at will and disseminated with great ease. This situation portends great dangers for societies.

Given the recent ugly developments resulting from fake news across many countries of the world, it is perhaps little wonder that fake new has become a major preoccupation for international organizations, policy-makers, the media and media actors, civil society, the academia and libraries in particular as the established custodians of authentic, reliable and verifiable information. However, considering the dangers posed by fake news to social stability, this chapter seeks to identify how libraries as social institutions whose main commodity is information, and library and information professionals can help combat fake news in this era of information explosion thereby alleviating its negative effects on social stability and peaceful co-existence in a multicultural and heterogeneous society like Nigeria that is increasingly becoming more diverse and complex as the population increases.

The Concept of the Information Age

Information is considered the oil that powers the society, a veritable tool and resource for result-oriented planning, decision-making and implementation among individuals, organizations, governments, and global society (Jimoh, 2011). Information is the commodity for development, a processed data to which meaning may be attached which is useful for planning, decision making and reduction of uncertainty (Oluwaseyi, Akanni and Busuyi, 2017).

The historical development of the advent of writing and information resources in general can be divided into three major distinct periods. The first period is the printing revolution in the fifteenth century, which led to changes in society's management of information and language itself. The second period is the emergence of complete writing systems in Mesopotamia in the fourth millennium BC, which then spread to Egypt, the Indus Valley, and further east (Angela 2020). The third period is the modern period which is the development of information technology in the later twentieth century, which has allowed for easy and rapid access to enormous amounts of information. There has also been an explosion of scientific and technical progress, leading to an unprecedented level of information volumes, possession and dissemination (Cumerma 2020). These three periods mark significant milestones in the spread and availability of information throughout history. However, the modern era's problem is information explosion, which includes the proliferation of information resources with attendant misinformation and disinformation, which have come to be known as fake news in journalism and mass communication.

The present era, characterized by the advent and existence of digital technologies, the internet and social media, which are used to manipulate and use information, is referred to as the information age or information explosion era. As a result of the vast sources and amount of information available at humankind's disposal, this has led to bigger challenges for information professionals and information users in terms of identifying, not just the right and accurate information, but the reliable sources of genuine information. Complexity and change are the two defining characteristics of the Information Age. Our successes as individuals, families, organizations, communities and societies will depend more than ever upon our abilities to, in near real time, deal with increasingly complex and dynamic situations which will be characteristic of the Information Age.

The Information Age, otherwise known to many as the Information Era, the Computer Age, or the Digital Revolution is a phenomenon that did not occur overnight. It is a concept directly linked to computing, electronics, engineering, science, and communications reformation from the analog system to the digital system, characterizing the current age by the ability of individuals to transfer information freely and to have instant access to information previously unavailable or difficult to locate (Borkovich 2011).

The Information Age is based on the widespread proliferation of emerging information and communication technologies and the capabilities that those technologies provide and will provide to humankind to overcome the barriers imposed on communications by time, distance, and location as well as the limits and constraints inherent in human capacities to process information and make decisions.

The overall concepts of digital technologies used to process, store, and transport information, has provided vast benefits to many, such as the ability to control, access, share and manage data, information, and knowledge transferring. It is equally important to note that the emergence of Web 2.0 made its appearance in the form of social media, blogs, wikis, images, photos, rich site summaries (RSS), aggregate feeds, and other social-cultural behavioral transmissions that quickly augmented and often surpassed email as the primary form of electronic communication. Providing the public with an interactive platform of participation, such as commercial online ordering, banking, shopping, bill-paying, and game-playing, in addition to alerting a select group of followers or the public in general of an individual's every move, became addictive and self-perpetuating (Borkovich 2011).

Concept of Fake News and its Characteristics

The phrase "fake news" is made up of two simple words "fake", meaning something not genuine, but meant to be taken as genuine and "news", meaning information about current events. Therefore, one can say that fake news is an information or news that is not genuine but which is presented as such. However, this definition may seem too simplistic and narrow, so we explore available literature for the opinions of other scholars from various backgrounds on the concept of fake news. According to McGongale (2017), fake news is information that has been deliberately fabricated and disseminated with the intention to deceive and mislead others into believing falsehoods or doubting verifiable facts. In this regard, it is disinformation that is presented as, or likely to be perceived as, news. Alawode, Olorede and Azeez (2018) view fake news as news articles that are intentionally and verifiably false and could mislead readers. The authors explain that fake news includes false information that can be verified as such, and is created with dishonest intention to mislead readers. Michailidis and Viotty (2017) describe fake news as hoax-based stories that perpetuate "hear-say" rumors, and misinformation. Another simple but well-accurate definition about the term of fake news has been provided by Leonhardt and Thompson (2017) in their article in *The New York Times* where they describe fake news as a type of propaganda or yellow and misleading journalism that consists of deliberate disinformation or hoaxes spread via traditional media outlets (their print or online editions) or online via social media. Tandoc et al (2018) elaborated more on this by providing an overview of the various types of fake news, which are news satire, news parody, fabrications manipulation, advertising and propaganda. In summary, fake news can be described as simply stories that contain misleading and fake facts which aim at spreading propaganda, influencing the audience's political views, producing a funny content or making a profit.

So far, the core deducible elements of fake news are the "falsity" of news content and the intent to deceive or mislead. Therefore, a false content that is created in error and circulated without the intention of misleading the consumers may be dismissible since no human system is immune to mistakes. With this in mind, Wardle and Derakhshan (2018) identify certain properties of fake news;

– False Context: when a genuine content is often being re-circulated out of its original context. For example, an image from Vietnam, captured in 2007, re-circulated seven years later, was shared under the guise that it was a photograph from Nepal in the aftermath of an earthquake there in 2015 (Pharm, 2018).
– False Connection: when headlines, visuals or captions do not support the content
– Misleading Content: misleading use of information to frame an issue or an individual in a certain way
– Imposter Content: when genuine sources are impersonated with false and made-up sources
– Manipulated Content: when genuine information or imagery is manipulated to deceive, as with a doctored photo.
– Fabricated Content: when news content is 100% false, designed to deceive and do harm.

From the foregoing, it is clear that fake news takes different forms. In an attempt to define fake news, Belloir et al. (2022) state the notions of "real" and "false" facts are at the heart of the fake news concept and should thus be considered. Also of note is the emotional and cognitive process which is exploited through the specific way the attacker influences the readers of the fake news through its implicit conclusion; according to Belloir et al. (2022), a fake news is false but verifiable news composed of false facts based on real ones. Drafted in a way to trigger an emotional load, it aims to deceive its readers and influence their opinion through an implicit conclusion.

On the other hand, most authors believe that fake news does not have to be 100% false or manufactured. Fake news can be totally or partially false (Tandoc et al., 2018; Gelfert 2018; Rini, 2017; Allcott and Gentzkow 2017), and follows the media agenda and tries to describe real events while also distorting and manipulating the truth (Rini, 2017). Even so, the concept of fake news has been widely contested in relation to the "deliberate intention to deceive" and some authors do not consider the intention of the creator of fake news to be fundamental (Pepp et al., 2019; Walters, 2018). Furthermore, Pepp et al. (2019) argue that fake news simply does not match editorial criteria and journalistic practices, regardless of the producer's intention. Fake news is related to the wide dissemination of content that seems to have been treated with journalistic rigor but that does not obey such practices.

In view of this problem imposed by the difficulty of defining fake news, it is our belief that we should have a clear-cut definition of the terminology that allows us to respond to the new challenges of journalism and communication. Baptista and Gradim (2020) consider fake news to be a type of online disinformation, with totally or partially false content, created intentionally to deceive and/or manipulate a specific audience, through a format that imitates a news or report (acquiring credibility), through false information that may or may not be associated with real events, with an opportunistic structure (title, image, content) to attract the readers' attention and to persuade them to believe in the falsehood in order to obtain more clicks and shares, therefore, leading to higher advertising revenue and/or ideological gain.

Library and Information Professionals Roles in the Information Age

It is no longer news that the world has moved from the industrial age into the information age. Humankind's quest for knowledge has led to the creation and accumulation of a tremendous amount of information. The world today boasts of a knowledge-based society with the quest for knowledge knowing no bounds and limits and is never satisfied. According to Issa (2003), there has come to be in today's world, a full realization of the fact that information remains the prime commodity of the present age with this continuing from the dawn of civilization to the modern age. Indeed, the availability and free flow of information bring about knowledge which has great potentials to provide impetus for the social, cultural, spiritual, political, economic, scientific and technological advancement of a nation. This hard-earned knowledge and information is valuable for all of humankind hence, the need for the preservation of knowledge and dissemination of information which led to the establishment of the Library and Information Profession.

According to Abel and Oxbro (2001), library and information professionals have to identify, acquire and evaluate internal and external sources of knowledge and integrate, organize and make relevant knowledge available to the right person at the right time. Librarians provide a base to research by collection, processing, storage and distribution of knowledge and information. Libraries make a link between information and the information seeker. Knowledge dissemination is another area where libraries can work as members of the knowledge management system, librarians can undertake the effective role of acquisition and analysis of documents which comprises of organization, classification and sorting and archiving for easy retrieval and faster dissemination, creating knowledge vocabulary and importing information literacy instruction to the users and as well as training for them to maximize use of the knowledge repository.

The changing role of library and information professions implies a set of modern skills needed for fighting the challenges of the Information Age. The Library Professionals need to focus on technical skills in the library environment. Although information technology presents the librarian with ethical challenges, the librarian has to be ready for the role of the information professional in the connected world and has to acquire modern skills that can contribute to success in this era.

In the present Information Age also known as the digital era professionals have to evolve as the information profession changes. Now information scientists have to work as e-information resources in which various professional groups are expected to map strategies that lead to generate, manage, maintain, disseminate and package and repackage of information. Thus, the library professional has to change themself and adopt new skills and additional roles in the era of Information Age.

Information professional has to adopt leadership position such as the role of information manager in the design, development, management, ethical management, and dissemination of knowledge-based information systems in order to meet the information needs and responsibilities of their patrons. Their roles as E-resource Managers and digital information disseminators are also essential.

2 Statement of the problem

The problem of fake news has become a significant threat to almost every aspect of life. We now live in an era where information can be used as a weapon that can cause chaos of unimaginable proportions. For instance, the effect of fake news has led to what the World Health Organisation described as an "infodemic", with the outbreak of coronavirus leding to significant amounts of misinformation about the pandemic on social media and websites that contained inaccurate information (Zoe 2020). The implications of fake news in the society includes electoral violence, ethno-religious conflict, jungle justice and leadership mistrust. However, acquiring analytical and critical thinking skills and information literacy are some of the tools of tackling fake news as outlined in previous studies (Durodolu and Ibenne, 2020). Apart from the recommended means of combating fake news highlighted earlier, the researchers are of the view that collaborative interventions by relevant stakeholders could be of immense benefits to solving the menace of fake news in the society since the required solution may not be achievable without synergy among stakeholders. Therefore, this chapter investigated the roles of Library and Information Professionals in facilitating collaborative intervention of relevant stakeholders in combating the menace of fake news in the society.

Objectives of the Study

The specific objectives of the study seek to:

1. describe fake news and its causes, sources and consequences in the information age;
2. ascertain the need for collaborative intervention in combating fake news in diverse societies;
3. examine tools and strategies for combating fake news in the information age; and
4. identify the role of Library and Information Professionals in combating fake news in the information age.

Literature Review

Fake News: Causes, Sources and Consequences

Driven by foreign actors, citizen journalism, and the proliferation of talk radio and cable news, many information systems have become more polarized and contentious, and there has been a precipitous decline in public trust in traditional journalism (Netanel 2020). Other reasons for the spread of fake news include:

1. Political Desperation: desperate political actors including politicians, political parties, interest groups, organizations, etc are often caught distorting information to gain public sympathy and support. Prior to the digital age, the German concept of "groBe Luge," a big lie, was used by Nazi politicians as a propaganda tool during the Second World War. It was adopted to tell a lie so colossal that no one would believe that someone could have the impudence to distort the truth so infamously. It was believed that when such lies are repeatedly told, they become accepted as the truth. Neri (2011) notes that democratic decay, the gradual deterioration of the institutions of democracy, is the consequences of fake news in the digital age and the objectives of much of digital disinformation. Banjo (2019) explains that social media poses no harm if the use and intent is to sensitize the public, spread harmony or improve the economy. But the purposes are often darker. In what is known as state-sponsored trolling, for instance, governments create digital hate mobs to smear critical activities or journalists, suppress dissent, undermine political opponents, spread lies and control public opinion. Nemr and Gangware (2019), in a study published by the United State Department of Defense, state that a growing number of states in the pursuit of geopolitical ends, are leveraging digital tools and social media networks to spread narratives, distortions, and falsehoods to shape public perceptions and undermine trust in the truth. The study refers to misinformation and disinformation as "weapons of mass distraction."

2. Poor Professional Ethics Among Information Professional and Information Users

Some media houses were established either by private owners, organizations or governments to drive specific agenda. Many websites are partisan, and as such are biased in their reports and opinion pages.

3. Poor Information/Digital Skills

In a study reported by Buckland (2020), fewer than 20% of middle school students were even able to distinguish between "sponsored content" and real news stories, let alone access underlying bias. Digital natives may be able to lift information from Facebook and put it on Twitter or WhatsApp but their ability to evaluate information is often questionable, with Waring (2018) in a study facilitated by the Stanford History Educational Group describing young people's ability to reason about information on the internet as bleak. About 80% of students in their study could not differentiate between a real news article and sponsored contents on social media despite the explicit appearance of "sponsored content" on the posts.

4. Lack of Adequate Laws to Regulate Social Media Use.

Very few democracies have established laws to combat fake news, especially on the internet. When such laws were muted in Nigeria, there was public outcry against them as citizens considered them ploys to gag the media and strangle the freedom of expression by the ruling political class. The Protection From Internet

Falsehood and Manipulation Bill 2019 was presented in the senate for a second reading in November 2019 but was met with public outcry (Ewang, 2019). Berger (2019) notes that the spread of fake news is made possible largely through social networks and social messaging, which begs the question of the extent of regulation and self-regulation of companies providing these services. In their character as intermediary platforms, rather than content creators, these businesses have to date generally been subject to only light-touch regulation (except in the area of copyright).

However, absence of information may be as dangerous as wrong information or even worse because socio-political, economic, personal, and career decisions are taken by individuals and groups based on the available information at their disposal. The ability of digital technologies to not just manipulate written texts and graphics but also to distort and create audio-visual information is already a phenomenon that is being recognized among information professionals and users as Deepfake. However, the dangers they pose to societal stability in all areas of human endeavors cannot be overemphasized.

According to Bharadwaj (2020), both disinformation and misinformation contribute to fake news, and, pose risks to brands and their audiences. Misinformation is false information that is given without malice while disinformation is false information with the intent to deceive, such as government propaganda. In Myanmar, misinformation like hate speech and propaganda, fueled violence against Rohingya Muslims. And while there were riots in Sri Lanka after stories that proved false set the country's Buddhist majority against Muslim (Bretfeld 2018). Elsewhere in the aftermath of the panic over fake news that followed the 2016 US presidential election, librarians and other information professionals were urged to "take leadership in the current crisis" (Sullivan, 2019).

At present, democracy is being threatened by the ease at which fake news about civic issues is being allowed to spread and flourish (Waring 2018). The spread of inaccurate information has been regarded as a severe threat to democracy (Brito 2022). As a lack of trust and accusations of partisanship keep trailing conventional media houses and their social media handles, Wardle and Derakhshan (2017) notes that libraries are one of the few institutions where trust has not declined. According to the Pew Research Center, 78% of U.S adults believe that libraries help them find information that is trustworthy and reliable.

Fake news is now viewed as one of the greatest threats to democracy, journalism, and freedom of expression. It has weakened public trust in governments (Zhou and Zafarani, 2020), with examples of its potential impact see with the contentious "Brexit" referendum and the equally divisive 2016 U.S. presidential election (Pogue 2017).

Fake News and Role of Library and Information Professionals

Most authors consider fake news to be an article that mimics the format of a news story or report, with fake content that was created with the intent to deceive (Rini 2017; Tandoc et al 2018b; Allcott and Matthew 2017). In fact, it is the intention to deceive that makes it possible to distinguish between fake news and false news (Meneses 2021). False news is not intended to mislead the reader. The false content of a report or piece of news may result from a journalistic error or the journalist's lack of professionalism in verifying its sources (Nielsen and Graves 2017; Meneses 2021). The very word "fake" refers us to the intention to deceive and to lie. "Fake" is associated with counterfeiting and imitating the real (Gelfert 2018). Fake news seeks to be credible and gain legitimacy by imitating the format of the news in order to manipulate and deceive the reader and make the fake content look real (Levy 2017). It is important to highlight that we often approach contemporary fake news, that is, in an online context, in which false statements are widely shared in the digital universe, namely in social media. The goal of contemporary fake news is to go viral (Rini 2017; Meneses 2021). For these reasons, fake news can take the form of a news feed post (in the case of Facebook) or tweet (in the case of Twitter), just like the real news is presented on these social media (headline, image, signature/source). In addition, fake news links to sites that mimic real news sites (Silverman 2017). Several investigations have recently sought to understand which factors are associated with the selection and consumption of fake news in an online environment. The belief, consumption and dissemination of fake news may be related to several aspects such as the growing distrust in media (Swift 2016; Nielsen and Graves 2017) and users' level of education and digital literacy (Flynn, Nyhan and Reifler 2017). However, information professionals are playing an increasing role in dealing with information in electronic formats by creating webpages to promote their services to external customers and choosing automated library management systems. Skills in information organization and verification are thus more necessary in this age of information explosion. Librarians and information professionals have a key role to play in this era. For example, librarians are well equipped to take intranet projects through the various stages of design and maintenance as they understand their users and their organizations information needs and have the range of skills to manage knowledge effectively. The role of librarians in this context is to help users find the information they require and then provide them with the tools to asses and use the resources for their individual needs. Information Age librarians and information professionals should be able to manage the Digital Information System (DIS) as this encompasses the overall competencies (knowledge, know-how, skills and attitudes) necessary to create, store, analyze, organize, retrieve and disseminate digital information (text, images, sounds) in digital libraries or any other types of information. Librarian and information professionals have as part of their professional competencies, expert knowledge of the content of information resources, including the ability to critically evaluate and filter information. One of the main tasks of libraries is to provide timely and reliable information to users. It is not just fabricated stories and propaganda that are threatening to users, pseudoscience, hoaxes and biases are also misinforming materials that need to be recognized by knowledge professionals. Most times library users may find it difficult to be able to distinguish reliable sources of information from unreliable ones. However, it is pertinent to state that librarian cannot entirely eradicate the origination and spread of fake news, and rather they can play a pivotal role in reducing it dissemination. Ebhonu and Onobrakpor (2021) highlighted some major roles of librarians including the 3Ds (Detection, Documentation and Debunking) Information Verification, Fact Checking, Media Literacy, and Counter Information.

Theoretical Underpinnings

The following theories were considered for this study;

1. Undeutsch hypotheses

The theory is known as news-related theory because it reveals the possible characteristics of fake news content compared to true news content. This theory was developed by forensic psychology targeting deceptive statements or testimonies (i.e disinformation). The easy way to combat fake news is to be able to verify the presence of some attributes such as information, sentiment and polarity in a statement. These attributes help to distinguish between the fake news and the truth. The undeutsch hypotheses states that a fake statement differs in writing style and quality from a true one, however, it also emphasizes that identified attributes can be used to automatically detect fake news (Zhou et al 2019). Therefore, deliberate and concerted effort should be taken by library and information professionals to curtail the spread of fake news in the society.

2. User-related theories

User-related theories include normative influence theory, social identity theory, and availability cascade (Kuran and Sustein 1999). These investigate the characteristics of users involved in fake news activities e.g posting, forwarding, liking and commenting.

This study examines the collaboration integration of normative influence, social identity and availability cascade (Kuran and Sustein 1999) which were built around the fact that the fake news, unlike information such as a fake review (Jindal and Liu 2008) can “attract” both malicious and normal users (Shao et al 2018). Malicious users (e.g some social bots (Ferara et al 2016) spread fake news often intentionally and are driven by benefits. (Kahneman and Tversky 2013) like Click-bait. Some normal users (which we denote as vulnerable normal users) can frequently and unintentionally spread fake news without recognizing the falsehood. Normal users are encouraged to engage in fake news activities when many users have done so (i.e peer pressure) or the bandwagon effect (Cheng-Jun et al. 2019). Meanwhile, it has been demonstrated that it is difficult to correct one’s cognition after fake news has gained their trust (i.e confirmation bias). However, effort to reduce the spread of fake news should not only be geared towards malicious users only but also the normal users.

3 Materials and Methods

A community-driven approach was used for building the article list used in this literature review. However, data for this chapter was collected through the literature review whereby all articles with information about combating fake news in the information age were collected and used in this study. The collection of papers was done using various online tools with relevant information including LibHub – a Library Gateway to Information Resources (http://libhub.kiox.org) and Google scholar (scholar.google.com). The literature searching was conducted after three major steps; (i) search formulation where the topic was analyzed to generate key concepts (Fake news, Libraries, Librarians, Information professionals, Information age): (ii) choosing the appropriate information sources, and (iii) using the identified keywords in step (i) to conduct a search in the LibHub and Google Scholar. The searches were restricted to recent publications, and only search results which were found relevant to the topic of Combating Fake News in the Information Age: The roles of Library and Information Professionals in Diverse Societies were selected and used.

Discussion and findings

Roles, Tools and Strategies for Combating Fake News in the Information Age by Libraries and Librarians

In the 1800s, the public library was considered a vital force for strengthening democracy. Today, librarians are poised to play no less critical a role – helping tomorrow's leaders navigate an ever-swelling sea of information, discerning the hard truth from convincing lies (Buckland 2020). There are several steps that can be taken to address the threats posed by fake news in the digital era. Since librarians and information scientists cannot fully control those who engage in the use of fake news to deliberately project their selfish objectives, they can engage in activities and programmes that can protect the information users, the public from fake news. Some of them include:

Information and Media Literacy campaign: according to Pinto et al. (2020), the fight against the dark side of the digital era is accomplished through investment in information and media literacy. Only the most enlightened and prepared are able to deal with the information and communication technologies and will be able to distinguish truthful information from disinformation, misinformation, or fake news, namely purposeful deception conveyed on any digital platform. This is where librarians and libraries have huge roles to play by engaging in information and media literacy activities through their websites and other onsite user education programs. Governments at local levels can partner with academic libraries to engage students, who constitute a greater percentage of internet users of digital literacy programs as part of user education programs

Textual Fact-Checkers: fact checking is an essential component of the process of news reporting which addresses the claim's logic, coherence, and context (Thorne and Vlanchos 2018). Fact checking can help detect misinformation but does not distinguish it from disinformation because it only establishes the veracity of information in textual form. Examples of fact-checking websites include Africa Check, Newschecker, PolitiFact, and Full Fact. Fact-checking can also inhibit political lying, thus making it more costly for political figures to distort the truth.

Multimodal Fact-Checkers: the findings of Hameless et al. (2024) in research funded by the Amsterdam School of Communication Research point to the importance of fact-checking as an emerging journalistic discipline. It shows that fact-checks can be effective in countering multimodal disinformation. In an era where citizens and other nonprofessional communicators join the digital newsfeed, journalists' role as truth crusaders becomes even more important.

Fake news is harmful to our community and makes the world less informed, erodes public trust (Mosseri 2017) and threatens national unity and security. Therefore, all relevant stakeholders-telecommunication companies, the media, librarians and other information professionals as well as teachers have a responsibility in addressing the menace of fake news. From the telecommunication perspective, it is believed that most fake news is financially motivated, and one of the most effective approaches to fighting it is by removing the economic incentives for traffickers of misinformation who make money by masquerading as legitimate news publishers and posting hoaxes that get people to visit their sites (Mosseri 2017).

Additionally, telecommunication companies should build new products to identify and limit the spread of fake news on their platforms. Also, relevant stakeholders should be supported in developing participatory and transparent initiatives for creating a better understanding of the impact of disinformation and propaganda on democracy, freedom of expression, journalism, and the civic space, as well finding a way to support core traditional media for acceptability over social media. Nigeria in particular must ensure that its national newspaper and television stations are sustainable to practice quality journalism that will be acceptable and preferable to the people (Okigba 2018). The need to rebuild trust in the traditional media is born out of the fact that political encroachment is having very negative impacts on most media houses in Nigeria, making most Nigerians believe that every piece of news is either out rightly fake or somewhat fake with only a touch of the real.

Furthermore, the International Specialized Mandate on Freedom of Expression and/or Media (2017) recommends that measures should be taken to promote media and digital literacy as part of regular school curricula and to engage with civil society and other stakeholders to raise awareness about the issue of fake news.

Finally, librarians as traditional information providers are at the very center of all information business. They have a significant role to play in curbing fake news, according to Denise (2018): libraries have a long history of working to strengthen the information literacy skills of their patrons, and one can easily view the fake news phenomenon as just the most recent iteration of the need to do this. The bigger landscape of fake news reveals a complicated issue that resonates with the indisputable importance of critical and evaluative skills.

In this regard, the intrusion of fake news has provided libraries and librarians with a challenging but unique opportunity to engage their user communities on a broad discussion on the subject of critical and analytical thinking as part of their information seeking behaviour in the current information environment. The emphasis must therefore be to continue promoting information literacy skills among user communities.

4 Tips on How to Identify Fake News

Separating fact from falsity can be tasking, but getting the truth is a worthy course no matter what it costs. To identify fake news, these tips can be followed; first, consider the source of information by clicking away from the story to investigate the site, its missions and its contact information: reading beyond the headlines: researching the author: investigating the article: looking out for fake images: developing a critical and evaluative mindset, examining the evidence; keeping in mind that credible news stories usually include sufficient facts and figures from credible sources: and searching for other sources (Raghavendra & Pragati, 2023). It is also important to ask the professionals when in doubt by visiting trustworthy fact-checking websites like FactCheck.org, International Fact-Checking Network (IFCN), and PoltiFact.com among others to verify information, as well as by using automated methods like Machine learning and deep learning models (Adline and Augasta, 2023). Stay away from unverifiable news as much as possible and do not share it.

Conclusion

It is ironic that as anyone in the world can easily get information at the tip of their fingers, it is all the more essential to be meticulous about the information at one's disposal by following the tips on how to verify the authenticity of information. However, librarians and information professionals in this information age will become the guardians of digital information and the vehicle to preserve democratic access to information. The role of librarians and information professionals in this era will be increasingly geared toward offering consultancy to users by providing assistance in activities such as accessing digital reference services and electronic information services, navigating, searching and retrieval of digitalized information through Web documents that span the Universal Digital Library or the Global Digital Library. Librarians and information professionals in this new era will be an embodiment of a digital information professional or digital knowledge worker, who will ensure that the information in the digital space is filtered. The knowledge that library and information professionals bring to this knowledge environment will make sense of a multiplicity of digital collections and resources provide access to a network of key contacts, identify cost-effective strategies for information retrieval, and assist users in the publication and creation of new knowledge. Several personal characteristics as well as the new skills and competencies are essential for library and information professionals in this Information Age to create and manage resources.

Demonstrated skills and sophisticated expertise competencies are needed for library and information professionals to manage the Digital Information System (DIS) in the important area of imaging technologies, optical character recognition, makeup language cataloguing, multimedia indexing, database technology, user interface design, programming and Web technologies.

Recommendations

In addition to the already suggested strategies for curbing the negative implications of fake news in this information explosion era, the following are recommended:

1. The mass media should make a deliberate effort in promoting media literacy to enable the general public to know how to differentiate facts from fiction.
2. Libraries, librarians, and the academia in general must work collaboratively in teaching information literacy skills to their students in particular and their user communities as a whole.
3. The Federal Government through its agencies like the Ministry of Information and the National Orientation Agency (NOA) should intensify campaigns against fake news in Nigeria.
4. The media regulating agencies like the Nigerian Press Commission should insist on media professionalism from media houses and professionals in order to avoid them being used as instruments for the spread of propaganda especially by the elite political class.

References

Abdusalami L.T., Okezie, Q. I. & Agbo, A.D. (2013). The Role of the Library in the promotion of knowledge societies in Nigeria. *Advances in Applied Science Research*, *4*(1), 58–70.

Abell, A. & Oxbro, N. (2001). Competing with knowledge: The informational professionals in the knowledge management age. Library Association Publishing,

Adline, R. M. & Augasta, G. M. (2023). A Novel Framework for Fake News Detection using Double Layer BI-LSTM. doi: 10.1109/ICSSIT55814.2023.10061026

Alawode, W., Olorede, J. O. & Azeez, L. D. (2018). Fake news and public perception of Nigerian's online media: implications for national security. A paper presented at the 1st national conference of the academic Staff Union of Polytechnics, Federal Polytechnic Offa, held

Alfredo, A. (2017). Origins of Writing. doi: 10.1007/978-981-10-6887-4_4

Allcott, H. & Matthew, G. (2017). Social Media and Fake News in the 2016 Election. *Journal of Economic Perspectives* (31), 211–36.

Angela, S. (2020). Historical development of definitions of information literacy: A literature review of selected resources. *The Journal of Academic Librarianship*, doi: 10.1016/J.ACALIB.2020.102116
Banjo, A. A (2019). The Actualization of Personal Data Protection in Nigerian Law: Law: An Analysis of Personal Data Protection in the Nigerian and European Union Legal Systems
Baptista, J. P. & Gradim, A. (2020). Understanding Fake News Consumption: A Review. MDPI, *Journal of Social Sciences*. *9*, 185. doi: 10.3390/socsci9100185
Belloir, N., Ouerdane, W., Pastor, O., Frugier, E., Louis-Antoine de Barmon. (2022). A Conceptual characterization of Fake News: A Positioning Paper. 16th International Conference on Research Challenges in Information Science (RCIS'22), Barcelone, Spain. hal-03679073.
Berger, G. (2019). Whither MIL: Thoughts for the road ahead. *Understanding media and information literacy (MIL) in the digital age: A question of democracy*, 25–36.
Bharadwaj, A. (2020). Assessing India's Preparedness. *Wild Blue Yonder*.
Bin, L. Hashin, W., Haliza, N. & Mokhtar, W. (2012). Preparing New Era Librarians and Information Professionals: Trends and Issues. *International Journal of Humanities and social science*, *2*(7),
Borkovich, D. J. (2011). The Social Science of data warehousing: Its ever-evolving corporate culture. *Issues in information systems*, *12*(1), 23–35.
Bretfeld, S. (2018). Buddhism Endangered by Hidden Enemies: Conspiracy Narratives in Sri Lankan Buddhist Present and Past. In *Handbook of Conspiracy Theory and Contemporary Religion* (pp. 255–278). Brill.
Brito, I. A. (2022). The world of shadows of disinformation: the emerging technological caves. *Revista da Faculdade de Direito da Universidade de Lisboa*, *63*, 365–406.
Buckland, M. K. (2020). *Ideology and Libraries: California, Diplomacy, and Occupied Japan, 1945–1952*. Rowman & Littlefield.
Cheng-Jun Wang, Jonathan J.H. Zhu, (2019). Jumping onto the bandwagon of collective gatekeepers: Testing the bandwagon effect of information diffusion on social news website, Telematics and Informatics, 41, 34–45
Cumerma, A. (2020). *Cuba's Nation Brand: War and Intelligence in the Fiction of Norberto Fuentes* (Doctoral dissertation, Johns Hopkins University).
Debra, J. B. & Philip, D. N. (2014). Big Data in Information Age: Exploring the Intellectual Foundation of Communication Theory. *Information Systems Education Journal*, *12*(1).
Denise, E. A. (2018). Information literacy and libraries in the age of fake news. California, Libraries Unlimited. Available at www.abc-clio.com Global Peace Index (2019). Nigeria is one of the least peaceful places on earth. Available at www.saharareporters.com. Accessed 25/6/2021.
Durodolu, O. O. & Ibenne S. K. (2020). The fake news infodemic Versus Information literacy. *Library HiTech News*, Emerald Publishing limited, 0741–9058. Dio:10.1108/LHTN-03-2020-0020
Ebhonu, S. I. & Onobrakpor, U. D. (2021). Combating Fake News in the Society: The Roles of Librarians. *Library Progress (International)*, *41*(2), 188–199.
Ferrara, E., Varol, O., Davis, C., Menezer, F. & Flammini. A. (2016). The rise of social bots. *Commun. ACM*, *59*(7), 96–104.
Flynn, D. J., Nyhan, B. & J. Reifler. (2017). The Nature and Origins of Misperceptions: Understanding False and Unsupported Beliefs about Politics. *Political Psychology*, *38*, 127–50.
Gelfert, A. (2018). Fake News: A Definition. Informal Logic *38*, 84–11
Hameleers, M., & van der Goot, E. (2024). Look at what the real facts and experts say! The use of expert references and objectivity claims in disinformation: A qualitative exploration and typology. *Journalism*, 14648849241257383.
Harvey, L. (1950). Bandwagon, snob, and Veblen effects in the theory of consumers' demand. *The quarterly journal of economics*, *64*(2), 183–207.

Issa, A. O. (2003). Libraries and librarians in the knowledge and information society (Ed). Centre for Continuing Education, the Federal Polytechnic, Offa, Kwara State.

Jessica, P., Michaelson, E. & Sterken, R. K. (2019). What's New about Fake News. *Journal of Ethics and Social Philosophy, 16*, 67.

Jindal, N. & Liu, B. (2008). Opinion spam analysis. *In Proceedings of the 2008 International Conference on Web Search and Data Mining. ACM*, 219–230.

Kahneman, D. & Tversky, A. (2013). Prospect theory: An analysis of decision under risk. In Handbook of the fundamentals of financial decision making: Part I. *World Scientific*, 99–127.

Kleis, N. R. & Graves, L. (2017). 'News You Don't Believe': Audience Perspectives on Fake News. Oxford: Reuters Institute for the Study of Journalism, Available online: https://ora.ox.ac.uk/objects/uuid:6eff4d14-bc72-404d-b78a-4c2573459ab8

Kuran, T. & Sunstein, C. R. (1999). Availability cascades and risk regulation. *Stanford Law Review*, 683–768.

Leonhardt, D. & Thompson, S. A. (2017). Trump's lies'. *New York Times*. Retrieved from https://www.nytimes.com/interactive/2017/06/23/Opinion/trumps-lies.html

Levy, N. (2017). The Bad News about Fake News. *Social Epistemology Review and Reply Collective, 6*, 20–36.

McGonagle, T. (2017). Fake news: False fears or real concerns? *Netherlands Quaterly of Human Rights, 35*(4), 203–209. www.journals.sagepub.com

Meneses, J. P. (2021). Deepfakes and the 2020 US elections: what (did not) happen. *arXiv preprint arXiv:2101.09092*.

Michalidis, P. & Viotty, S. (2017). Spreadable spectacle in digital culture: Civic expression, fake news, and the role of media literacies in "post-fact" society. *American Behavioral Scientist, 61*(4), 447–454.

Mihai-Ionut, P. & Ene, I. (2019). Influence of the Educational Level on the Spreading of Fake News Regarding the Energy Field in the Online Environment. *In Proceedings of the International Conference on Business Excellence. Warsaw: Sciendo, 13*, 1108–17.

Mosseri, A. (2017). Working to stop misinformation and false news. Available at www.facebook.com. Accessed 2/6/2021.

Nemr, C., & Gangware, W. (2019). Weapons of Mass Distraction: Foreign State Sponsored Disinformation in the Digital Age. U.S. Department of State – United 107 States Department of State. https://www.state.gov/wpcontent/uploads/2019/05/Weapons-of-Mass-Distraction-Foreign-State-SponsoredDisinformation-in-the-Digital-Age.pdf

NERI, Marcelo.(2011). *A Nova Classe Média*. São Paulo: Saraiva, 2011.

Netanel, N. W. (2020). Mandating digital platform support for quality journalism. *Harv. JL & Tech., 34*, 473.

Nielsen, Rasmus Kleis, and Lucas Graves. 2017. 'News You Don't Believe': Audience Perspectives on Fake News. Reuters Institute for the Study of Journalism. http://reutersinstitute.politics.ox.ac.uk/our-research/news-you-dont-believe-audience-perspectives-fake-news.

Okigba, E. (2018). 2019 polls: BBC raises concern over fake news. Available at www.vanguardngr.com. Accessed 25/6/2021.

Oluwaseye, A. J., Akanni, M. J., & Busuyi, A. O. (2017). Information needs and seeking behaviour of medical students at the college of medicine, university of Ibadan, Nigeria. *Journal of Applied Information Science and Technology, 10*(2), 49–62.

Paulo, M. J. (2018). Sobre a Necessidade de Conceptualizar o Fenomeno Das Fake News. *Observatorio (OBS)*, 12.

Pepp, J., Michaelson, E., & Sterken, R. K. (2019). What's new about fake news. *J. Ethics & Soc. Phil.*, *16*, 67.
Pharm, N. (2018). Haunting 'Nepal quake victims' photo' from Vietnam. Retrieved from http://www.bbc.co.uk/news/worldasia-32579598
Pinto, C., Braga, I., Martins, S., & Carvalho, M. (2020). How disinformation affects the organizations' performance in the digital era. *Encontros da Licenciatura em Ciências e Tecnologias da Documentação e Informação*.
Pogue, D. (2017). How to stamp out fake news. *Scientific American*, *316*(2), 24–24.
Pričins, M. (2022). Misinformation and Information Manipulation in Pre-Election Messages of Candidates from Latvian Political Parties. Media i Społeczeństwo, doi: 10.22364/ms22.11
Raghavendra, K. & Pragati, A. (2023). Fake News Identification: An Effective Combined Approach using ML and DL Techniques. doi: 10.1109/PCEMS58491.2023.10136087
Rini, R. (2017). Fake News and Partisan Epistemology. *Kennedy Institute of Ethics Journal 27*, E-43-E-64.
Shao, C., Ciampaglia, G. L., Varol, O., Yang, K., Flammini, A. & Menezer, F. (2018). The spread of low-credibility content by social bots. *Nature communications*, 9(1), 4787.
Silverman, Craig. 2017. "The Rise and Fall of Fake News." WNYC, January 27. http://www.wnyc.org/story/rise-and-fall-fake-news/.
Sullivan, M. C. (2019). Why librarians can't fight fake news: *Journal of Librarianship and Information Science*, doi: 10.1177/0961000618764258
Swift, A. (2016). Americans' Trust in Mass Media Sinks to New Low. Gallup News *14*, Available online: https://news.gallup.com/poll/195542/americans-trust-mass-media-sinks-new-low.aspx
Tandoc, E., Lim, Z. C. Lim, Z. & Ling, R. (2018). Defining "fake news": A typology of Scholarly definitions. *Digital Journalism*, *6*(2), 137–153.
Thorne, J., & Vlachos, A. (2018). Automated fact checking: Task formulations, methods and future directions. *arXiv preprint arXiv:1806.07687*.
Undeutsch, U. (1967). Beurteilung der glaubhaftigkeit von aussagen. Handbuch der psychologie *11*, 26–181.
Walters, R. M. (2018). How to Tell a Fake: Fighting Back against Fake News on the Front Lines of social media. *Texas Review of Law and Politics 23*,111.
Wardle, C. & Derakhshan, H. (2018). Thinking about information disorder: formats of misinformation, disinformation and mal-information. In Ireton, C., Possetti, J. (Eds.), Journalism, fake news and disinformation: *Handbook for journalism education and training*, 43–54. UNESCO.
Waring, A. (2018). *The new authoritarianism*. BoD–Books on Demand.
Zhou, X. & Zafarani, R. (2020). A Survey of Fake News: Fundamental Theories, Detection Methods, and Opportunities. *ACM Comput. Surv. 1*(1), Article1, 37.
Zhou, X., Jain, A., Phoha, V. & Zafarani. R. (2019). Fake News Early Detection: A Theory-driven Model. *arXiv preprint arXiv*: 1904. 11679.
Zoe, T. (2020). WHO says fake Coronavirus claims causing 'infodemic' {Interview} British Broadcasting Corporation, available at: www.bbc.com/news/technology-51497800

Akanji Comfort Adebukola and Fashina Abiola Yetunde

The Effect of Scientific Misconception on Biology Students' Conceptual Understanding in Senior Secondary Schools in Lagos State Nigeria

Abstract: Scientific misconceptions can be referred as a view or an idea that is inappropriate based on incorrect thinking. It can be also known as alternative concepts or erroneous understanding about sciences. Biology is an interesting and exciting school subject that incorporates in its study laboratory and field skills in the acquisition of meaningful and relevant knowledge. This study investigated the effectiveness of misconception on biology students' conceptual understanding. Some of the reasons responsible for inconsistencies in students' performance in biology are: misconception of the content, incompetent mode of teacher's delivery of the subject, and misuse of instructional materials. A total of 257 (111 male, 146 female) senior secondary school Year II (SS2) biology students from 102 government owned secondary schools in district II and 59 government owned secondary schools in district IV in Lagos state in Nigeria were purposively selected. Four schools from Education districts II and IV also participated in the study. Using a pre-test and post-test non-equivalent control group quasi experimental research design, the study examined the effect of misconception on biology students' conceptual understanding, with the Biology Conceptual Understanding Test (BCUT) used to collect data. The data collected were analyzed using mean, standard deviation, and analysis of covariance (ANCOVA) where pre-test scores served as covariates, with results showing that misconception was effective in students' conceptual understanding in biology. Finally, the implications of the findings for teaching and learning were drawn, with recommendations also made.

Keywords: misinformation, scientific misconception, biology students and conceptual understanding, Lagos State

Akanji Comfort Adebukola, University Of Lagos, Nigeria, Distance Learning Institute (Science Education), e-mail: drcomfortomotara2019@gmail.com
Fashina Abiola Yetunde, Distance Learning Institute, University of Lagos, e-mail: yfashina@unilag.edu.ng

https://doi.org/10.1515/9783111176345-012

Introduction

Misconceptions can be referred to as misunderstanding a factual idea or being given contrary information from credible sources such as parents and teachers. The major problem is that once a misconception has been formed, it may be difficult to change (Eggen and Kauchak 2012) and that possessing misconceptions can have serious impacts on learning (Hanuscin n.d.).

Biology is concerned with the study of life whether in the context of human beings, animals, lower animals, or plants. It is an interesting and exciting school subject in that it incorporates a study laboratory and field skills in the acquisition of meaningful and relevant knowledge. Biology has many branches of its discipline such as: anatomy, biochemistry, bio-informatics, botany, cell biology and genetics, ecology, evolution, histology, microbiology, molecular biology, morphology, physiology, and zoology. As noted by Osuafor and Okonkwo (2013), biology, apart from being closely related with these branches, is also closely related with other science subjects like agricultural science, chemistry, geography, mathematics, and physics. It is little wonder then that biology finds application in many specialized areas like medicine, pharmacy, food production and processing industries, biotechnology, genetic engineering, agriculture and horticulture, environmental protection, tourism industry (biological gardens), among others. The importance and the direction of teaching science are captured and in the biology secondary school curriculum Durodolu (2018) identified the objectives of teaching science, including knowledge f the science academic discipline; acquiring the skills of the scientific method; having clear explanations for societal issues through an increasing interest in science literacy and societal goals; personal needs; and career awareness.

In Nigeria, the sciences are taught in school subjects as biology, chemistry, and physics. The National Policy on Education in the National curriculum for Senior Secondary Schools volume 3 Science states specific objectives to be achieved. The cardinal objectives for biology include adequate laboratory and field skills in biology, meaningful and relevant knowledge in biology, the ability to apply scientific knowledge to everyday life in matters of personal and community health and agriculture, and reasonable and functional scientific attitudes.

There are extant literatures on reasons responsible for poor performance in biology by students in secondary school. Misconception of the content (Durodolu 2021), incompetent mode of the teacher's delivery of the subject, and misuse of instructional materials were identified by Nwagbo (2009). In addition, teachers in many Nigerian secondary schools do not apply appropriate science strategies as identified and recommended to be effective for science instruction (Norom 2009). Others include parents/ family background influence (Essien 2002; Okeke 2009;

Osuala and Okonkwo 2013) and cognitive style (Okoye and Okeke 2007). Perhaps what is missing is that meaningful learning has actually not taken place.

Learning in a meaningful context occurs when concepts are combined to form statements or propositions and the extent to which new concepts are integrated into cognitive structure which is a function of effort as well as the quantity and quality of existing relevant cognitive structures (Novak 2002; McNeil et al. 2006; Okebukola 2012). The way students learn within the schema tradition posits that when new information is received as sensory data, the learner assigns it to a new organizing framework known as schema (Ausubel 2000; Novak 2002). In exposition of meaningful learning, Okebukola (2012) notes that it is best described by contrasting it with diverse-rote learning because the arrangement of both rote and meaningful learning takes different turns in one's cognitive structure. Meaningful learning occurs when concepts are linked to similar ones, correlated, and applied in order to generate new ideas (Oluwole, Okiki and Durodolu, 2018). On the other hand, rote learning is anchored on knowledge that is isolated which is not relatable, which cannot be spontaneously constructed when given a similar task and is very prone to forgetfulness. In addition, Artino (2008) points out that educational literature confirms that learners' success is contingent on their flowing, comprehensive cognitive inventory with good utilization and application in solving problems; therefore, the learner is not static and should acquire concepts and knowledge which would help in their development in order to remain continual and interactive with others, so that they will be able to solve real problems in meaningful tasks.

Many researchers have developed models and strategies based on constructivist theory, including the learning cycle, concept maps, the V model, constructive analysis model, realistic model, cooperative learning strategies, and other teaching approaches grounded in constructivism. Despite these efforts, students enter the learning environment with pre-existing knowledge (or conceptions) that are continuously built upon throughout their education. This ongoing process also involves addressing common student misconceptions in science. In order to teach science effectively, it is vital to ensure that existing conceptions are sound and to modify any misconceptions that will compromise them, following the logic that misconceptions themselves can be considered to be consecutive and therefore lead to ever increasing issues with learning as students continue to build their knowledge on current understandings (Durodolu 2021). There are many strategies available to help teachers modify misconceptions, but before this can be achieved the teacher needs to have strategies for identifying exactly what misconceptions a student may have.

This study was observed as a case among biology students in senior secondary schools in Lagos state, Nigeria, as part of their due courses. As such it was more of a learning exercise than a true research project. The objectives of the study were to define three scientific concepts and identify for each some of the misconceptions that students commonly have in biomes, ecology of population, and succession, in an attempt to discover what the students' misconceptions were, how they acquired them, and whether the exercises, combined with discussion, have helped to modify any of these occurrences. The study indicated that the experimental group evidenced more learning achievements, science process skills in general and in each subscale, and critical thinking in general and in the interpretation subscale than the control group. As a consequence of this, the study examined conceptual understanding of students' concepts in biology as a dependent variable.

Tanner and Allen (2005) note that underpinning science education reform movements in the last 20 years – at all levels and within all disciplines – is an explicit shift in the goals of teaching science from students simply creating a knowledge base of scientific facts to students developing deeper understanding of major concepts within a scientific discipline. Conceptual understanding is a function of the internal knowledge structures of an individual in which knowledge structures can be conceptualized as organized networks of concepts with their interrelations (Bunde 2007). Rich, interrelated, and integrated knowledge are characteristic features of conceptual understanding. Superficial understanding is characterized by isolated and fragmented knowledge (De Jong and Ferguson-Hassler 1998). Measuring conceptual understanding can also be achieved by assessing the characteristics of the knowledge structure of the individual. This requires assessment strategies that capture rich, dynamic, interrelated, and integrated domains of concepts (Anderson and Schunn 2000). Integration reflects a deeper level of understanding which tends to approximate new knowledge elements to existing mental structures as well as the connection between them. On the other hand, correct knowledge with the co-existence of alternate conceptions reflects isolated, fragmented, and low understanding (Novak 2002). Another characteristic feature of determining conceptual understanding is the appearance of not explicitly presenting knowledge which implies a surface representation. According to Novak (2002), a deeper level of representation is characterized by an organization of knowledge that includes implicit concepts with their relationship and the principle. When students exhibit conceptual understanding, there is a possibility that retention may be enhanced.

Statement of the Problem

Students' performance in biology in external examinations such as the Senior Secondary Certificate Examination (SSCE) has not significantly improved in the last decade, with the average percentage of A1–C6 in biology still below 50.0%, which implies that students taking the subject generally do not exhibit conceptual understanding. A lack of conceptual understanding tends to also facilitate poor retention of the concepts. Many reasons have been adduced as being responsible for students' poor performance in biology: misconception of the content, incompetent mode of teacher's delivery of the subject, and inadequate use of instructional materials and ill-equipped biology laboratories. In addition, teachers in many Nigerian secondary schools do not apply appropriate instructional strategies for teaching the science subjects as identified and recommended to be effective for science instruction (Norom 2009).

Despite many interventions in terms of innovative teaching strategies to improve students' performance, these have not yielded significant improvement. Therefore, learning strategies that will help students to enrich their information, develop their mental abilities, and train them to be innovative and novel are desirable. However, this cannot be done without giving the students the opportunity to express, share, and test their ideas by providing them with appropriate sources, arousing their interests, and inciting their deep thinking with the help of guided approaches. Thus, this study proposes employing a 7E-learning cycle that can enable the students to elicit, engage, explore, explain, elaborate, evaluate, and extend to construct meaningful concepts in biology which could erase the effect of misconception on students' conceptual understanding in learning biology.

Purpose of the Study

The purpose of the study is to investigate the effect of misconception on biology students' conceptual understanding. Specifically, the objectives of the study are to:

1. Ascertain the effect of misconception on students' conceptual understanding
2. Assess how misconception influences students' conceptual understanding
3. Investigate the influence of students' attitude to biology on their conceptual understanding

Significance of the Study

This study would be of immense benefit to biology teachers who could avail of the opportunity to learn and also understand the instructional processes involved in misconception. It is expected to provide information on how misconception influences conceptual understanding of students as the basis for planning as well as on executing worthwhile instructional/learning strategies that could be geared towards improving students' learning outcomes in biology.

The curriculum experts, administrators, authors, parents, peers, and employers of teachers in secondary schools will also profit from this research. Curriculum experts are expected to give more provoking considerations to other concepts in biology that students can learn when exposed to misconception and thus improve on out of the box thinking. Administrators of schools can also look for ways to create more time for biology teachers in order to specify misconception. The findings can encourage the government to organize seminars, conferences, and workshops in which misconception will be discussed and taught in order to improve the teaching and learning of biology. Finally, the findings of this study will provide enrichment in literature in the area of conceptual understanding.

Methodology

This study adopts the use of quasi-experimental design which is suitable for this type of research. The sample for this study comprises 257 students studying biology at the senior secondary school 2 level. The composition was made up of 122 (47.4%) students in the intervention group which represents the misconception learning with 135 (52.6%) students being in the control group which represents the traditional way teachers' present biology to students with no misconception. The sample was also was made up of 111 (43.2%) males and 146 (56.2%) females. Their age ranged from 13 to 18 with the mean age being 16.59 years. With regard to the field of study, humanities students had the highest composition followed by science, while students in the business group had the smallest sample.

The instruments used to collect data were the Biology Conceptual Understanding Test (BCUT) and lesson notes. BCUT comprises sixty 60 multiple choice objective questions with four options to be selected by the students. Students are also to provide reasons for each of the options selected. Conceptual understanding is measured based on non-significant differences between the objective score and the reasoning score. Each correctly scored objective question will attract a score of 1 while an incorrect option will not attract any score. A scientific reason to the option selected will also attract a score of 1. The maximum score obtained for the objective and reasoning score will be 60 marks each. Two separate lesson notes were prepared, one for the control and the other for the experimental group based on misconception. The topics of the lesson were biomes, ecology, and succession.

Result

Data were analyzed based on research questions and hypotheses. Research questions were analyzed quantitatively using descriptive statistics of mean and standard deviation, with qualitative data analyses of transcribing, coding, and content analysis also used. The hypotheses were analyzed using a paired sample t-test, Analysis of Covariance (ANCOVA), and Scheffe post-hoc. The level of significance was put at α=.05 and the data were analyzed using Statistical Package for Social Scientist (SPSS) Version 20.0.

Table 1: Descriptive statistics of respondents by treatment groups, gender, field of study, and age.

S/N	Variable	Levels	N	%
1	Treatment	Misconception	122	47.4
		Traditional Method (control)	135	52.6
2	Gender	Male	111	43.2
		Female	146	56.8
3	Age	13–15 years	39	13.6
		16–18 years	247	86.4
4	Field of Study	Science	99	34.6
		Humanities	101	35.3
		Business	86	30.1

Table 1 indicates that the field of study comprises 71 (%=27.6) science students, 99 (%=38.6) students in the humanities as well as 87 (%=33.8) in the business class.

Research Question 1

What is the impact of misconception and the conventional technique on students' conceptual understanding?

Table 2: Descriptive data on objective score and reasoning score for measuring conceptual understanding in biology.

Group	Pretest			Post-test		
	N	Mean	Sth	Mean	Sth	Mean% Diff
7E-Learning (Exp.) Obj. score	122	27.47	6.94	46.23	3.76	18.76
Reasoning score	122	24.83	7.93	45.59	3.97	20.76
Control Obj. score	135	28.37	4.62	29.29	2.74	0.92
Reasoning score	135	25.67	5.57	26.52	3.17	0.85

Table 2 shows that the mean pretest of students' objective score in biology in the control group ($\bar{x}$=28.37, SD=4.62) was greater than that of the experimental group ($\bar{x}$=27.47, SD=6.94). Similarly, the mean pretest of students' reasoning score in biology in the control group ($\bar{x}$=25.67, SD=5.57) was greater than that of the experimental group ($\bar{x}$=24.83, SD=7.93). On exposure to the intervention which is misconception, students' post-test objective score improved by a mean difference of 18.76 while that of the reasoning score improved by a greater mean difference of 20.76. Thus, the misconception is more effective in declining the reasoning score than the objective score. The conventional technique of teaching biology could only increase the objective score by 0.92 while that of the reasoning score improved by 0.85. The conventional method is more effective in improving the objective score than improving the reasoning score. Thus, conceptual understanding of students in biology with misconception considerably increased when compared to students in the control group.

Hypothesis 1

There will be no significant main effect of misconception and the conventional technique on students' conceptual understanding.

Variable	Mean	N	SD	t-value	Df	Sig. of P
Objective Score	37.12	286	9.06			
Reasoning Score	35.32	286	10.16	8.903	285	.000

Table 3 shows that students exhibited limited conceptual understanding in biology as evidenced in the significant difference between students' objective score and reasoning score in biology (t=8.903, df:285, p=.000). Thus, the null hypothesis stated was rejected as the level of significance of probability was less than 0.05.

Research Question 2: How do the categories of the field of study influence students' conceptual understanding?

Table 3: Descriptive statistics of mean and standard deviation of effect of field of study on students' conceptual understanding.

Field of study	N	Mean ($\bar{x}$)	SD
Science	99	75.18	23.08
Humanities	101	63.03	22.58
Business	86	66.67	20.90

A cursory look at Table 3 shows that the science field of study influenced biology students' conceptual understanding most ($\bar{x}$=75.18, SD=23.08). This was followed by students in the business field ($\bar{x}$=66.67, SD=20.90). The humanities field of study had the least influence on students' conceptual understanding in biology ($\bar{x}$=63.03, SD=22.08).

Hypothesis 2

There will be no significant main effect of the field of study (science, business, and humanities) on students' conceptual understanding, retention, and biology self-efficacy.

It can be seen from Table 4 that there was a significant main effect of the field of study on students' conceptual understanding scores in biology ($F_{(2,\ 279)}$ =11.855, p=.000, η^2_p=.078). Thus, the null hypothesis stated was rejected as the field of study significantly influenced students' conceptual understanding in biology, which was because the significance of probability was less than 0.05. The effect size as measured by Partial Eta Square (η^2_p) reveals that the field of study accounted for 7.8% of the variances to the dependent variable measure of students' conceptual understanding in biology.

Table 4: Pair-wise Comparison of Effect of the Field of Study on Students' Conceptual Understanding in Biology.

(I) Field of Study	(J) Field of Study	Mean Difference (I-J)	Std. Error	Sig.[b]	95% Confidence Interval for Difference[b]	
					Lower Bound	Upper Bound
Science	Humanities	.708	.922	1.000	−1.513	2.928
	Business	4.305*	.948	.000	2.021	6.589
Humanities	Science	−.708	.922	1.000	−2.928	1.513
	Business	3.598*	.932	.000	1.353	5.842
Business	Science	−4.305*	.948	.000	−6.589	−2.021
	Humanities	−3.598*	.932	.000	−5.842	−1.353

Research Question 3

To what extent do the levels of students' attitude to biology influence students' conceptual understanding?

Preamble

Percentile was used to classify students into three categories of attitude: low, medium, and high. Attitude was measured on a scale of 1.0–4.0, with attitude scores of up to 3.27 classified as low. those of 3.28 to 3.40 classified as moderate, and scores greater than 3.40 classified as high.

Table 5: Descriptive statistics of mean and standard deviation of levels of students' attitude to biology on conceptual understanding.

Levels of Attitude	N	Mean ($\bar{x}$)	SD
Low	103	67.41	21.62
Moderate	93	64.67	22.27
High	90	73.28	23.61

Students with a high attitude demonstrated the highest level of conceptual understanding as evidenced in their scores in biology concepts ($\bar{x}$=73.28, SD=23.61) as shown in Table 5. This was followed by students with a low attitude ($\bar{x}$=67.41, SD=21.62). Students with a moderate attitude had the lowest score in conceptual understanding in biology ($\bar{x}$=64.67, SD=22.27).

Hypothesis 3

There will be no significant main effect of students' attitude to biology on students' conceptual understanding.

Field of Study	N	Mean ($\bar{x}$)	SD
Science	99	75.18	23.08
Humanities	101	63.03	22.58
Business	86	66.67	20.90

It can be observed from the above table that there was a non-significant main effect of attitude on students' conceptual understanding in biology ($F_{(2, 279)}$ =.687 p=.504 η^2_p=.0005). This is as a result of the fact that the level of significance of probability was greater than 0.05. Thus, the null hypothesis was not rejected. It can be inferred that students' levels of attitude do not vary significantly to cause a change in students' conceptual understanding in biology.

Discussion

The findings of this study revealed among others that students with misconceptions performed significantly less than those taught without misconceptions. The calculated t-test of difference in the students' gain score of the two experiments is significant at a 0.05 level of significance. This finding is in agreement with the study of Mecit (2006) that investigated the effect of misconception on inquiry based learning and observed a significant effect of the intervention on the critical thinking skills of Grade 5 learners upon attitude towards achievement in science and biology The study suggests that living displays in the classroom serve to stimulate interest and curiosity thereby increasing students' performance in biology.

Similarly, a related study by Marek (2008) noted poor performance results in children, students, or learners through a scientific misconception by having them first explore materials and then construct a concept. The findings of their study revealed that the students with misconception performed at a lower standard than students taught with instructional strategy. The second hypothesis stated that there will be no significant main effect of the field of study on students' conceptual understanding in biology.

The result in Table 3 shows that there was a significant main effect of the field of study on students' conceptual understanding scores in biology ($F_{(2, 279)}$ =11.855, p=.000, η^2_p=.078). Thus, the null hypothesis stated was rejected as the field of study significantly influenced students' conceptual understanding in biology. Some researchers support this finding, for example Anderson and Schonborn (2008) whose first article on conceptual understanding (mini-series) described how cognitive skills are essential for the construction of expert knowledge and argued that conceptual understanding and the associated cognitive skills are multifaceted in nature, as well as essential to explicitly develop such knowledge and cognitive abilities in students in order to scaffold them along the novice-expert continuum.

The third hypothesis stated that there will be no significant main effect of students' attitude to biology on students' conceptual understanding, with the result showing that there was a non-significant main effect of attitude on students' conceptual understanding in biology ($F_{(2, 279)}$ =.687 p=.504 η^2_p=.0005). This is as a result of level of significance of probability being greater than 0.05, meaning the null hypothesis was not rejected. This is in agreement with the findings of Okebukola (2011) who revealed that students' attitude is a major barrier to their underachievement in science.

Conclusion

Based on the findings of this study it is evident that misconception is directly linked to students' conceptual understanding. When students are exposed to misconceptions, they tend to have lower conceptual understanding than they would have with instructional strategy. There is thus the need for teachers to be very careful in introducing new topics in such a way as to prevent students from developing misconceptions in the teaching of science and integrate new knowledge into older understandings in such a way that links are maintained and correct concepts are maintained.

Recommendations

1. Biology teachers should endeavor to relate scientific concepts to misconceptions.
2. The Ministry of Education should regularly organize counseling seminars and workshops on scientific misconceptions for teachers.
3. The school should have a reward for teachers that adequately integrate new knowledge into older understandings to facilitate learning.

References

Anderson, C.W., & Schonborn, K.J (2008). Teaching and learning in science: An overview of research on the integration of inquiry-based learning and scientific reasoning. Journal of Research in Science Teaching, 45,(8), 925–950.

Anderson, J.R., and Schunn, C.D. (2000). Implications of the ACT-R learning theory: no magic bullets. In R. Glasser (Ed.), Advances in instructional psychology (Vol.5), Eribaum.

Anderson et., (2006).Investigating the impact of a practicum experience in an aquarium on pre-service teachers. Teaching Education, 17(4), 341–353

Artino R (2008). A Brief Analysis of Research on ProblemBased Learning, University of Connecticut. June Eric, 6:111–123.

Ausubel, D.P. (2000). The acquisition and retention of knowledge.Dondrech t: Kluvert

Bunde, L. (2007). On the improvement of students' conceptual understanding in statistics education. A Masters Dissertation of the University of Maastricht, University of Germany

De Jong et al., (1996).Types and qualities of knowledge. EducationalPsychology 31: 105–113.

De Jong, T., & Ferguson-Hessler, M.G.M. (2005). Types and qualities of knowledge: A cognitive taxonomy for the design of learning environments. Educational Psychologist, 40(3), 167–178.

Durodolu, O.O (2018). Information Literacy, Self-Concept and Metacognitive Ability of Teacher-Librarians at the University of Zululand. Library Philosophy and Practice (e-journal)

Durodolu, O.O (2021) Flattering the curve of fake news in the epoch of infodemic: an epistemic challenge. Handbook of research on Deep Fakes, Fake News, and Misinformation in online Teaching and Learning Technologies.

Eggen, P. and Kauchak, D. (2012) Strategies and Models for Teachers: Windows, teaching content and thinking skills (6th ed). Boston: Pearson.

Essien, I. T. (2002). Influence of home environment on secondary school students' achievement in Geography in Akwa-Ibom state. Journal of the Nigerian Society for Educational Psychologists (NISEP). 1(1), 109–116.

Fisika, J. &Alam, F.M.D. (2013). Learning Cycle 7E-Model to increase students' critical thinking skills in science.JournalPendidikan Indonesia, 9, 58–66. ISSN: 1693–1246.

Hanuscin, D. (n.d.) Misconceptions in Science E328: Elementary Methods. [Online] http://www.indi ana.edu/~w505a/studwork/deborah/. [20/7/2005].

Marek, E.A. (2008). The role of inquiry-based learning in science education. Journal of Research in Science Teaching, 45(2), 161–185.

McNeill, K.L., Lizotte, D.J., Krajcik, J., & Marx, R.W. (2006). Supporting students' construction of scientific explanations by fading scaffolds in instructional materials. The Journal of the Learning Sciences, 15(2), 153–191.

Mecit, O. (2006). The effect of 7E-learning cycle model on the improvement of fifth grade students critical thinking skills.A Thesis of the Middle East Technical University.

Norom, A.B. (2009). Innovative approaches to learning in secondary education. Journal of Educational Methods, 15(4), 45–58.

Novak, J. D. (2002). Meaningful learning.The essential factor for conceptual change in limited or inappropriate propositional hierarchies leading to empowerment of learners'.Science Education, 86: 548–571.

Nwagbo, C.R. (2009). Effective strategies for teaching and learning biology in secondary schools in Nigeria. International Journal of Educational Research, 7(2), 23–36

Okebukola, P.A.O. (2011). Status of science in Nigerian schools. Lagos: Okebukola Science Foundation

Okebukola, P.A.O. (2012). Breaking barriers to meaningful learning f science. A paper delivered at the international conference of institute of science and technology education, South Africa from 22–25 October, 2012, 29–56.

Okeke, E. A. C. (2009). Attracting women into science-based occupations: Problems and Prospects. In Nnaka, C. and Njoku, Z. (eds). Gender and STM Education Series, No 3.89–108. STAN.

Okoye, B. E & Okeke, O. C, (2007). Efficacy of eliminating superstitious beliefs strategy of achievement and knowledge retention in Genetics among Secondary School Students.Journal of the Science Teachers Association of Nigeria, 42(1&2), 73–77.

Oluwole, O.C.O & Durodolu, O.O (2018). Enhancing job performance of Librarians in Academic Libraries through Technological change. University of Ibadan Journal of Library and information science

Osuafor, A. & Okonkwo, I. (2013). Influence of family background on academic achievement of secondary school biology students in Anambra state. African Research Review. An International Multidisciplinary Journal, 7(3), 156–167.

Tanner, K., & Allen, D. (2005). Approaches to teaching and learning: Implementing active learning techniques. Journal of College Science Teaching, 34(4), 32–37

Petros Nhlavu Dlamini, Esther Oluwayemi Jatto and Adeyinka Tella

Disinformation and Indigenous Knowledge Systems: A Literature Review

Abstract: This study aims to examine disinformation and indigenous knowledge systems through a literature review. This is to define disinformation and indigenous knowledge systems; define the link between disinformation and indigenous knowledge systems; determine the disinformation against indigenous knowledge systems; discuss the indigenous peoples and innovative strategies for resisting disinformation; and describe what the future holds for Indigenous people in light of disinformation. The method adopted involves a review of literature from various sources. The findings reveal the link between disinformation and indigenous knowledge systems, and the implications of this relationship for indigenous peoples and their communities. Disinformation against indigenous systems includes false claims about the effectiveness of traditional healing practices, misrepresentation of indigenous environmental knowledge, denial of indigenous history and cultural heritage, stereotyping of indigenous people as primitive or uncivilized, misrepresentation of indigenous knowledge as static or unchanging, appropriation of indigenous knowledge by non-indigenous people, among other issues. The innovative strategies for resisting disinformation by indigenous people revealed are using social media to tell their own stories, building networks of trusted sources, developing media literacy programs, and collaborating with allies. The findings report that the future of indigenous peoples in light of disinformation is uncertain and complex. While the rise of social media and other digital technologies has created new opportunities for Indigenous peoples to share their stories and perspectives with a global audience, as well as to develop innovative strategies for resisting disinformation, the study concludes that the future for Indigenous peoples in light of disinformation will depend on the collective efforts of Indigenous communities, institutions, and societies as a whole to promote truth, reconciliation, and healing. This will require ongoing dialogue, education, and advocacy to address

Petros Nhlavu Dlamini, Department of Information Studies, Faculty of Arts, University of Zululand, e-mail: DlaminiP@unizulu.ac.za
Esther Oluwayemi Jatto, Department of Library and Information Science, Ajayi Crowther University, Oyo, Oyo State, Nigeria, e-mail: eo.jatto@acu.edu.ng, estherbanjo86@gmail.com, Orcid Number: 0000-0001-5398-1066
Tella Adeyinka, Dept of Library and Information Science, University of Ilorin, Nigeria, Dept of Information Science, University of South Africa, Pretoria, e-mail: tellayinkaedu@yahoo.com, tella.a@unilorin.edu.ng, https://orcid.org/0000-0002-5382-4471

https://doi.org/10.1515/9783111176345-013

the root causes of disinformation and promote greater understanding, empathy, and respect for indigenous cultures and perspectives.

Keywords: disinformation, Indigenous knowledge, Indigenous people, Indigenous knowledge system, literature review

Introduction

Disinformation is the deliberate spread of false or misleading information with the intention of deceiving people (College of Staten Island Library 2023). It can take many forms, including rumours, hoaxes, propaganda, and fake news, and can be spread through various channels such as social media, news outlets, and government agencies. It can be spread through various channels, including social media, news outlets, and even government propaganda. Disinformation can have serious consequences, including damaging reputations, undermining trust in institutions, and even inciting violence. Indigenous knowledge systems refer to the knowledge, practices, and beliefs that are passed down through generations in traditional societies. This knowledge is often based on the interactions between people and the natural environment, and can be essential for their survival and well-being. Indigenous knowledge systems can include traditional healing practices, agricultural techniques, and spiritual beliefs.

Disinformation can have a significant impact on Indigenous knowledge systems. For example, false information about the effectiveness of Western medicine can lead to a loss of confidence in traditional healing practices. Elsewhere, disinformation about the benefits of modern agriculture can lead to a decline in traditional farming techniques, which can be more sustainable and better adapted to local conditions (OECD 2021). Additionally, disinformation can be used to discredit indigenous people and their cultures, perpetuating harmful stereotypes and erasing their contributions to society.

Overall, the spread of disinformation can have a detrimental effect on indigenous knowledge systems and the people who rely on them (Colomina, Sánchez Margalef, and Youngs 2021). It is important to be aware of the sources of information we consume and to critically evaluate the information we receive, especially when it pertains to traditional knowledge and cultural practices.

Disinformation can have serious consequences, such as damaging reputations, undermining trust in institutions, inciting violence, and even affecting elections or public policy decisions. Disinformation can be created and disseminated by individuals, organizations, or governments, and it can be used for various purposes, including political gain, financial profit, or ideological reasons. The spread of disinformation has become a major challenge in the digital age, and it is essential to be aware of its existence and to critically evaluate the information we consume.

Examples of Disinformation

Examples of disinformation abound in the literature, some of which are considered here:

False claims about the COVID-19 vaccine: throughout the pandemic, there have been numerous false claims about the COVID-19 vaccine, including that it can alter DNA or cause infertility. These claims have been debunked by health experts, but they continue to be circulated on social media and other platforms. According to a study by the Center for Countering Digital Hate, 65% of anti-vaccine content on social media platforms can be traced back to just 12 accounts (Bond, 2021).

Russian disinformation during the 2016 U.S. presidential election: in 2016, Russian operatives used social media platforms to spread disinformation and sow discord during the U.S. presidential election. According to a report by the Senate Intelligence Committee, the Russian Internet Research Agency "sought to influence the 2016 U.S. presidential election by harming Hillary Clinton's chances of success and supporting Donald Trump at the direction of the Kremlin" (Wu, 2019).

Disinformation about climate change: there have been numerous false claims about climate change, including that it is a hoax or that it is not caused by human activity. These claims have been debunked by climate scientists but they continue to be promoted by some politicians and media outlets. According to a study published in the journal *Global Environmental Change*, the fossil fuel industry has been a major source of disinformation about climate change (The Guardian 2024) These are just a few examples of disinformation, but they illustrate the wide range of topics and sources that can be involved. It is important to critically evaluate the information we consume and to seek out reliable sources.

Other examples might include but are not limited to:

Election fraud claims in the 2020 U.S. presidential election: following the 2020 U.S. presidential election, then-President Donald Trump and some of his supporters claimed that the election had been stolen through widespread voter fraud. These claims were repeatedly debunked by election officials and courts, and there was no evidence of widespread fraud. Nevertheless, the claims were widely circulated on social media and other platforms, contributing to distrust in the election process and culminating in the January 6, 2021 attack on the U.S. Capitol (NPR 2023).

Disinformation about the origins of COVID-19: since the outbreak of COVID-19, there have been various theories about the origin of the virus, including that it was intentionally created in a lab or that it was transmitted from animals to humans through a wet market in Wuhan, China. While the exact origins of the virus are still under investigation, there is no credible evidence to support the lab leak theory, and most experts believe that the virus likely originated from animals. However, the lab leak theory has been promoted by some politicians and media outlets, contributing to anti-Asian sentiment and damaging international cooperation in responding to the pandemic (Wade, 2021).

Disinformation about the safety and efficacy of vaccines: disinformation about vaccines is not new, but it has become particularly prevalent during the COVID-19 pandemic. False claims about the safety and efficacy of vaccines have been spread on social media and other platforms, contributing to vaccine hesitancy and hindering efforts to control the pandemic. For example, a study published in the journal *Vaccine* found that Twitter accounts that spread vaccine misinformation were more likely to be followed by real people than accounts that shared accurate information (Institute for Data, Democracy & Politics Columbian College of Arts & Sciences, 2020).

These examples show how disinformation can have serious consequences, including damaging public health, undermining democracy, and promoting hatred and violence. It is important to be aware of the sources of information we consume and to critically evaluate their credibility.

Indigenous Knowledge Systems with Examples

Indigenous knowledge systems refer to the complex knowledge systems developed by Indigenous peoples over generations, through their experiences, observations, and interactions with the environment. These knowledge systems include a wide range of practices, beliefs, values, and traditions, and are closely tied to indigenous cultures and ways of life.

Examples of Indigenous Knowledge Systems

Traditional ecological knowledge: Indigenous peoples have developed intricate knowledge systems about the environment, including plants, animals, and ecosystems. This knowledge is often based on direct observation and experience, and is transmitted orally from generation to generation. For example, the Tla-o-qui-aht First Nation in British Columbia, Canada, has developed a detailed understanding of the local ecosystem, including the life cycles of salmon and other fish and the relationships between species in the ecosystem. This knowledge has been instrumental in managing local fisheries and preserving the ecosystem for future generations.

Medicinal plants and traditional healing: Indigenous peoples have developed a rich knowledge of medicinal plants and traditional healing practices. Many of these plants have been used for centuries to treat a wide range of illnesses and ailments, and are often viewed as sacred. For example, the Mayan people in Central America have used the bark of the chuchuhuasi tree for centuries to treat various ailments, including arthritis, fever, and respiratory problems. The bark is also used in spiritual ceremonies, and is believed to have protective powers.

Oral traditions and storytelling: Indigenous peoples have a rich tradition of oral storytelling, which is used to pass down knowledge and traditions from generation to generation. These stories often contain important lessons about history, culture, and the relationship between humans and the environment. For example, the Navajo people in the South-western United States have a tradition of oral storytelling, which includes stories about the creation of the world and the origins of their culture. These stories are viewed as sacred and are an important part of Navajo identity and culture.

Indigenous knowledge systems are often tied to specific cultural and spiritual practices, and are deeply rooted in Indigenous ways of life. These knowledge systems are increasingly recognized as valuable sources of information and perspectives, particularly in the areas of environmental conservation, public health, and cultural preservation. However, they have also been threatened by colonization, globalization, and other forces that have disrupted Indigenous cultures and ways of life.

Astronomy and cosmology: Indigenous peoples have developed sophisticated knowledge systems related to astronomy and cosmology. Many indigenous cultures view the stars, planets, and other celestial bodies as sacred, and have developed intricate systems for tracking their movements and using them to navigate and mark important events. For example, the Inca people in the Andes mountains of South America developed a complex system of astronomy which they used to predict seasonal changes, mark agricultural cycles, and plan religious ceremonies.

Traditional knowledge of architecture and construction: Indigenous peoples have developed unique forms of architecture and construction that reflect their cultural and environmental contexts. These building techniques often rely on locally available materials and are designed to maximize comfort, sustainability, and cultural expression. For example, the igloo, a traditional form of housing used by Inuit peoples in the Arctic, is built using blocks of snow or ice and is designed to retain heat and protect against cold winds. The igloo's design allows it to be quickly constructed and easily moved, making it well-suited to the nomadic lifestyle of many Inuit communities.

Traditional food systems: Indigenous peoples have developed complex systems for gathering, cultivating, and preparing food, based on their knowledge of local ecosystems and cultural practices. These food systems often emphasize the importance of sustainable harvesting practices and the use of locally available ingredients. For example, the Mapuche people in Chile have developed a traditional food system based on the cultivation of native crops, such as quinoa, and the gathering of wild foods, such as mushrooms and berries. These foods are often prepared using traditional techniques and are an important part of Mapuche culture and identity.

These examples demonstrate the diversity and richness of indigenous knowledge systems, and their importance for understanding the complex relationships between humans, the environment, and culture. However, these knowledge systems have also been threatened by historical and ongoing colonization, environmental degradation, and cultural assimilation. It is important to recognize and support the preservation and revitalization of indigenous knowledge systems, as a means of promoting cultural diversity, sustainability, and social justice.

The Link Between Disinformation and Indigenous Knowledge Systems

The relationship between disinformation and indigenous knowledge systems is an area of growing concern and research. Indigenous peoples have historically been subject to disinformation campaigns aimed at undermining their cultures, erasing their histories, and suppressing their knowledge systems. This literature review explores the link between disinformation and indigenous knowledge systems, and the implications of this relationship for indigenous peoples and their communities.

Several studies have highlighted the ways in which disinformation campaigns have targeted indigenous communities and knowledge systems. For example, a study by Burt (2018) examined the role of disinformation in the Dakota Access Pipeline (DAPL) protests, which were led by Indigenous activists in the United States. The study found that social media platforms were used to spread false and misleading information about the protests, and that this disinformation campaign was aimed at discrediting Indigenous activists and their claims to land and sovereignty.

Other studies have focused on the impact of disinformation on Indigenous knowledge systems, particularly in the context of environmental conservation. For example, a study by Ratuva et al 2020) examined the role of disinformation in shaping public perceptions of Indigenous conservation practices in Fiji. The study found that disinformation campaigns, which portrayed Indigenous conservation practices as backward and ineffective, had undermined the credibility of Indigenous knowledge systems and eroded support for Indigenous-led conservation initiatives.

Despite these challenges, there is growing recognition of the importance of Indigenous knowledge systems for addressing global challenges, such as climate change and environmental degradation. Several studies have highlighted the potential of Indigenous knowledge systems for informing sustainable and culturally appropriate solutions to these challenges. For example, a study by Ens et al. (2015) examined the role of Indigenous knowledge systems in shaping ecosystem-based management strategies in Canada. The study found that Indigenous knowledge systems had contributed to more holistic and adaptive management approaches, and had helped to bridge gaps between scientific and traditional knowledge systems.

In conclusion, the link between disinformation and Indigenous knowledge systems is a complex and multifaceted issue. Disinformation campaigns have historically targeted Indigenous peoples and their knowledge systems, and continue to pose a threat to Indigenous cultures, histories, and ways of life. However, there is also growing recognition of the value and importance of Indigenous knowledge systems for addressing global challenges and promoting sustainability and social justice. Future research in this area will be crucial for understanding the relationship between disinformation and indigenous knowledge systems, and for developing strategies to protect and promote indigenous cultures and knowledge systems.

Furthermore, the impact of disinformation on Indigenous knowledge systems has implications for both Indigenous peoples and the wider society. Indigenous knowledge systems are often closely tied to the local ecosystems and are developed through generations of experience and observation. They represent a unique and valuable source of knowledge that is often overlooked or dismissed by mainstream scientific and political institutions. When Indigenous knowledge systems are discredited or ignored, it can lead to the loss of valuable knowledge and practices that could be used to address contemporary environmental and social challenges.

Moreover, disinformation campaigns can have detrimental effects on Indigenous peoples and their communities. Discrediting Indigenous knowledge systems can undermine the cultural identity and sovereignty of indigenous communities, and can perpetuate historical injustices and marginalization. Disinformation campaigns can also lead to social and political conflict, as evidenced by the DAPL protests and other Indigenous-led movements around the world.

However, there are also opportunities for Indigenous knowledge systems to challenge and resist disinformation. Indigenous peoples and their communities have developed innovative strategies for resisting disinformation and asserting their cultural identities and knowledge systems. For example, many Indigenous communities have turned to social media and other digital platforms to share their stories, challenge disinformation, and connect with allies and supporters around the world.

The link between disinformation and Indigenous knowledge systems is a complex and evolving issue that requires further research and engagement from multiple stakeholders. Recognizing the value and importance of Indigenous knowledge systems, and supporting their preservation and revitalization, is crucial for promoting cultural diversity, sustainability, and social justice.

Disinformation Against Indigenous Knowledge

Disinformation against Indigenous knowledge refers to the spread of false or misleading information about the beliefs, practices, and knowledge systems of Indigenous communities. This type of disinformation is often used to discredit Indigenous people and their cultural traditions, and to advance political or economic interests. Here are some examples and citations of disinformation against Indigenous knowledge:

False claims about the effectiveness of traditional healing practices: Indigenous communities around the world have developed their own traditional healing practices, which are often based on the use of plants and other natural remedies.

However, disinformation campaigns have sought to discredit these practices and promote Western-style medicine as the only legitimate form of healthcare. For example, in South Africa, the government has promoted a campaign called "Operation Khanyisa" which seeks to "eradicate harmful traditional practices" and promote Western medicine instead (SAPA News 2019).

Misrepresentation of indigenous environmental knowledge: Indigenous communities have long been stewards of their land and have developed sophisticated knowledge systems about their environment. However, disinformation campaigns have sought to portray Indigenous people as "primitive" and lacking in scientific knowledge. For example, in the Amazon rainforest, there have been efforts to discredit Indigenous knowledge about sustainable forestry practices and promote large-scale industrial logging instead (The Guardian 2020).

Denial of indigenous history and cultural heritage: Disinformation campaigns have also sought to deny the history and cultural heritage of indigenous communities, often in the service of political or economic interests. For example, in Australia, there has been a long-standing campaign to deny the history and ongoing impact of colonization on Indigenous Australians, with some politicians even claiming that "there was no slavery in Australia" (The Guardian, 2020b).

These examples demonstrate how disinformation campaigns can be used to undermine the knowledge, culture, and rights of Indigenous communities. It is important to recognize and challenge these campaigns in order to support the preservation and advancement of Indigenous knowledge systems.

Stereotyping Indigenous people as primitive or uncivilized: disinformation campaigns have often portrayed Indigenous people as primitive, uncivilized, and backwards, which can perpetuate harmful stereotypes and justify discriminatory policies. For example, in the United States, there has been a long history of stereotyping Native Americans as "savages" or "redskins," which has contributed to systemic racism and violence against indigenous people (NPR 2019).

Misrepresentation of indigenous knowledge as static or unchanging: another form of disinformation against Indigenous knowledge is the idea that it is static or unchanging, and therefore irrelevant to modern society. This ignores the fact that Indigenous knowledge systems have always evolved and adapted to new circumstances. For example, in Canada, there has been a tendency to treat indigenous knowledge as a relic of the past rather than a living and evolving tradition that has much to offer in terms of sustainable resource management (The Conversation 2018).

Appropriation of Indigenous knowledge by non-indigenous people: finally, disinformation campaigns can also involve the appropriation of Indigenous knowledge by non-Indigenous people, often without proper acknowledgement or compensation. This can lead to a distortion or erasure of Indigenous knowledge, and can contribute to the marginalization of Indigenous communities. For example, in New Zealand, there has been controversy over the use of Maori traditional knowledge in the development of new pharmaceuticals, with some arguing that this represents a form of exploitation and theft (Radio New Zealand 2021).

These are just a few examples of the many ways in which disinformation can be used to undermine Indigenous knowledge and rights. It is important to remain vigilant and critical of such campaigns, and to support the ongoing preservation and advancement of Indigenous knowledge systems.

Indigenous Peoples and Innovative Strategies for Resisting Disinformation

Indigenous peoples have long been the targets of disinformation campaigns aimed at erasing their histories, cultures, and land rights. In recent years, these campaigns have taken on new urgency as the rise of social media has enabled disinformation to spread quickly and easily. To combat this problem, Indigenous peoples have developed innovative strategies for resisting disinformation. Here are some examples:

Using Social Media to Tell Their Own Stories

One of the most effective strategies for combating disinformation is for Indigenous peoples to use social media to tell their own stories. This allows them to bypass traditional media outlets that may be biased against them and to share their perspectives directly with a global audience. For example, the Standing Rock Sioux Tribe used Facebook and Twitter to document their protests against the Dakota Access Pipeline in 2016 and 2017, drawing international attention to their cause (LaDuke 2017).

Building Networks of Trusted Sources

Another strategy is for Indigenous communities to build networks of trusted sources who can help them verify information and identify disinformation. For example, the Indigenous Environmental Network has developed partnerships with scientists, journalists, and other experts who can help them fact-check claims made by corporations and government agencies (Klein 2019).

Developing Media Literacy Programs

Indigenous organizations are also developing media literacy programs to teach their communities how to identify and resist disinformation. For example, the Native American Journalists Association has created a guide to media literacy that includes tips for identifying fake news and propaganda (NAJA 2019).

Collaborating with Allies

Indigenous peoples are also collaborating with allies to resist disinformation. For example, the Maori community in New Zealand has worked with the tech industry to develop a tool called “Fight Fake News” that helps users identify and report disinformation on social media (Buchanan 2019).

Generally, these innovative strategies demonstrate that Indigenous communities are not passive victims of disinformation but are actively working to protect their cultures and land rights in the face of misinformation campaigns.

What Does the Future Hold for Indigenous People in Light of Disinformation?

The future for Indigenous peoples in light of disinformation is uncertain and complex. On one hand, the rise of social media and other digital technologies has created new opportunities for Indigenous peoples to share their stories and perspectives with a global audience, as well as to develop innovative strategies for resisting disinformation (Bidwell 2019). However, these same technologies also make it easier for disinformation to spread quickly and widely, potentially undermining Indigenous cultures, histories, and land rights.

In order to address these challenges, it will be important for Indigenous communities to continue to develop and refine their strategies for resisting disinformation (SAPA News 2019). This may involve building stronger networks of trusted sources, engaging in media literacy education, and collaborating with allies from other communities and sectors.

At the same time, governments, tech companies, and other institutions will need to take responsibility for addressing the broader social and political contexts that enable disinformation to thrive (Castel 2012).This may involve regulating social media platforms, supporting independent journalism, and recognizing and respecting Indigenous land rights and sovereignty. Ultimately, the future for Indigenous peoples in light of disinformation will depend on the ability of communities, institutions, and societies as a whole to address the underlying causes of disinformation and to promote greater understanding, empathy, and respect for diverse cultures and perspectives (Christen 2012).

In addition to the strategies mentioned earlier, Indigenous peoples can also explore the use of emerging technologies such as blockchain, artificial intelligence (AI), and machine learning (ML) to combat disinformation (Davis 2020). For example, blockchain can be used to create decentralized systems for storing and sharing information, which can help prevent the spread of false or manipulated information. Elsewhere, AI and ML can be used to analyze large volumes of data and detect patterns of disinformation, which can help Indigenous communities to identify and respond to threats more quickly and effectively (Davis 2020).

Another important issue to consider is the role of traditional media in shaping public perceptions of Indigenous peoples. While social media has created new opportunities for Indigenous peoples to share their stories and perspectives, traditional media outlets such as newspapers, television, and radio continue to play a significant role in shaping public opinion. Unfortunately, these media outlets have a long history of perpetuating harmful stereotypes and erasing Indigenous cultures and histories (Frontera News 2021).To address this issue, Indigenous peoples can work to build relationships with traditional media outlets and advocate for more balanced and accurate coverage of Indigenous issues (Kovach 2009). They can also develop their own media outlets and platforms, such as community radio stations and newspapers, to share their perspectives and counteract disinformation.

Overall, the future for Indigenous peoples in light of disinformation will depend on the ability of communities, institutions, and societies as a whole to address the underlying causes of disinformation and to promote greater understanding, empathy, and respect for diverse cultures and perspectives. This will require ongoing collaboration and innovation, as well as a commitment to upholding Indigenous rights and sovereignty (McKenna and Mason 2019).

It is also important to recognize that the impacts of disinformation on Indigenous peoples are not uniform and can vary depending on factors such as geographic location, cultural context, and political climate. For example, Indigenous peoples in countries with strong legal protections for land rights and self-determination may be better equipped to resist disinformation and defend their cultures and identities (Paradies 2016).However, Indigenous peoples in countries with weak legal protections and ongoing conflicts over land and resource rights may face greater challenges in resisting disinformation and asserting their sovereignty. In these contexts, it may be necessary to work with international organizations, human rights groups, and other allies to raise awareness of Indigenous issues and advocate for stronger protections for Indigenous rights (UN, 2007).

In addition to addressing the broader social and political contexts that enable disinformation to thrive, it is also important to support the mental health and well-being of Indigenous peoples who may be impacted by disinformation. Disinformation campaigns can cause significant psychological distress and trauma, particularly for those who are targeted by hate speech and other forms of online harassment. This can have long-term impacts on individuals, families, and communities, and may require specialized mental health services and support.

Ultimately, the future for Indigenous peoples in light of disinformation will depend on the collective efforts of Indigenous communities, institutions, and societies as a whole to promote truth, reconciliation, and healing. This will require ongoing dialogue, education, and advocacy to address the root causes of disinformation and promote greater understanding, empathy, and respect for Indigenous cultures and perspectives.

Likely Challenges Misinformation May Create for Indigenous Knowledge

Misinformation poses significant challenges to indigenous knowledge, which encompasses the rich cultural, traditional, and ecological wisdom of indigenous communities. Misinformation can distort, devalue, and erode indigenous knowledge in various ways, undermining its preservation, transmission, and rightful recognition. Here is a detailed discussion of the challenges that misinformation can create for indigenous knowledge:

Cultural Erosion and Loss of Authenticity: misinformation can misrepresent or misinterpret indigenous cultural practices, beliefs, and traditions, leading to a distortion of their authentic heritage (Fafunwa 2018).Communities might lose the true essence and significance of their knowledge due to the spread of inaccurate information.

Undermining Traditional Ecological Wisdom: indigenous knowledge often includes ecological wisdom and sustainable practices that have been honed over generations (Martin and Mirraboopa 2003). Misinformation can undermine these practices by promoting harmful practices or misrepresenting indigenous environmental stewardship.

Violation of Cultural Rights: misinformation can perpetuate stereotypes and inaccuracies about indigenous communities, contributing to the violation of their cultural rights and dignity. Cultural appropriation and offensive portrayals can arise from misinformation, further marginalizing and disrespecting indigenous knowledge (Grimwood and McManus 2016).

Threat to Self-Determination: misinformation may misrepresent indigenous history, land rights, and narratives, weakening claims to self-determination and autonomy (Wright 2016).Distorted information can impact legal, political, and social discussions about indigenous rights and sovereignty.

Interference with Transmission: misinformation can disrupt the intergenerational transmission of indigenous knowledge, as inaccurate information is passed down to younger generations. Traditional teachings and values may become diluted or lost due to the infusion of misinformation.

Diminished Cultural Preservation: misinformation can divert attention and resources away from efforts to document, preserve, and revitalize indigenous languages, traditions, and practices. This diversion can hamper cultural preservation initiatives and the revitalization of indigenous knowledge systems.

Lack of Recognition and Respect: misinformation can contribute to the lack of recognition and respect for indigenous knowledge systems by portraying them as outdated, inferior, or irrelevant. This can perpetuate a hierarchy of knowledge that marginalizes indigenous perspectives.

Environmental and Health Risks: misinformation can lead to the spread of incorrect indigenous knowledge related to medicinal plants, natural resource management, and environmental practices. This can have adverse effects on both ecosystems and community health.

Research Ethics and Exploitation: misinformation in research about indigenous communities can result in unethical practices, including the unethical use of indigenous knowledge without proper consent or acknowledgement.

Lack of Trust and Cultural Appropriation: misinformation can erode trust between indigenous communities and outsiders, as outsiders may rely on false information to engage with indigenous knowledge. Cultural appropriation can also arise when outsiders adopt, adapt, or misuse indigenous knowledge based on misinformation.

Addressing these challenges requires a comprehensive approach that involves Indigenous communities, policymakers, and educators, researchers, and media outlets. Promoting accurate, respectful, and ethical representations of Indigenous knowledge while empowering Indigenous communities to share their own narratives is crucial to countering the negative impacts of misinformation.

Strategies for Curbing Misinformation to Protect Indigenous Knowledge

Curtailing misinformation to protect indigenous knowledge requires a multifaceted approach that involves indigenous communities, educators, researchers, policymakers, media, and society at large. Here are strategies that can be employed to curb misinformation and safeguard indigenous knowledge:

Community Empowerment and Engagement: there is a need to involve indigenous communities as active participants in information dissemination, fact-checking, and content creation related to their knowledge. It is important to also support community-led initiatives to document, preserve, and share authentic indigenous knowledge through various media, including oral traditions, videos, and websites.

Cultivate Media Literacy and Critical Thinking: with this, it is important to develop media literacy programs that educate indigenous communities and the broader public about discerning credible sources and verifying information. Foster critical thinking skills to enable individuals to critically assess information and recognize potential misinformation.

Culturally Relevant Education: there is a need to incorporate indigenous knowledge and cultural context into educational curricula, promoting accurate and respectful representations (Smith 2012; Broudy and Hunt 2017). We need to teach students about the importance of validating information from indigenous sources and engaging in culturally sensitive research.

Digital Storytelling and Social Media: We should encourage indigenous communities to use digital platforms, such as social media and websites, to share their stories, knowledge, and perspectives directly with the public. There is also the need to provide training on digital storytelling techniques to ensure effective and culturally appropriate communication.

Collaborative Research and Ethical Practices: We can engage in collaborative research partnerships with indigenous communities that prioritize informed consent, equitable participation, and respectful use of indigenous knowledge. We can also establish ethical guidelines that ensure indigenous communities are acknowledged, consulted, and appropriately compensated for their contributions.

Fact-Checking and Verification: There is a need to establish fact-checking mechanisms specific to indigenous knowledge to verify information before dissemination (Martin 2003). Encouraging the use of reputable sources and experts from indigenous communities to validate information can also work out.

Open Access and Digital Archives: It is important to create open-access repositories and digital archives of indigenous knowledge, curated and validated by indigenous experts. We can also ensure that such repositories are user-friendly, accessible, and searchable for both indigenous communities and the general public.

Media Representation and Cultural Sensitivity: there is a need to encourage media outlets to consult with indigenous experts when covering stories related to indigenous knowledge and culture. Cultural sensitivity and accuracy in media portrayals of indigenous communities should also be promoted (UNESCO 2021).

Legal Protection and Intellectual Property Rights: We can strengthen legal frameworks to protect indigenous intellectual property rights and prevent the unauthorized use or misrepresentation of indigenous knowledge (Indigenous Peoples and Sustainable Development Knowledge Platformn.d.). We can also facilitate the documentation of traditional knowledge and the registration of indigenous cultural expressions.

Public Awareness Campaigns: We can launch public awareness campaigns that highlight the value of indigenous knowledge, debunk myths, and emphasize the importance of accurate representation (Martin 2003). We can also foster a broader understanding of indigenous cultures and knowledge systems.

Collaboration with Technology Companies: we can collaborate with technology companies to implement fact-checking tools and algorithms that detect and counter misinformation related to indigenous knowledge.

Community Elders and Cultural Authorities: there is a need to empower indigenous elders, cultural authorities, and knowledge holders to play a central role in combating misinformation by sharing accurate information and correcting misconceptions.

Combating misinformation to protect Indigenous knowledge requires a comprehensive effort that respects cultural diversity, promotes inclusivity, and empowers indigenous communities to share their knowledge on their terms. It also necessitates ongoing collaboration, education, and advocacy to ensure that indigenous knowledge is preserved, respected, and accurately represented.

Conclusion

The study concludes that the future for Indigenous peoples in light of disinformation will depend on the collective efforts of Indigenous communities, institutions, and societies as a whole to promote truth, reconciliation, and healing. This will require ongoing dialogue, education, and advocacy to address the root causes of disinformation and promote greater understanding, empathy, and respect for Indigenous cultures and perspectives.

Recommendations

Based on the foregoing discussion, it is important to recognize the value of Indigenous Knowledge and work towards integrating it into decision-making processes. Additionally, it is important to work towards decolonizing knowledge systems and recognizing Indigenous Knowledge as a valid form of knowledge.

References

Bidwell, P. (2019). Disinformation and Indigenous Peoples: A Global Overview. Centre for International Governance Innovation.

Bond, S. (2021). Just 12 people are behind most vaccine hoaxes on social media, research shows. Special Series Untangling Disinformation – Retrieved from https://www.npr.org/2021/05/13/996570855/disinformation-dozen-test-facebooks-twitters-ability-to-curb-vaccine-hoaxes (access 15 August 2024).

Broudy, D., & Hunt, D. (2017). Cultural bias, misinformation, and conservation policy: Challenges to Indigenous knowledge and sovereignty in the Arctic. *Cultural Anthropology*, 32(3), 327–355.

Buchanan, E. (2019). Indigenous communities are fighting fake news with local journalism and Facebook. MIT Technology Review. Retrieved from https://www.technologyreview.com/2019/06/07/239336/indigenous-communities-are-fighting-fake-news-with-local-journalism-and-facebook/

Burt, B.Q. (2018). Indigenous-led environmentalism: news framing of pipeline protest in the harper and trudeau eras. A thesis, Department of Sociology Memorial University of Newfoundland. Retrieved from https://research.library.mun.ca/13414/1/Burt_BenjaminQuinn_master.pdf (access 15 August 2024).

Castells, M. (2012). Networks of Outrage and Hope: Social Movements in the Internet Age. Polity Press.

Christen, K. (2012). Does Information Really Want to be Free?: Indigenous Knowledge Systems and the Question of Openness. *International Journal of Communication*, 6, 2870–2893.

College of Staten Island Library (2023). Misinformation and disinformation: thinking critically about information sources. Retrieved https://library.csi.cuny.edu/misinformation.

COLOMINA, C, SÁNCHEZ MARGALEF, H. & YOUNGS, R. (2021). The impact of disinformation on democratic processes and human rights in the world. https://www.europarl.europa.eu/RegData/etudes/STUD/2021/653635/EXPO_STU(2021)653635_EN.pdf

Davis, J. (2020). Blockchain and Indigenous Land Claims: Indigenous Land Titling and Blockchain Technology. *International Indigenous Policy Journal*, 11(4), 1–15.

Ens, E. J., Pert, P., Clarke, P. A., Budden, M., Clubb, L., Doran, B., Douras, C., Gaikwad, J., Gott, B., Leonard, S., Locke, J., Packer, J., Turpin, G., & Wason, S. (2015). Indigenous biocultural knowledge in ecosystem science and management: Review and insight from Australia. *Biological Conservation*, *181*, 133149. https://doi.org/10.1016/j.biocon.2014.11.008

Fafunwa, A. A. (2018). Indigenous knowledge and epistemicide: Challenges and prospects. *Journal of African Cultural Studies*, 30(3), 312–324.

Frontera News. (2021). AI and Indigenous Peoples: A Strategic Partnership to Fight Misinformation. https://fronteranews.com/indigenous-people-and-ai-a-strategic-partnership-to-fight-misinformation/

Grimwood, B. S., & McManus, R. J. (2016). Indigenous Knowledge Systems, Intellectual Property and Misinformation in the Digital Age. *The International Indigenous Policy Journal*, 7(4).

Indigenous Peoples and Sustainable Development Knowledge Platform. (n.d.). Guidelines for the Use of the Intellectual Property System to Protect Traditional Knowledge, Traditional Cultural Expressions and Genetic Resources.

Institute for Data, Democracy & Politics Columbian College of Arts & Sciences (2020). Vaccine hoaxes are rampant on social media. Here's how to spot them. https://iddp.gwu.edu/vaccine-hoaxes-are-rampant-social-media-heres-how-spot-them (access 15 August 2024).

Klein, N. (2019). How Indigenous people are using technology to protect the planet. The Guardian. Retrieved from https://www.theguardian.com/environment/2019/may/17/how-indigenous-people-are-using-technology-to-protect-the-planet

Kovach, M. (2009). Indigenous Methodologies: Characteristics, Conversations, and Contexts. University of Toronto Press.

LaDuke, W. (2017). Standing Rock Sioux Tribe wins a major victory in court. The Nation. Retrieved from https://www.thenation.com/article/archive/standing-rock-sioux-tribe-wins-major-victory-court/

Lindgren, S. (2020). Countering Misinformation about Indigenous Peoples: What Can We Learn from the Current Crisis? *Journal of Indigenous Wellbeing*, 5(1), 1–12.

Martin, K. (2003). Ways of knowing, ways of being and ways of doing: A theoretical framework and methods for Indigenous re-search and Indigenist research. *Journal of Australian Studies*, 76(203), 203–214.

Martin, K., & Mirraboopa, B. (2003). Ways of knowing, ways of being and ways of doing: A theoretical framework and methods for Indigenous re-search and Indigenist research. *Journal of Australian Studies*, 76(203), 203–214.

McKenna, B., & Mason, L. (2019). Reimagining Traditional Media through a First Nations Lens. *Canadian Journal of Communication*, 44(1), 65–83.

Native American Journalists Association (NAJA). (2019). Media literacy guide. Retrieved from https://najanewsroom.com/2019/03/20/media-literacy-guide/

NPR. (2019). Why Native American mascots have become so controversial. Retrieved from https://www.npr.org/2019/02/08/692665253/why-native-american-mascots-have-become-so-controversial

NPR (2023). Social media's role in Jan. 6 was left out of the final report. https://www.npr.org/2023/01/26/1151360750/social-medias-role-in-jan-6-was-left-out-of-the-final-report (access 15 August 2024).

OECD (2021). Adoption of technologies for sustainable farming systems wageningen workshop proceedings. https://www.oecd.org/greengrowth/sustainable-agriculture/2739771.pdf

Paradies, Y. (2016). Colonisation, Racism and Indigenous Health. *Journal of Population Research*, 33(1), 83–96.

Radio New Zealand. (2021). Use of Māori traditional knowledge in drug development 'exploitative'. Retrieved from https://www.rnz.co.nz/news/national/446209/use-of-maori-traditional-knowledge-in-drug-development-exploitative

Ratuva, S., Crichton-Hill, Y., Ross, T., Basu, A., Vakaoti, P. & Martin-Neuninger, R. (2021). Integrated social protection and COVID-19: rethinking Pacific community responses in Aotearoa. *Journal of the Royal Society of New Zealand*, DOI: 10.1080/03036758.2020.1861033

SAPA News. (2019). SA Government to eradicate harmful traditional practices. Retrieved from https://www.sanews.gov.za/south-africa/sa-government-eradicate-harmful-traditional-practices

Smith, L. T. (2012). Decolonizing methodologies: Research and Indigenous peoples (2nd ed.). Zed Books.

The Conversation. (2018). Indigenous knowledge has been warning us about climate change for centuries. Retrieved from https://theconversation.com/indigenous-knowledge-has-been-warning-us-about-climate-change-for-centuries-105305

The Guardian (2024). Smoking gun proof': fossil fuel industry knew of climate danger as early as 1954, documents show. https://www.theguardian.com/us-news/2024/jan/30/fossil-fuel-industry-air-pollution-fund-research-caltech-climate-change-denial (access 15 August 2024).

The Guardian. (2020a). Brazil's Amazon rainforest under siege from Covid-19, loggers and miners. Retrieved from https://www.theguardian.com/world/2020/jul/09/brazil-amazon-rainforest-covid-19-loggers-miners

The Guardian. (2020b). Australian politicians reject slavery claims, call for 'honest conversation' about history. Retrieved from https://www.theguardian.com/australia-news/2020/jun/12/australian-politicians-reject-slavery-claims-call-for-honest-conversation-about-history

UN (2007). The United Nations Declaration on the Rights of Indigenous Peoples. A Manual for National Human Rights Institutions. Retrieved from https://www.ohchr.org/sites/default/files/Documents/Issues/IPeoples/UNDRIPManualForNHRIs.pdf (accessed 15 August 2024).

UNESCO. (2021). Media and Information Literacy: Policy and Strategy Guidelines.

United Nations. (2007). United Nations Declaration on the Rights of Indigenous Peoples. https://www.un.org/development/desa/indigenouspeoples/declaration-on-the-rights-of-indigenous-peoples.html

Wade, N. (2021). The origin of COVID: Did people or nature open Pandora's box at Wuhan? *Bulleting of the Atomic Scientist*, May 5, https://thebulletin.org/2021/05/the-origin-of-covid-did-people-or-nature-open-pandoras-box-at-wuhan/ (access 15 August 2024).

Wright, S. (2016). Misinformation and its consequences in the Myanmar context: How misinformation undermines indigenous knowledge and exacerbates tension and violence. Devpolicy Blog, 28.

Wu, N. (2019). Senate report calls for action against foreign election interference amid Trump-Ukraine controversy. US Today, https://www.usatoday.com/story/news/politics/2019/10/08/senate-russia-helped-trump-against-clinton-2016/3909680002/ (access 15 August 2024).

Oluwole O. Durodolu, Olatokunbo Christopher Okiki
and Stella C. Nduka

Information Detergent: a Panacea to the Mounting Threat of Information Pollution

Abstract: Information detergent seeks to promote a healthier information environment and reduce the negative impacts of fake news and misinformation. The chapter discusses using the following concepts to explain information detergent: confirmation bias, cognitive bias, echo chamber, and post-truth. In combatting information disorder due to the proliferation of social media, information disorder strategies were canvassed in tackling fake news that permeates the entire social media space, such as verifying the source of information, fast-checking, seeking multiple sources, avoiding echo chambers, applying critical thinking, and reporting misinformation to the appropriate authority. The chapter recommended that librarians should carry out information literacy training programs for their users to reduce the incidence of information disorder and promote information hygiene.

Keywords: Fake news, Information Detergent, Information Pollution, Information Disorder, misinformation, Propaganda, Cognitive Bias, Cognitive Bias, Post-Truth, Echo Chamber

Introduction

Fake news and misinformation are referred to as fictitious and deceptive information content which confuses the recipient, and have become a topical issue in the contemporary world because of their impact on the political, economic, and social well-being of any nation. Even though the occurrence is not new, the proliferation of social media and satellite television network channels have contributed to the transparency and precision of how information flows openly with attendant problems. In contrast, this has also elicited the escalation of fake news. These days fake news has become an instrument of manipulation, revealing the ignorance of the

Oluwole O. Durodolu, Department of Information Science, University of South Africa, Pretoria, South Africa, e-mail: woledurodolu@gmail.com
Olatokunbo Christopher Okiki, University of Lagos, Akoka, Lagos, Nigeria, e-mail: cokiki@unilag.edu.ng
Stella C. Nduka, University of Lagos, Akoka, Lagos, Nigeria, e-mail: stellacnduka@gmail.com

https://doi.org/10.1515/9783111176345-014

sender's information literacy (Rozanov, Kharlamova, and Shirshikov 2021). In circulating news and delicate information, the essential guidelines are rooted in the notions of truth as well as the principles of accuracy in communication. Nevertheless, recently, social media has redefined the complication of public information. The impact of social media, particularly on socio-political matters, has been drawing significant attention, because it enables broad distribution and sharing of information and news. Allied to information provision, another notable purpose of social media is to facilitate people in engaging in social interaction, communication, and entertainment. But sadly the social media platform is largely unsupervised and leaves room for manipulation, propaganda, distortion of fact, and misinformation (Rozanov, Kharlamova, and Shirshikov 2021).

Fake news refers to deliberately misleading or fabricated information that is camouflaged as real news. It can be spread using traditional media outlets, social media, or other online outlets (Pennycook and Rand 2021). The spread of fake news can have substantial negative effects, including the erosion of trust in the media, the amplification and increase of misinformation, and the manipulation of public opinion (Domenico and Visentin, 2020). To combat fake news, it is imperative to fact-check information before dissemination and to be cautious of sources that seem defective or deceitful. Moreover, media literacy education can help to better fathom how to evaluate sources of information and distinguish between real news and fake news (Durodolu, Ibenne, and Dube 2021) and (Durodolu and Adekanye, 2017).

Misinformation is false or inexact information that is spread unintentionally through various means, including social media, word-of-mouth, and traditional media. Misinformation can lead to misunderstandings, confusion, and incorrect beliefs or actions (Jevin and Bergstrom 2021). Misinformation can be predominantly precarious when it concerns essential public health issues or volatile political matters. Recently, the spread of misinformation has become a major concern, mostly in the context of social media and online platforms where information can be speedily and effortlessly shared (Kozyreva et al. 2023).

Information hygiene refers to the practice of upholding a healthy and non-toxic connection with the information that is consumed, shared, and interacted with on a daily basis. It comprises being mindful of and alert to the sources of information, critically appraising the content and accuracy of information, and being responsible in sharing information with others. Practicing good information hygiene can aid individuals to avoid undesirable impacts of misinformation and fake news. By adopting these practices, individuals can help promote a healthier information environment and avoid the spread of false or misleading information (Chisita et al. 2022).

Propaganda can be a dominant tool for manipulating public opinion which can contribute to social unrest in certain circumstances. Propaganda can be defined as information or ideas that are spread with the intention of promoting a particular cause or influencing people's beliefs or actions. It can be used by governments, political groups, corporations, and other organizations to shape public opinion and advance their own interests (Morris 2023). The problem with relying solely on social media for news is that the sources are often unverified and there is no editorial oversight, which means that false information can be spread quickly and it can be challenging to separate fact from fiction. Additionally, social media algorithms often show users content that aligns with their existing beliefs and opinions, which can create a "filter bubble" where users are only exposed to information that reinforces their existing views (Srba et al. 2023). Consumers of news should also be aware that social media platforms may have a commercial or political interest in the content that is shared on their platforms which may affect the accuracy and reliability of the news that is shared. Therefore, it is crucial to verify information from multiple sources and not rely solely on social media for news. It is also essential to choose reputable news sources and maintain a critical perspective when consuming news (Clever et al. 2023).

When propaganda is used in a situation of socio-political pressure, it can give room to social unrest by stimulating prevailing divisions and generating new ones. For example, propaganda that demonizes certain groups or encourages a particular ideology can lead to increased divergence and conflict between different groups within society (Hook and Verdeja 2022). In extreme cases, propaganda can add to violence and civil disorder, such as with the Rwandan genocide in 1994. In the preceding years leading up to the genocide, propaganda was used by some groups to demonize the Tutsi ethnic group and encourage the indication that they were a threat to the Hutu majority. In turn, the propaganda enabled and created a fertile ground for a state of fear and distrust, which ultimately led to the massacre of hundreds of thousands of Tutsis. This illustrates how propaganda can lead to communal unrest by stimulating acrimonious ideas, demonizing certain groups, and creating an atmosphere of fear and suspicion (Yanagizawa-Drott, 2014).

The authors, in finding a solution to the problem of information pollution, came up with the new concept of "information detergent" which is used to describe the process of removing misinformation or fake news from the internet and other information outlets. This comprises recognizing false or deceptive information and taking steps and precautions to remove or correct it. Information detergent can take many procedures, including fact-checking articles, debunking false claims, and removing misleading content from social media platforms or other online forums. It may comprise a combination of manual and computerized processes to detect and address misinformation. The aim of information detergent is to uphold and stimulate a healthier information environment and moderate the negative impacts of fake news and misinformation. By removing false or misleading information and promoting accurate, reliable sources of information, individuals and organizations can help prevent the spread of harmful information and support a more informed public discourse.

Statement of Problem

Information pollution is one of the serious contemporary problems, associated with the problem of excessive, deceptive, or inappropriate information disseminated through various mediums such as social media, news outlets, and other online platforms (Baptista and Gradim 2020). Information overload and the volume of information accessible on the internet and other media is concerning and the ease and comfort of dissemination of this information is gradually leading to a diminished integrity and control as well as the spread of deceitful or distorted information.

Information pollution also poses a serious danger with consequences, including a deterioration in trust in reliable sources of information, increased confusion and anxiety among people, and the promotion of harmful beliefs and behaviors. This problem is particularly acute in the age of social media, where anyone can share information without proper vetting or fact-checking, leading to the proliferation of fake news, conspiracy theories, and other forms of disinformation.

The problem of information overload is a common issue in today's digital age, where we are constantly bombarded with vast amounts of information from various sources, including emails, social media, news, and other online platforms. The problem arises when we are exposed to more information than we can process, leading to stress and cognitive overloads. Information overload can have several negative consequences, including reduced productivity, decreased focus, and difficulty in retaining information. It can also lead to decision paralysis, where we become overwhelmed with options and are unable to make choices effectively. To overcome information overload, it is essential to manage the flow of information by prioritizing and filtering the most relevant and important information. This can be achieved by setting specific goals and priorities, organizing and categorizing information, and using tools such as email filters, news aggregators, and productivity apps. Another effective way to combat information overload is to take regular breaks from technology and practice mindfulness and meditation to reduce stress and improve focus. Learning how to manage our attention and limit distractions is also important, as it can help us maintain our focus and attention on the most important tasks and information. Overall, managing information overload requires a combination of organization, focus, and mindfulness, as well as the adoption of effective strategies and tools to help filter and prioritize the information we receive.

Factors Leading to Information Disorder

The increasing use of social media platforms has led to the issue of fake news. The wide spread of fake news or misinformation often arises when a message or news is partially true, has false statements to back it up, making the information look sensible, is and sent out through any social media platforms, radio, or television. This message tends to bring to mind certain emotions and opinions regarding the story. Also, the publication of fake news has caused information disorder to people due to the present information ecosystems which arose from individuals' subjective reading of news and their spreading of this to other people who believe fake news and in turn further spread such misinformation (Thornhill et al. 2019). However, the discernment of the credibility of information is often influenced by some factors such as confirmation bias, cognitive bias, echo chambers, post-truth, and echo chambers, which are briefly discussed below.

Confirmation Bias

Confirmation bias, which is also referred to as "myside" bias, is described as the inclination for an individual to understand information in such a way that conforms with their existing principle. Confirmation bias according to Cherry (2020) will have an effect on how individuals gather information and also influence how they interpret and recollect information, namely in a biased manner. In addition, Grohol (2020) avowed that confirmation bias is manifest in daily assessments, that is, people can primarily depend on evidence in such situations that backs their beliefs and pay no attention to anything opposing to those beliefs. This bias can impact how people search for information and have an impact on what they discover. Other ways bias can occur is the way a person interprets information or news or remembers information, namely, they only commit to memories with information that validates their existing convictions (Grohol 2020), which is known as selective recall. In addition, the different ways or types that confirmation bias can take place, as identified by Cherry (2020), are: (i) biased attention, when an individual selectively focuses on information that confirms their views and ignores information data that doesn't conform to their beliefs; (ii) biased interpretation, when an individual interprets information intentionally in a way that authenticates their views; and (iii) biased memory, when a person selectively remembers information that supports their views and forgets any information that is contrary. Confirmation bias can have an effect on a person's daily lives, such as the way one interprets current matters, the way one chooses their friends and associates, that is, their personal relationships, and their decision making. However, the impact of confirmation bias on a person cannot be downplayed due to the fact that this bias can prevent a person from taking a second opinion of a situation objectively, which can influence their subsequent decision, thereby leading them to poor options.

Cognitive Bias

Cognitive bias can be defined as a logical, that is, non-random and predictable deviation from rationality in decision-making (Vonk and Shackelford 2017). Cognitive bias can also be referred to as an organized mistake that occurs when people are processing and interpreting information in the world around them and which can affect their decisions and lead to an inaccurate conclusion. This is because one often analyzes information using the complex cognitive machine; due to being flooded with information from millions of sources throughout an average day, the brain builds up ranking systems by deciding which information deserves attention, is significant, and thus should be stored in memory (Stanborough 2020a).

Cognitive biases have been proposed to underlie many beliefs and behaviors that are dangerous or problematic to individuals: superstitions, pseudoscience, prejudice, poor consumer choices, and so on (Vonk and Shackelford, 2017). Cherry (2020) posits that biases such as the cognitive version regularly act as regulations that help a person make sense of situations and arrive at decisions with great speed. Subsequently, some of these biases can be related to memory, meaning that the way a person recalls an event may be biased for a number of reasons, which in turn may lead to bias in reason and decision-making. Furthermore, some other cognitive biases could be related to problems with attention as it can be a scarce resource, meaning one may have to be selective about what to pay to attention to in the world around them.

Post-Truth

The term post-truth, according to Kalpokas (2019), has been around for some time but genuine interest in the term came about in 2016. Keyes (2004) defined post-truth as the vague impression of boundaries between lying and truth-telling and, likewise, fact and fiction. Kalpokas (2019) reported that post-truth is characterized by "qualitatively new dishonesty on the part of politicians," especially in making up facts to support whatever they are promoting instead of the truth. Statements that can be verified are not as common as they once were; rather many are made up and presented to the public because this suits a particular story or a broader agenda. In addition, Cooke (2017) submitted that the post-truth era is one where arguments and information are more likely to be accepted if they appeal to emotion and conform to existing views. The author further stated that news or stories does not exist within an emptiness due to end users of information allowing emotion and motivation to lead their information behaviour, thereby making them vulnerable to falsehood.

Echo Chamber

An echo chamber is a situation whereby a person only comes across information or opinions that echo and reinforce their own views. It can also lead to misinformation and twist a person's point of view, making them have difficulty considering opposing viewpoints as well as discussing complicated topics. This is caused in part by confirmation bias, that is, the ability to favor news that reinforces existing beliefs (Gcf Global 2022). An echo-chamber is also described as a set of users differentiated by two properties, opinion and network polarization; opinion polarization means that users are more inclined to share similar views while network polarization means that they are more closely connected with each other than with the outside network (Törnberg 2018). Subsequently, an echo chamber is a tightly connected set of users more inclined to share a common view on a given narrative (Törnberg 2018). It must be pointed out that echo chambers can take place anywhere information is carried out, be it online or real life. However, people can quickly connect with like-minded people and perspectives via social media and countless news sources due to the Internet. In order to avoid echo chambers or being an echo chamber, one has to take the steps such as checking multiple news sources to confirm that one is getting complete and genuine news, making time to interact with people of diverse viewpoints, and remembering that because you want something to be true does not make it fact (Gcf Global 2022).

Using Information Literacy to Combat Fake News

Globally, the world has seen a dramatic increase in access to information and communication technologies. It is well known that the knowledge society is loaded with information across platforms such as print, broadcast, and digital content, while others are information starved, however, the quality of information people access primarily determines their perceptions, beliefs, attitudes, and judgement. Fake news is a type of news that is intentionally misleading. In the age of social media, spreading fake news is an increasingly global problem; according to Anyaoku and Osuchukwu (2022), fake news is false information presented as news that can potentially harm the recipient. It is also intentionally and verifiably incorrect information that can mislead readers or listeners (Allcott and Gentzkow 2017). Various authors have described fake news as information, media manipulation, mis-information, dis-information, mal-information or information warfare (Warwick and Lewis 2017; Khaldarova and Pantti 2016; Wardle and Derakshan 2017.

Wardle and Derakhshan (2017) conducted research on fake news and formed the following three groups:

- Misinformation is when false information is shared through various formats but with no intended harm or liability.
- Disinformation is when incorrect information is willfully shared within society to cause harm.
- Malinformation is when genuine information is transmitted to cause damage and disaffection among the populace, especially from private individuals into the public sphere.

In Nigeria, fake news as a concept has assumed the same dimensions as expressed above, with many different meanings and interpretations. Social media sites continuously allow the widespread and unregulated spread of false news, fake images, false audio messages, edited videos, and old and unrelated videos which permeate the social sphere. Anderson (2019) averred that the spread of false and misleading information has become a social malaise causing panic, tension, and fear that affects people's wellbeing, health and social cohesion. This social ill is common because it cuts across cultures, hence the need for intervention.

Competence Needed to Combating Fake News

The Internet and social media are significant sources of information for active citizens who are mainly young. These are simple tools used for social interaction, sharing information, communicating, playing music, viewing videos, playing games, and entertainment (Anyaoku 2019). Stein-Smith (2017) reported that it is tempting to believe that fake news is easy to recognize, yet, active youths find it difficult to analyze the credibility of information that permeates the digital space and often lack the requisite knowledge to identify authentic sources and biased, false, and misleading information. Abegglen, Burns, and Sinfield (2016) asserted that a citizenry needs the requisite skills and capacity to assess sources and content to navigate the Internet and select reliable information. Burns and Sinfield (2016) in turn averred that critical thinking is the art of making clear, reasoned judgments based on interpreting, understanding, applying, and synthesizing evidence gathered from observation, reading, and experimentation. According to the Council of Europe (2022), "Critical thinking is germane for citizens and especially for teaming youth identity."

Library and education centers need to raise awareness and provide literacy training on safer internet habits so individuals can learn how to effectively use information literacy skills when searching for new knowledge or answers on the web. Instilling these skills early is essential for developing future informed citizens who can make sound decisions based on available facts instead of rumors or false stories masquerading as fact-based knowledge. It is also crucial for instructors to support users in learning to evaluate sources before sharing any personal opinion online or offline since this can easily undermine personal privacy and lead individuals down dangerous pathways.

It is misleading to assume that fake news is obvious and easy to dictate or recognize. A study by Donald (2016) revealed the opposite and how there is difficulty judging the credibility of information online. Donald (2016) and Durodolu, Nduka and Dube (2022) recommend that librarians assume a leadership role in educating users to have mastery of information and utilize the same information literacy skills that have driven library/bibliographic instruction for many years in their daily use of social media and other online news/entertainment sites. Within the context of active information literacy and library clinic programs, research has shown that libraries and librarians now offer on-site and online instruction sessions on effectively determining what fake news is (Benson 2019; De Paor and Heravi 2020; Ayoung, Naazi-Ale Baada, and Bugre 2022).

Information Detergent Strategy

Information disorder is a term used to describe the phenomenon of misinformation, disinformation, and malinformation spreading rapidly in today's digital age. Filtering information disorder can be challenging, but there are several strategies that can help.

Verify the source: one of the most critical steps to filtering information disorder is to verify the source of the information. Check if the source is reputable, has a track record of providing accurate information, and is trustworthy.

Fact-check the information: before sharing or using information, fact-check it by comparing it with reliable sources or expert opinions. Alternatively, use fact-checking websites or tools to verify the accuracy of the information.

Look for multiple sources: check for multiple sources that provide the same information. If a single source is providing a particular piece of information, verify it from other sources to ensure its accuracy.

Avoid echo chambers: echo chambers are groups or platforms where people only interact with others who share the same beliefs or opinions, which can lead to misinformation spreading quickly. Try to broaden your information sources and interact with people with different perspectives.

Use critical thinking: develop a critical mindset towards the information you receive; be skeptical of information that sounds too good to be true, and think critically and analyze the information before accepting it as fact.

Report misinformation: report any misinformation you come across on social media or other platforms. This can help stop the spread of false information and prevent it from causing harm.

By following these strategies, you can filter information disorder and stay informed with accurate and reliable information.

Conclusion

The proliferation of social media has resulted in the spread of fake news among many citizenries, especially youths, which has led to a negative impact on the use of social media in the society. One of the negative effects of social media usage is the spread of fake news which can lead to information disorder. From the review in this discussion, it was observed that information detergent should be applied to contents or information received through social media channels. The study concluded that in order to combat the malaise of information disorder, certain strategies must be applied such as verifying the source, fast-checking the information, seeking multiple sources, avoiding echo chambers, applying critical thinking, and reporting misformation as this would help individuals filter information disorder, thereby making information accurate and reliable. The study also confirmed that librarians should assume a leadership role by conducting information literacy training programs for their users as this will equip them with the necessary skills needed to safely use the internet effectively and also form good internet habits when searching social media searching disseminating information.

The study recommends that equipping users with sufficient information literacy abilities at a young age is essential for creating future informed citizens who may make wise judgments based on the available facts rather than rumors or misleading stories disguising as fact-based knowledge.

References

Abegglen, S., Burns, T., and Sinfield, S. (2016). Utilizing critical writing exercises to foster critical thinking in diverse first-year undergraduate students and prepare them for life outside university. *Double Helix*, *4*, 266–277.

Alexander Rozanov, Julia Kharlamova, VladislavShirshikov (2022). The role of fake news in conflict escalation: a theoretical overview. Information Systems Frontiers. https://doi.org/10.1007/s10796-022-10242-z

Allcott, H., and Gentzkow, M. (2017). Social media and fake news in the 2016 election. *Journal of economic perspectives*, 31(2), 211–36.

Anderson P. (2019). Tackling fake news: the case of Nigeria. https://www.ispionline.it/it/pubblicazione/tackling-fake-news-case-nigeria-23151

Anyaoku, E. N. (2019). Internet Use and Online Safety for Children and Young People in the Library Platform. In Library and Information Services for Children and Young People in the Digital Age: Festschrift for Professor Virginia W. Dike 290–306 Nnewi: Greeno Publishers

Anyaoku, E. N., and Osuchukwu, N. P. (2022). Librarians Use of Information Literacy Strategic Tools for Teaching Students in Nigeria Universities to Combat Fake News and Misinformation. Retrived on 19th August, 2022 at https://repository.ifla.org/bitstream/123456789/1961/1/058-anyaoku-en.pdf

Ayoung, D. A., Naazi-Ale Baada, F., & Bugre, C. (2022). Curbing Fake News: A Qualitative Study of the Readiness of Academic Librarians in Ghana. *International Information & Library Review*, 1–14.

Baptista JP, Gradim A. (2020). Understanding Fake News Consumption: A Review. Social Sciences. Vol. 9(10), Pp.185. https://doi.org/10.3390/socsci9100185

Benson, M. (2019). Teaching information literacy: Combating fake news through library-faculty collaboration. *Journal of New Librarianship*, *4*(2), 538–541.

Burns, T., and Sinfield, S. (2016) Essential Study Skills: The Complete Guide to Success at University (4th ed.) London: SAGE,

Cherry, K. (2020) What Is the Confirmation Bias? Accessed 20 Aug. From https://www.verywellmind.com/what-is-a-confirmation-bias-2795024

Chisita, C. T., Kahakatshi, N., Rusero, A. Marutha, N., Chigwada J., & Durodolu, O. O (2022). Evaluating the Socio-Economic Impact of Digital Information Manipulation and Weaponization of Information. In Chisita, C. T., Kahakatshi, N., Rusero, A. Marutha, N., Chigwada J., & Durodolu, O. (Ed), Library and Media Roles in Information Hygiene and Managing Information. IGI Global. Pp. 300. DOI: 10.4018/978-1-7998-8713-3. ISBN13: 9781799887133.https://www.igi-global.com/book/library-media-roles-information-hygiene/272677

Clever, L., Schatto-Eckrodt, T., Clever, N. C., &Frischlich, L. (2023). Behind Blue Skies: A Multimodal Automated Content Analysis of Islamic Extremist Propaganda on Instagram. Social Media + Society, 9(1). https://doi.org/10.1177/20563051221150404

Council of Europe (2022). The Observatory on History Teaching in Europe – The adopted texts of the Parliamentary Assembly of the Council of Europe. The Observatory On History Teaching In Europe. Pp. 1–37

Cooke, N. A. (2017) 'Posttruth, Truthiness, and Alternative Facts: Information Behavior and Critical Information Consumption for a New Age', The Library Quarterly, 87(3), pp. 211–221. doi: 10.1086/692298.

De Paor, S., and Heravi, B. (2020). Information literacy and fake news: How the field of librarianship can help combat the epidemic of fake news. *The Journal of Academic Librarianship*, *46*(5), 102218.

Domenico, G and Visentin M. (2020). Fake news or true lies? Reflections about problematic contents in marketing. International Journal of Market Research, Vol. 62(4) Pp.409–417.

Donald, B. (2016, November 22). Stanford researchers find students have trouble judging the credibility of information online. Stanford University School of Education. Retrieved from https://ed.stanford.edu/news/stanford-researchers-find-students-have-trouble-judging-credibility-information-online

Durodolu, O.O & Adekanye, E. A. (2017). The Perception of Information Literacy Skill among the Librarians: A Survey of University of Lagos Library. International Journal of Processing and Communication. Vol. 5, No.1 & 2. Pp. 148–159.

Durodolu, O.O, Ibenne, S.K and Dube, TV (2021). New mental costumes: Leveraging ACRL's value of information frame in dealing with infodemics in a post-truth era. College & Research Libraries News (College & Research Libraries News). Vol. 82 (4). Pp. 175

Durodolu, O. O., Nduka, S. C. and Dube, T.V (2022). Media Literacy and Fake News: Evaluating the Roles and Responsibilities of Radio Stations in Combating Fake News in the COVID-19 Era. International Journal of Information Sciences Management (IJIMS). Vol. 6 (1), pp. 32–49

Fedeli, G. (2020). 'Fake news' meets tourism: a proposed research agenda. Annals of Tourism Research, 80, 102684.

GCF Global (2022) Digital Media Literacy – What is an Echo Chamber? https://edu.gcfglobal.org/en/digital-media-literacy/what-is-an-echo-chamber/1/

Grohol, J.M. (2020) The Psychology of Confirmation Bias. The PsychCentral newsletter. Accessed 20 Aug. From https://psychcentral.com/blog/the-psychology-of-confirmation-bias

Hook, K. Verdeja, E (2022). Social Media Misinformation and the Prevention of Political Instability and Mass Atrocities. In Human Rights & IHL.

Jevin D, and Carl T. Bergstrom (2021). Misinformation in and about science. Proceedings of the National Academy of the United States of America PNAS). Vol. 118 (15). https://doi.org/10.1073/pnas.1912444117

Kalpokas, I (2019) Post-truth: The Condition of Our Times. DOI: In book: A Political Theory of Post-Truth 10.1007/978-3-319-97713-3_2

Keyes, R. (2004). *The Post-Truth Era: Dishonesty and Deception in Contemporary Life*. New York: St. Martin's Press

Khaldarova, I., &Pantti, M. (2016). Fake news: The narrative battle over the Ukrainian conflict. *Journalism practice*, 10(7), 891–901. Retrived on 19th August, 2022 at. https://doi.org/10.1080/1369118X.2015.1064461

Kozyreva, A., Herzog, S. M., Lewandowsky, S. and Reifler Jason. (2023). Resolving content moderation dilemmas between free speech and harmful misinformation. Vol. 120 (7). https://doi.org/10.1073/pnas.2210666120

Morris, J. (2023). How homophobic propaganda produces vernacular prejudice in authoritarian states. Sexualities, 13634607221144624.

Pennycook, G and Rand, D.G (2021). The Psychology of Fake News. Journal Trends in Cognitive Science. Volume 25, Issue 5, Pp. 388–402

Rozanov, A., Kharlamova, J. and Shirshikov, V. (2021). The Role of Fake News in Conflict Escalation: a Theoretical Overview. Available at SSRN: https://ssrn.com/abstract=3857007

Srba, I., Moro, R., Tomlein, M., Pecher, B., Simko, J., Stefancova, E., &Bielikova, M. (2023). Auditing YouTube's recommendation algorithm for misinformation filter bubbles. ACM Transactions on Recommender Systems, Vol. 1(1), Pp.1–33

Stanborough, R.J. (2020) Is Cognitive Bias Affecting Your Decisions? https://www.healthline.com/health/mental-health/cognitive-bias

Stanborough, R.J (2020) Is Cognitive Bias Affecting Your Decisions? https://www.healthline.com/health/mental-health/cognitive-bias

Stein-Smith, K. (2017). Librarians, information literacy, and fake news. *Strategic Library*, *37*(1–4).

Thornhill, C., Meeus, Q., Peperkamp, J., &Berendt, B. (2019). A digital nudge to counter confirmation bias. *Frontiers in big data*, *2*, 11. https://doi.org/10.3389/fdata.2019.00011

Törnberg, P. (2018). Echo chambers and viral misinformation: Modeling fake news as complex contagion. *PLoS one*, *13*(9), e0203958.

UNESCO. (n.d.). Media and Information Literacy. Retrived on 19th August, 2022 at https://en.unesco.org/themes/media-pluralism-and-diversity/media-information-literacy

Vonk, J and Shackelford, T.K. (2017) Encyclopedia of Animal Cognition and Behavior, DOI 10.1007/978–3-319-47829-6_1244- Accessed 21 Aug from https://www.researchgate.net/profile/Fernando Blanco11/publication/317344882_Cognitive_bias/links/59cbafeba6fdcc451d5cde62/Cognitive-bias.pdf

Wardle, C., and Derakhshan, H. (2017). Information disorder: Toward an interdisciplinary framework for research and policy making. Report presented to Council of Europe, 27. Council of Europe. Retrived on 19th August, 2022 at https://tverezo.info/wp-content/uploads/2017/11/PREMS-162317-GBR-2018-Report-desinformation-A4-BAT.pdf

Warwick, A., and Lewis, R. (2017). Media manipulation and disinformation online. Data & Society. Retrived on 19th August, 2022 at https://www.chinhnghia.com/DataAndSociety_MediaManipulationAndDisinformationOnline.pdf

Yilmaz I, Erturk O. (2023). Religious Necropolitical Propaganda in Educational Materials for Children. Religions. Vol. 14(1):67. https://doi.org/10.3390/rel14010067

Yanagizawa-Drott, D (2014). Propaganda and Conflict. The Quarterly Journal of Economics. Vol. 129, No. 4. pp. 1947–1994 (48 pages)

Fatimah Jibril Abduldayan, Samson Okpanachi Abu,
Lukman Oyewobi and Idris Mohammed Yelwa

Disinformation, Fake News, and the Prolonged Academic Staff Union of Universities' (ASUU) Strike in Nigeria: The Role of Librarians

Abstract: Information literacy skills are essential for everyone, especially in this age of information overload where it is often difficult to sieve the truth out of the numerous information dissemination channels available on the Internet. Librarians and libraries are at the forefront of ensuring that only accurate, timely, and reliable information is offered to their users, however, the scourge of disinformation and fake news has created a huge gap which libraries and librarians are expected to fill. This study examined the impact of disinformation and fake news on the prolonged Academic Staff Union of Universities (ASUU) strike in Nigerian universities. Secondary data was gathered from the report by ASUU National executives and government officials from the Ministry of Education and Ministry of Labour, while reactions from Nigerian students to some of the information on the Internet were mined from the Twitter (also known as X) platform. The findings of the study provided a framework that can be adapted by librarians to fill the information and communication gap between the university union, the government, and Nigerian students in order to limit the effect of disinformation and fake news on the public and the Nation at large. The study recommended that libraries should have official social media handles where current information can be disseminated regularly.

Keywords: ASUU, disinformation, fake news, Nigeria, labour strike, social media

Fatimah Jibril Abduldayan, Federal University of Technology Minna Niger State,
e-mail: fj.dayan@futminna.edu.ng
Samson Okpanachi Abu, Federal University of Health Sciences Otukpo, Benue State
Lukman Oyewobi, Federal University of Technology Minna Niger State
Idris Mohammed YELWA, National Information Technology Development Agency (NITDA)

https://doi.org/10.1515/9783111176345-015

Introduction

Disinformation episodes have been well documented before the invention of the Internet, with the term "fake news" being commonly used from the end of the nineteenth century. Even if these phenomena are nothing new, these issues have been recently taken into serious consideration at both scientific and political levels, such that many national and supranational institutions have considered the related technical and ethical problems. For example, the Council of Europe released a report in 2017 (Wardle and Derakhshan 2017) pointing out that information pollution (and its variants such as misinformation, disinformation, and malinformation) is seen as a serious risk for democracies, and its role in manipulating the public discourse is a problem worth investigating. We need to understand, among other things, that it can be linked to declining trust in evidence-based journalism, to the growing popularity of news outlets that promote low-quality information, and to a measurable impact of computer-assisted devices, such as bots, that contribute significantly to a self-feeding diffusion process that could amplify the virility of fabricated content. This has serious implications for public discussions regarding different topics, from the Academic Staff Union of Nigerian Universities (ASUU) strikes to political party propaganda.

The Academic Staff Union of Universities (ASUU) has been on indefinite strike because Nigeria's higher education system has been plagued by persistent unresolved issues. The problem includes a lack of funding and infrastructural facilities, poor and unfair remuneration, non-compliance to agreements by the government, a delay in reimbursements of university staff salaries, and political meddling in education. On February 14, 2022, ASUU initiated a nationwide strike at government-owned universities scheduled for one month to implement the 2009 deal with the government. People's thoughts and feelings about the ongoing strikes have been flowing freely into social media platforms as a direct consequence of the proliferation of these events. Information is sent digitally from user to user in today's fast-paced environment which might affect how other users interpret a specific occurrence (McGregor 2019).

The impact of disinformation and fake news on the prolonged ASUU strike cannot be over-emphasized as different media outlets and social media platforms have disseminated different information on the same subject matter which in turn has created confusion in the mind of the society. For instance, the Minister of Labour said the Federal Government had brought ASUU to the industrial court for proceedings, while the Federal Government said they did not do so. In the same vein, the Minister of Education had said that ASUU is not cooperating with the Federal Government in resolving the lingering issue, thereby painting the union as "selfish" to the public, whereas the 2009 agreement that contains improved salary structure and payment of backlog of allowances has yet to be implemented.

Our case study leans towards understanding the role of librarians in selecting and filtering disinformation, even though we are aware that applying a single label to such complex phenomena can be misleading. Fake news-related issues encompass the learning ecosystem, educational industrial action, and the participation of libraries in those processes (Rapti et al. 2022). Librarians cannot be neutral information doctors prescribing a mere method but rather be interventionists in the media ecology for democratic purposes. It is not a simple one-to-one correspondence of good information that equals better democracy; rather the signals we send with our spaces and services (physical and virtual) are important. Libraries will also need to revisit and re-evaluate their position of neutrality regarding information access and collection development to a position that reflects their strong opposition to fake news (De Paor and Heravi 2020). Libraries should help to counter fake news both through specific educative actions aimed at it and by being broadly educative institutions with a coherent notion of their role and relationship to informational discernment in a democratic society. Hence, the aim of the study was to examined misinformation, fake news, and the prolonged Academic Staff Union of Universities (ASUU) strike in Nigeria, as well as the role of librarians in the management of information in media spaces.

Conceptual Review

ASUU: Historical Appraisal

The acronym "ASUU" stands for the Academic Staff Union of Universities, which was founded in 1978 and is a prominent trade union of Nigerian University Academic Staff which is affiliated with the Nigeria Labour Congress (NLC). Historically, ASUU is born from the then "NAUT," that is, the Nigerian Association of University Teachers, which was formed in 1965 including academic staff in all federal and state universities in the country. From its inception, ASUU has embarked on a number of industrial actions in an attempt to mitigate against unfair treatment occasioned by the government on its members. In actuality, its active struggle began in the 1980s during the time when the nation suffered intense pressure from the then-military regime. In 1988, the union organized a National Strike to obtain fair wages and university autonomy and, as a result, the ASUU was proscribed on August 7, 1988 with all its properties seized. It was allowed to resume in 1990, but after another strike was banned on August 23, 1992. However, an agreement was reached on September 3, 1992 that met several of the union's demands including the right of workers to collective bargaining. The ASUU organized further strikes in 1994 and 1996, protesting against the dismissal of staff by the Sani Abacha military regime. This realistic provision stemming out of the nation's third republic is essentially the foundational basis of ASUU.

The new democratic dispensation that began in 1999 saw further protests against FGN's oppressiveness on some of the rights and privileges of university workers. As a result, from 1999 until today, the union has embarked on five major strikes; 2007, 2013, 2018, 2020, and 2022 respectively. In all these years, however, the union has recorded series of profitable results which are in effect beneficial to members of the union, university establishments all over the nation, the educational sector in general, and society at large.

Disinformation and Fake news

The term "disinformation and fake news" has become a buzzword, with the concept going back a long time in history even before the time of the printing press. False and fictional stories have always been part of society, whether for the purpose of entertainment, politics, malice, or business, which has continued to the internet age (Burkhardt 2017). Further examples of "fake news" include satirical and fictional accounts produced by individuals to discredit one another or simply for the purpose of amusement. This may indicate that the concept around the phrase "fake news" has not changed in its function but has changed in its nature and how it is now circulated. The definition of "fake news" is complex, transitional, and often misconstrued. However, the term itself is often used to represent a broader landscape of false or warped information, which could be intentional (disinformation) or un-intentional (misinformation). Lim (2020) proposed a typography of definitions for the classification of terms under the broad umbrella of "fake news." These definitions are divided into two categories for the purpose of this article: misinformation and disinformation.

Misinformation is positioned at one end of the scale and defined as the inadvertent sharing of false information, while disinformation sits at the opposing end and constitutes the deliberate creation and sharing of information known to be false (Lim 2020). Seven definitions of fake news exist within these categories, with this scale helping differentiate the various definitions of information that constitute this term: satire or parody, misleading content, false connection, false context, imposter content, manipulated content and fabricated content (Lim 2020). Rochlin (2017) argued that news is being upsold on the notion of opinion as opposed to fact. The author further discussed the emergence of "selected exposure" and "confirmation bias" which presents people as tending to search for information that reaffirms and compliments their pre-existing views. Therefore, if fake news resembles an ideology or popular worldview, it will be shared and circulated without being verified.

Social Media

The term "social media" has been defined in different ways by its users. For instance, Kaplan (2015, n.p.) defines social media as "a group of internet-based applications that build on the ideological and technological foundations of Web 2.0 and that allow the creation and exchange of user-generated content." It includes web-based and mobile-based technologies that are used to turn communication into interactive dialogue among individuals, organizations, and communities. Typical examples of social media platforms include websites such as Facebook, Twitter, Flickr, YouTube, and the interactive options on these websites, such as the "retweeting" option on Twitter. These instruments are referred to as media because they are tools which can also be used for the storage and dissemination of information. However, unlike the traditional media like television and radio, most of the social media tools allow their users to interact through the likes of "re-tweeting" on Twitter and "comment" options on Facebook.

Spohr (2017) discussed the influence of social media on fake news by its facilitation of echo chambers and filter bubbles. This means online users are exposed to content based on algorithmic technology that allows individuals to customize their newsfeed so that they are exposed to news and content that they are in agreement with or content shared within their network of friends. This method of news circulation creates a "bubble" or "chamber" where content is filtered down by personal preference while thoughts, beliefs, and opinions are echoed within the user's online environment. Not only does this method reinforce polarizing views but it eliminates entry barriers, allowing creators and producers of fake news to manipulate the algorithmic curation to attract users and spread misinformation and fake news stories for political and financial purposes (Spohr 2017). Further examples in Burkhardt's (2017) paper showed the influence of technology on the circulation of fake news stories, and how its effects have magnified due to lack of regulation and restrictions in online technology. Burkhardt (2017) discussed how the dissemination of news had always been thoroughly regulated and reviewed before reaching the public, whether through newspapers, the radio or television. However, at present, the freedom and unregulated nature of the Internet has contributed to the large masses of fake news stories being created and circulated online, causing confusion among society. The author also discussed the financial component of fake news that has made it a profitable and effective tool for advertisers and businesses (Burkhardt 2017).

Librarians and Information Literacy

Information literacy has traditionally been associated with the field of librarianship as it initially emerged from the concept of bibliographic instruction which assisted information consumers with locating and retrieving information. The concept received great attention when the American Library Association (ALA) officially defined the term in 1989, claiming that an information-literate individual should recognize when information is needed and have the ability to locate, evaluate, and use effectively the needed information (Tewell, 2015). In 2004, one of the leading library and information associations in the United Kingdom, the Chartered Institute of Library and Information Professionals (CILIP), developed its own definition for information which centred on knowing when and why you need information, where to find it, and how to evaluate, use, and communicate it in an ethical manner. However, in 2018, this definition was reviewed to promote the broader function of information literacy which was applicable to various contexts, from education and the workplace to everyday life. CILIP agreed upon the revised definition as "the ability to think critically and make balanced judgements about any information we find and use" (Secker 2018). It thus empowers individuals as citizens to reach and express informed views and to engage fully with society. The concept has gradually transformed and adapted providing criteria for individuals to assess information and to ensure credibility and reliability. On a broader scale, information literacy has also been identified as a vital component of proactive citizenship as it enables one to locate, access, retrieve, evaluate, interpret, and act on information, which allows one to participate in community affairs, develop community involvement, and to have an informed opinion about problems occurring locally, nationally, and internationally. Information literacy is therefore recognized as a tool of empowerment which provides individuals with the necessary skills and competencies to become informed citizens who actively contribute and engage in their communities and in society.

A multitude of information literacy frameworks and models have emerged within the field of librarianship that are based on a criterion of skills used to inform and educate information consumers. Many of the frameworks were created to educate and teach students of higher education how to locate, evaluate, and utilize information ethically. However, broader definitions have arisen due to issues emerging from the information age including fake news. An example is CILIP's Information Literacy Model which contains eight competencies that an information-literate individual should have an understanding of:

(1) A need for information,
(2) The resources available,
(3) How to find information,

(4) Need to evaluate results,
(5) How to work with or exploit results,
(6) Ethics and responsibility of use,
(7) How to communicate or share your findings,
(8) How to manage your findings (Secker 2018).

The CILIP model forms a one-size-fits-all approach requiring an information literate person to demonstrate each competency. While the model presents the basic skills associated with information literacy, it offers a more generalized overview that fails to elaborate on how these skills may be applied to different situations and contexts.

Fake News and Library Initiatives

In light of the ASUU prolonged 2022 strike and the increasing exposure to fake news stories and misleading clickbait, information professionals and librarians have to be positioned at the front lines of the information war. As the literature on libraries and their efforts to combat fake news is at an embryonic stage, initiatives and strategies are gradually emerging within the field. Using methods of library guides, fact-checking websites, and information literacy programs. and advocacy, librarians are using all resources available to inform and educate the public about fake news in the form of both misinformation and disinformation to help empower people on how to navigate successfully through the online world. A research study conducted in 2016 by The Stanford History Education Group illustrated how students' ability to evaluate information was extremely weak (De Paor and Heravi 2020). The students were required to distinguish between a news item and an advertisement which many failed to do (Fullerton and Kendrick 2022). In addition, academic librarians have been some of the first in the field to combat these issues through the implementation of instructional programs, workshops, and guidelines. Librarians at Indiana University East of Richmond also developed a library known as "LibGuide" which offers students criteria of questions to ask when investigating the reliability and credibility of a source or piece of information (Banks 2016). Fake News LibGuides have also been developed in other parts of the world, including Newcastle University in the United Kingdom and University College Dublin and Maynooth University in Ireland. In addition to library guides, some universities have included fact-checking resources and websites to help students verify information accurately (Eva and Shea 2018). Some of these would include websites such as FactCheck, Politifact, Snopes, and The Washington Post Factchecker which have been awarded on their usefulness and effectiveness (Batchelor 2017).

One of the roles of librarians is to provide instructions and to inform communities and the public about fake news. Using information and media literacy, Dallas Public Library held an eight-week training course in community journalism for high-school students (Banks 2016). Librarians collaborated with reporters from the Dallas Morning News as well as local journalists to inform students about good and bad journalism. As part of the project titled "Storytellers with Boarders," journalists taught students how to ask focused questions while librarians described how to use databases to find reliable information (Banks 2016). Oakland Public Library currently hosts interactive workshops within and outside of the library, with their staff travelling to local schools and organizations to help students discern real news from fake news (Eva and Shea 2018). Other methods that have originally emerged from academic libraries is the CARS and CRAAP tests which present a criterion which people of all educational backgrounds can utilize to verify information. The CARS method stands for credibility, accuracy, reasonableness, and support and has been used not only in higher-level institutions but in schools (Jacobson 2017). The CRAAP test, developed at Miriam Library at Chico in 2010, is an invaluable resource for patrons looking to evaluate and identify credible and reliable information (Batchelor 2017). The literature also highlights how librarians recognize the epidemic of fake news as an opportunity for change as well as for effective action.

Librarians have to develop different guidelines, infographics, workshops and programs to help inform and educate their patrons while also promoting their professional role as the original fact-checkers and information experts (Eva and Shea 2018). The American Library Association (ALA) have currently capitalized on the librarian as an information shepherd brand in their libraries. Similarly, Finley, McGowan, and Kluever (2017) discussed fake news as an opportunity for "real librarianship," listing how libraries can actively participate in the global effort to combat the phenomenon. The authors further stated that the role of librarians is to accurately inform the general public about fake news and equip them with tools to navigate information sources which focus on collection development, programming and instruction, and advocacy (Finley, McGowan, and Kluever 2017). Regarding collection development, there must be access to print resources for news and current events to ensure that patrons are informed appropriately. Therefore, a multitude of credible news sources should be made available at both ends of the political spectrum. While quality control is central, librarians must navigate the metaphorical line between selection and censorship following their traditional values.

Theoretical Framework

The Conflict Theory

This work is anchored on conflict theory as a basis for analysis. Conflict theory is premised on disorder, disagreement, and open hostility among individuals and groups and a lack of harmony in system parts. In the perspective of conflict theory, normal and harmonious cooperation is unusual. Social structure is temporary and often fragile combination of competing social forces. In conflict perspective, the task of the social scientist is to examine the sources of tension among people and groups with different amount of powers, techniques of conflict control, and the ways in which the powerful maintain and enlarge their influence on all aspects of the social structure. In the conflict view, the social order at any particular moment is the outcome of struggles among groups of unequal power. The conflict perspective is concerned with who benefits from any given social arrangement. Conflict within society is expressed through public protests and industrial conflicts (strikes and lockouts) as in the 2022 FGN/ASUU industrial conflicts.

This theory is pertinent and relevant to this work in that it focuses on open disagreement between the Federal Government and the Academic Staff Union of Universities (ASUU) based on unresolved issues contained in the terms of the 2009 FGN/ASUU agreement and the 2012 Memorandum of Understanding (MOU). While ASUU maintained that the FGN had breached the terms of the agreement, the Federal Government saw nothing wrong with repudiating some of the terms of the agreement. The relevance of the theory to this work hinged on disharmony and disequilibrium created among the two factions and the suspension of academic programs in the Nigerian University system for over seven months. We shall further explore the relevant application of this theory to the study as we progress throughout this paper. Another theory that is relevant to this discourse is the authoritarian media theory.

The Authoritarian Media Theory

This theory, which was propounded in 1956 by Siebert Peterson and Schuman, postulates that the media should do nothing that would undermine established authority or disturb order; the media should always be subordinate to established authority; the media should avoid offending majority or dominating moral and political values; and censorship should be justified to enforce these principles; unacceptable attacks on authority, deviation from official policy or offences against moral codes should be criminal offences; and journalists or other media professionals should have no independence within their moral organization. In the authoritarian media theory, whether the ownership is public or private, the system exists to serve the interest of the government in power. Under this system, the government exercises control over the media in various ways.

In Nigeria, during the 2022 ASUU/FGN industrial conflict, the government used cruel and uncivilized methods to ensure that pro-ASUU protests in the country were suppressed and all such pro-ASUU protests were blacklisted from media coverage, especially government-owned media. In some cases, security personnel were drafted to the scheduled venues where ASUU members were to publicize their grievances to allow the Nigerian public to understand why its members were on strike. The theory is relevant to this discourse in that the government exercised control and curtailed ASUU activities that were designed to express their views to the public, and there was not only restriction to members by the use of instrumentality of the State, with freedom of speech and expression by ASUU members highly curtailed.

Proposed Conceptual Framework

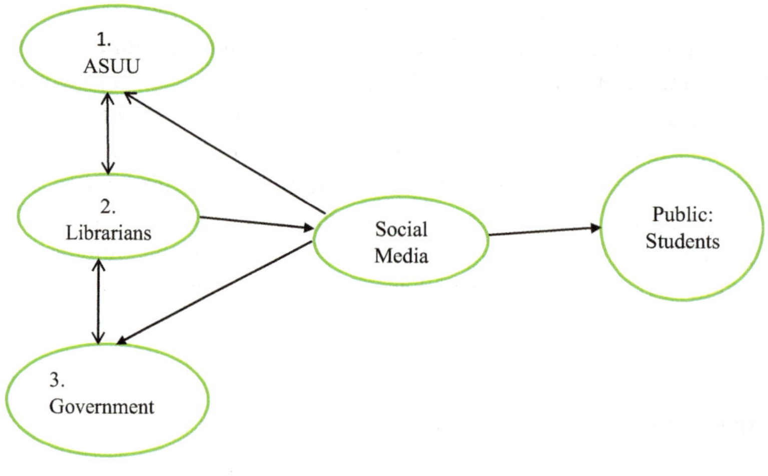

Figure 1: Conceptual framework.

Figure 1 showed the conceptual framework for handling disinformation by the librarians. The librarians play a key role in filtering and mediating the sources of information before releasing it to the larger society. Put differently, fake news is generated for certain intent and purpose and thereafter is circulated on different media platforms. The effect of disinformation on society if not managed by the information professionals may lead to crises such as protracted industrial strikes as in the case of ASUU in Nigeria. The librarians access current and accurate information directly from the government and ASUU and present it on the official social media handle of the library where the government officials, members of ASUU, and the students can access it. This framework will help to mitigate the spread of fake information regarding the dispute between the union and the government, thereby reducing recurrent protracted strike.

Methodology

The study adopted the exploratory research design and secondary data was gathered using reports from ASUU executives and the government officials involved in the ASUU-related issues. Reactions from Nigerian students to some of the information on the Internet was also mined from the Twitter platform. Thematic analysis was done on the gathered data using the Braun and Clarke thematic analysis approach.

Results and Discussion

Information from Government Perspective

Ume-Ezeoke (August 18, 2022) reported that in the face of lingering industrial action by the Academic Staff Union of Universities (ASUU), the Minister of Education denied receiving a directive from the President to resolve the protracted strike in two weeks. During the weekly ministerial briefing on Thursday in Abuja, the Minister told State House Correspondents that the President had instead asked to resolve the issues within the shortest possible time, contrary to what the Minister of Labour and Employment told the media. The Minister insisted that they would not succumb to ASUU's demand for their members to be paid the backlog of salaries withheld within the period of six months that they were on strike, noting that it was the penalty for their action. The Minister also disclosed that five of the university-based unions would likely call off their strike within the next week while that of ASUU remained uncertain.

In an interview on Channels Television, Oluwafemi (September 7, 2022) reported that the Minister of Education said Nigerians did not have "good reasons" to be disappointed with the Federal Government over the lingering strike of the Academic Staff Union of Universities (ASUU). ASUU had been on strike since February 14 over improved funding for universities, review of salaries for lecturers and other issues. Several meetings between ASUU and the Federal Government had ended in deadlock, as the strike entered its two hundred and fifth day. On Tuesday, the federal government said it could only afford a 23.5 percent salary increase for lecturers of all categories, as well as a 35 percent increase for professors. Speaking on the effort of the federal government, the minister said the federal government had made the "best" offer to ASUU in order to resolve the ongoing strike: "The government has already made an offer. As the minister, I know that the offer the government made is probably the best it can make. You cannot do better than your best." The Minister said it was not the fault of the federal government that ASUU had not suspended the strike: "If Nigerians are disappointed, I think they don't have very good reasons to be disappointed with the government on this. Why should they be disappointed? Just tell me in your opinion. How is it the fault of the government and not that of the union? You can only blame the federal government if it refuses to do what it is supposed to do and make the offer to satisfy their demands. There is no demand that can be satisfied 100 percent by any government."

Abdullah (August 18, 2022) reported that the ongoing industrial action by the Academic Union of Universities may linger. According to the Minister, ASUU had agreed with the government's recent offer to them and also agreed to call off the strike if the Federal Government paid them for the period they had been on strike. The Minister, however, said the government would not pay them for those months, saying that it was simply applying the law relating to their action. He said that apart from ASUU, ASUP, NASU, and SSANU, other striking unions would be called off by next month.

Ukwu (May 13, 2022) reported that: "ASUU Strike: we have reached a reasonable agreement with lecturers. We have reached some agreements and we hope that by next week, those agreements will be maturing and the different unions will have something to tell their members so that they can call off the strike." This was according to the Minister of Labour.

The Nation (September 19, 2022) reported that the ongoing strike by the Academic Staff Union of Universities (ASUU) would have dragged on for two years because of a lack of progress in negotiations, with the Minister of Labour and Employment saying: "If I leave them in education they will stay there two years without progress." The Minister addressed the audience at the public presentation of a book which chronicled the labour movement's struggles over 40 years in the country on Monday. He said he referred the dispute to the National Industrial Court of Nigeria (NICN) after negotiations between the union and the Federal Ministry of Education had broken down and shared that the case of ASUU was referred to the NICN in line with section 17 of the Trade Dispute Act 2004 after seven months of protracted discussions and negotiations with the union had failed. The Minister noted that the leadership of ASUU did not understand the importance of negotiation because they lacked the necessities of labour unionism: "We have to counsel our brothers on negotiation. No negotiation is forced. You cannot say it is either you give me 200 per cent or I will continue my strike. There are laws guiding strikes." He said either of the parties involved, the Federal Ministry of Education and ASUU could approach the NICN for an out-of-court settlement and maintained that Nigeria must be guided by laws and nobody should use the dispute to harangue anybody. He recalled that ASUU was at the stage of collective bargaining agreement (CBA) negotiation with their employer, the Federal Ministry of Education, when they embarked on strike.

Olisah (April 4, 2022) reported that the Minister of Labour and Employment had accused the members of the Academic Staff Union of Universities (ASUU) of intimidation and bullying and threatened to withdraw the certificate of the Director-General of the National Information Technology Development Agency (NITDA) and revoke the professorship of the Minister of Communications and Digital Economy, calling him a fake professor. This was made known when he was a guest on a Channels Television program, Politics Today, where he described such action as against labor negotiations. The Minister said that calling off of the strike depended on ASUU and urged them to come down from their high horse as well as to be considerate.

ASUU's Perspective

The ASUU NEC meeting in University of Lagos (2022) concluded from the meeting that the government had failed to fully implement the Memorandum of Action it signed with the Academic Staff Union of Universities (ASUU) on December 23, 2020, given that the draft report of the Renegotiated 2009 FGN/ASUU Agreement was submitted for finalization more than nine months previously, and noted that the forceful payment of ASUU members' salaries and emoluments with the Integrated Payroll and Personnel Information System (IPPIS) and non-adoption of the University Transparency and Accountability Solution (UTAS) continued to short-change their members. NEC thus resolved to embark on a four-week roll-over total and comprehensive strike action beginning from Monday, February 14, 2022.

Similarly, ASUU strike bulletin (3) asserted that the pronouncement by government that N92 billion was paid to ASUU as earned allowance was misleading, aimed at deceiving the public and causing confusion. The truth was that there is a balance of two tranches amounting to N40 billion yet to be paid.

The ASUU strike bulletin (4) proved that members should completely disregard the press release by NITDA. The report was completely different from the report of its Technical Team that tested UTAS. The union asserted that UTAS was subjected to a series of tests in the presence of other observers from agencies of government as stakeholders. The NITDA Technical Team scored UTAS 99.3% before and after remediation, however, there was false information on the social media negating the actual score.

Finally, the ASUU strike bulletin (5) asserted that the agents of government were redoubling their efforts at misinforming, misleading, and twisting facts relating to the union struggle to the Nigerian public, particularly the concerned students and their parents. The union emphatically stressed that this was aimed at dividing the union and sowing the seed of hatred against the patriotic struggle.

Tolu-Kolawole (August 18, 2022) reported that the Academic Staff Union of Universities on Thursday said it "had no words" for the Minister of Education. The PUNCH reported that the education minister, during the forty-seventh Session of the State House Ministerial Briefing organized by the Presidential Communications Team at the Aso Rock Villa, Abuja, said that ASUU should compensate students for the period wasted. The Minister also said the government would not pay the university workers for the time lost.

Daily Trust (August 27, 2022) reported that the Academic Staff Union of Universities (ASUU) had called on Nigerians to hold the Minister of Education and the Federal Government responsible for the prolonged strike. The union said that the government was not sincere in dealing with them, and more so, what the Federal Government was telling the people about their negotiations was not the truth. It was further said that the Minister had also said that whatever was written about the ASUU was believable: "You may recall that when asked to make comments on ASUU's submission to the President on Tuesday, the 9th of January, 2020, the Minister agreed with what ASUU presented upon which the President handed him ASUU's document and directed him to come up with a proposal for an amicable solution. For the same Minister to now lead his colleagues and the other ministers to misrepresent facts and mislead the good people of Nigeria against ASUU is rather unfortunate. It is the highest level of unpatriotic disservice a minister would do particularly in a sector like education which is the backbone of the development of any country."

Olisah (June 23, 2022) reported that the Academic Staff Union of Universities (ASUU) had maintained that they had not received any official invitation from the Federal Government for the earlier announced meeting with the government team aimed at resolving the areas of disagreements and calling off the strike.

Students' Comments on ASUU Strike on Twitter

ASUU: "Speaker of the House of Representative Deceived Us to Call Off Strike."
Anonymous_1: "This handsome man cannot deceive someone because he has human feelings and he understands it was affecting the students . . .. Talk another story again 😊😊."
Anonymous_2: "Lol look for another story, we all know y'all are trying to get engagement."
Anonymous_3: "Let them go for indefinite strike. Because there is increment of school fees too much."
Anonymous_4: "Be like say God won punish all you these ASUU people . . ."
Anonymous_5: "So with this information are we meant to start crying? Ahbi you people want to start what won't be easily settled."

#ASUU_StrikeUpdate
Reps: "For the record, at no point did the Speaker of the House of Representatives commit to offset the arrears of salaries owed to union members for the time they were on strike."
Unnamed source_1 (in response to another post): "I am sorry but this is shameless. Are you saying ASUU or number 4 man is going back on their words?"

The Academic Staff Union of Universities (ASUU), of the Federal University Lokoja (FUL) Branch, has called for the sacking of the Minister of Labour and Employment as well as the Minister of Education over their incompetence (November 30, 2022).
Unnamed source_2: "I swear, it's insane what our government has ruined our education to. I'm still learning the 1982 policy for some subjects in school and maybe it's grandchildren (not my own) that will use 2022's new scheme and policies, I want to rant more but make I just end it here."
Unnamed source_3: "ASUU and government when will this strike stop, keep hearing of this since I was born, education doesn't make any sense in Nigeria again."

The National Executive Council (NEC) of the Academic Staff Union of Universities #ASUU will be meeting on the August 1 2022. They are considering an indefinite strike this time around, and not the usual roll-over warning strike.
July 2022 comment by unnamed source_4: "But let the federal government also understand that if nothing happens to end the ASS STRIKE we the Nigeria student will also stand by our grounds and there will be no elections comes 2023."
Unnamed source_5: "Inside strike another strike go [sic] dey."
Unnamed source_6: "This is terrible huh indefinite strike? How about the five months of no studies? FG are not helping matters at all. Kept everyone at home. This is horrible, cruel and any word to define wickedness."

Impact of Disinformation and Fake News on the ASUU Strike

ASUU is an organized trade union of intellectuals seeking the socio-political and economic welfare of its members and promoting the cause of university education in the country as well as the entire good of Nigeria and all Nigerians therein. Since its birth in 1978, ASUU has never halted in its struggle for the reforming of education in Nigeria which is in dire need of such benevolent and indispensable effort. ASUU's main efforts have been to secure adequate funding, improved salary packages, autonomy, and academic freedom to curb the brain drain and ensure the survival of the university system. Consistent failings have, however, been occasioned by the government's frustrating attitude towards the growth of the academic sector. Notwithstanding this barrier, ASUU has, however, had a social impact which has come to effect through its relentless efforts.

The impact of disinformation and fake news on the prolonged academic staff union strikes in Nigerian universities is a complex and multifaceted issue. Throughout the strike periods, disinformation and fake news have played a significant role in shaping public perception, exacerbating tensions, and hindering the resolution of disputes. One of the key consequences of disinformation and fake news is the distortion of facts and narratives surrounding the strikes. False or misleading information, often spread through social media platforms, can easily fuel misinformation and misinterpretations among the public. This can lead to a polarized environment where different stakeholders hold divergent views and are less willing to engage in constructive dialogue.

Furthermore, disinformation and fake news can create a sense of distrust and skepticism towards the academic staff union and the university administration. The spread of false information can undermine the credibility and legitimacy of the parties involved, making it even more challenging to find common ground and negotiate fair resolutions. It can also contribute to the erosion of public confidence in the higher education system as a whole. The use of disinformation and fake news as a tool for manipulation and influence is a concerning trend in the context of university strikes. Various actors, including external entities, political interests, or even internal factions, may exploit disinformation to advance their agendas, further complicating the resolution process and impeding the restoration of normalcy in Nigerian universities.

Addressing the issue of disinformation and fake news requires a multifaceted approach. It involves promoting media literacy and critical thinking skills among the public to enable them to distinguish between reliable sources and misinformation by librarians. Educational campaigns and awareness programs can play a vital role in empowering individuals to become responsible consumers and sharers of information. Collaboration between stakeholders, including academic staff, university administrators, librarians, students, and the government, is crucial in countering the effects of disinformation. Open and transparent communication channels should be established, allowing for accurate information dissemination and proactive debunking of false narratives. Fact-checking organizations and independent media can also play a vital role in verifying information and providing accurate reporting.

The prolonged academic staff union strikes in Nigerian universities have been significantly impacted by the spread of disinformation and fake news. These phenomena have hindered the resolution of the strikes, perpetuated mistrust among stakeholders, and distorted public perception. Combating disinformation requires a concerted effort from various actors to promote media literacy, foster open communication, and ensure the availability of reliable information sources. By addressing the issue of disinformation with librarians playing their roles as expected, Nigerian universities can work towards more constructive dialogue, effective negotiation processes, and a stronger higher education system.

Conclusion

The issue of disinformation, fake news, and the prolonged Academic Staff Union of Universities (ASUU) strike in Nigeria highlights the critical role that librarians can play in promoting reliable information, combating misinformation, and fostering an informed society. Disinformation and fake news have become significant challenges in today's digital age, where information spreads rapidly and often without proper verification. The ASUU strike in Nigeria, which has caused significant disruptions in the educational system, has been subject to various forms of misinformation and propaganda. False narratives, distorted facts, and misleading content have contributed to the confusion surrounding the strike, exacerbating the negative impacts on students, parents, and the broader society.

Librarians, as information professionals, are uniquely positioned to address these challenges. They possess the expertise in information literacy, critical thinking, and research skills necessary to navigate the complex landscape of information. Librarians can serve as gatekeepers of reliable information by curating trustworthy resources, providing guidance on fact-checking and source evaluation, and promoting media literacy among students and the wider community.

Furthermore, librarians can actively engage in debunking false information and disseminating accurate information related to the ASUU strike. They can collaborate with relevant stakeholders, such as educators, journalists, and policymakers, to counter disinformation campaigns and promote evidence-based discussions. Librarians can leverage their skills in information organization and retrieval to create comprehensive databases, digital repositories, and online platforms that offer verified and up-to-date information about the strike, ensuring transparency and accountability.

In addition, librarians can play a vital role in educating individuals about the dangers of disinformation and fake news. They can conduct workshops, seminars, and awareness campaigns to teach students and the general public about the importance of critical thinking, media literacy, and responsible information consumption. By empowering individuals with the necessary skills to discern credible information from falsehoods, librarians contribute to the development of an informed citizenry that can actively participate in shaping public discourse and making well-informed decisions.

Recommendations

1. There should be an increased information literacy campaign by libraries.
2. The libraries should endeavour to access information from government officials and ASUU documents and disseminate this information fairly and responsibly.
3. The libraries should provide official handles where government officials, ASUU members, and students can access reliable and accurate information resources.

References

Abdullah, B. (August 18, 2022). *ASUU strike may linger – AdamuAdamu*. RadioNigeria. https://radionigeria.gov.ng/2022/08/18/asuu-strike-may-linger-adamu-adamu/

Banks, M. (2016). *Fighting fake news. American libraries*. December 27th. Retrieved from https://americanlibrariesmagazine.org/2016/12/27/fighting-fake-news/.

Batchelor, O. (2017). Getting out the truth: The role of libraries in the fight against fake news. *Reference Services Review*, 45(2), 143–148. https://doi.org/10.1108/RSR-03-2017-0006.

Burkhardt, J. M. (2017). *Combating fake news in the digital age*. Chicago, IL: American Library Association.

Daily trust (august 27, 2022). *Hold Education Minister, FG Responsible For Prolonged Strike – ASUU*. Daily Trust. https://dailytrust.com/hold-education-minister-fg-responsible-for-prolonged-strike-Asuu/

De Paor, S., & Heravi, B. (2020). Information literacy and fake news: How the field of librarianship can help combat the epidemic of fake news. *The Journal of Academic Librarianship*, *46*(5), 102218.

Eva, N., & Shea, E. (2018). Marketing libraries in an era of "fake news". *Reference & User Services Quarterly*, 57(3), 168–171.

Finley, W., McGowan, B., & Kluever, J. (2017). Fake news: An opportunity for real librarianship. *The Illinois Library Association Reporter*, 35(3), 8–11.

Fullerton, J., & Kendrick, A. (2022). A Test of PR Students' Ability to Differentiate Native Advertising from Editorial Content in Online Media. *Public Relations Journal*, 15(1).

Ineji, P. U., Bassey-Duke, V., & Brown, N. J. (2014). Application of political propaganda by government in the resolution of ASUU/FGN industrial Conflict of 2013. *Advances in Journalism and Communication*, *2*(03), 101.

Jacobson, L. (2017). *The smell test: Educators can counteract fake news with information literacy. Here's how*. Retrieved from https://www.slj.com/?detailStory=the-smell-test-educators-can-counter-fake-news-withinformation-literacy-heres-how.

Kaplan, A. M. (2015). Social media, the digital revolution, and the business of media. *International Journal on Media Management*, 17(4), 197–199.

Lim, S. (2020). Academic library guides for tackling fake news: A content analysis. *The Journal of Academic Librarianship*, *46*(5), 102195.

McGregor, S. C. (2019). Social media as public opinion: How journalists use social media to represent public opinion. *Journalism*, 20, 1070–1086.

Oberschall, A. (2010). Conflict theory. In *Handbook of Politics*. New York: Springer.

Olisah, C. (April 4, 2022). *Ngige accuses ASUU of bullying and intimidation, says suspension of strike depends on union*.Nairametrics. https://nairametrics.com/2022/04/22/ngige-accuse-asuu-of-bullying-and-intimidation-says-suspension-of-strike-depends-on-union/

Olisah, C. (June 23 2022). *ASUU disagrees with Labour Minister, Chris Ngige, denies receiving invitation for a meeting*. Nairametrics. https://nairametrics.com/2022/06/23/

Olisah, C. (June 23, 2022). *ASUU President accuses Ngige of insulting lecturers and spreading fake news over strike*. Nairametrics. https://nairametrics.com/2022/07/14/asuu-president-accuse-ngige-of-insulting-lecturers-and-spreading-fake-news-over-strike/.

Oluwafemi A. (September 7, 2022). *ASUU strike: Nigerians have no good reason to be disappointed with FG, says Minister*. The cable. https://www.thecable.ng/asuu-strike-nigerians-have-no-good-reason-to-be-disappointed-with-fg-says-minister/amp

Rapti, M., Tsakalidis, G., Petridou, S., & Vergidis, K. (2022). Fake News Incidents through the Lens of the DCAM Disinformation Blueprint. *Information*, 13(7), 306.

Rochlin, N. (2017). Fake news: belief in post-truth. *Library HiTech*, 35(3), 386–392.

Secker, J. (2018). The revised CILIP definition of information literacy. *Journal of Information Literacy*, *12*(1), 156–158.

Spohr, D. (2017). Fake news and ideological polarization: Filter bubbles and selective exposure on social media. *Business Information Review*, 34(3), 150–160.

Tewell, E. (2015). A decade of critical information literacy: A review of the literature. *Communications in Information Literacy*, *9*(1), 2.

The Nation (September 19, 2022). *ASUU strike would have dragged for two years – Ngige*.TheNation Newspaper. https://thenationonlineng.net/asuu-strike-would-have-dragged-for-two-years-ngige/

Tolu-Kolawole, D. (August 18, 2022). *ASUU replies Adamu on students compensation comment*. Punch Newspaper.

Ukwu, J. (May 13, 2022). *ASUU Strike: We've Reached Reasonable Agreement with Lecturers says Ngige*. Legit. https://www.legit.ng/education/1469737-asuu-strike-weve-reached-reasonable-lecturers-ngige/

Ume-Ezeoke, G. (August 18, 2022). *Adamu Denies Getting Presidential Order To Resolve ASUU Strike In Two Weeks.* ChannelTV. https://www.channelstv.com/2022/08/18/adamu-denies-getting-presidential-order-to-resolve-asuu-strike-in-two-weeks/

Wardle, C., & Derakhshan, H. (2017). Information disorder: Toward an interdisciplinary framework for research and policy making. *Council of Europe Report*, 27, 1–107.

Yerbury, H., & Henninger, M. (2020). Libraries and democracy: complementarity in a regime of truth. *Informed Societies*, 165.

Sowemimo R. Adekunmisi, Adebambo A. Oduwole
and Adedokun A. Adekunmisi

Roles of University Libraries in Curbing Information Disorder Among Undergraduates in Nigerian Universities

Abstract: Individuals, particularly undergraduates, now suffer from information disorder (ID), an act or condition where false or wrong information is created and shared either with or without the aim of harming the audience or the receiver. The syndrome is becoming more infectious as a result of revolutions of the social web, the abundance of media devices that aid the fluidity of information, and a gap in media and information literacy and other twenty-first century skills among undergraduates. Presently, the rate of proliferation of scam websites, communities, and fake social media accounts with fake recommendations, hashtags, rumors, videos, social feeds, and memes in particular is becoming worrisome and disturbing to individuals, societies, universities, and governmental organizations. Consequently, groups and individuals now have vested interests and gains at the expense of harmful impacts of ID on the public. This chapter examines the concept of ID while discussing its types, causes, syndromes, benefits (if any), challenges, and means to ameliorate its spread among undergraduates in a developing country like Nigeria.

Keywords: fake news, information disorder, librarians' roles in curbing misinformation

Introduction

Information disorder (ID), basically in the form of propaganda, fake news (FNs), and misinformation (MI), is as old as human beings but has taken on new dimensions and become more prominent with constant revolutions of the web and the high abundance of sophisticated media that allow online activities among neti-

Sowemimo R. Adekunmisi, Olabisi Onabanjo University, Nigeria,
e-mail: ibironkeadekunmisi@gmail.com
Adebambo A. Oduwole, Olabisi Onabanjo University, Nigeria, e-mail: bambose26@gmail.com
Adedokun A. Adekunmisi, Tai Solarin University of Education, Nigeria,
e-mail: adedokunadekunmisi@gmail.com

https://doi.org/10.1515/9783111176345-016

zens, particularly undergraduates. Undergraduates in particular share e-text, e-images, videos, and other e-content online (Adekunmisi 2021) to the extent that such content becomes popularized within a few hours (Allcott and Gentzkow 2017; Evason and Spousel 2019). The potential harm from this – hate speech, disorderliness, confusion, bullying, defamation of characters, riots, and violence – at times results in the loss of properties and lives, making this discourse timely, particularly in developing countries.

Nigeria being a typical example of countries where citizens post any content without minding the effects of this on other individuals makes this article imperative to broach awareness and educate undergraduates and, possibly, university management and members on steps to curb, if not totally eliminate, ID and its syndrome among the Nigerian populace. Educating students, particularly undergraduates, has implications for Library and Information Science (LIS) professionals in improving information and media literacy skills as well as improving the educational performances of undergraduates in the country.

Concept of Information Disorder

ID is a concept that emerged with the fluidity of information occasioned by revolutions of the Internet, social media platforms, search engines, proliferation and high penetrations of media devices, and a wide knowledge gap of twenty-first century skills such as information literacy and critical thinking. ID has been defined by various researchers, journalists, groups of researchers as well as media organizations. Wardle (2022) referred to ID as the various ways in which an information environment is contaminated or made morally impure, noting that most contents are often genuine but are frequently used out of context and created to be believed and shared by certain people. Peŕez-Escoda (2022) regarded ID as a sociological and communicative disease caused by overabundance of media (information). The World Health Organization (2021) and Casero-Ripollès (2021) regarded the concept as being an infodemic.

Kandel (2020) considered ID as a syndrome peculiar to developing and sharing of untrue or misleading information with or without the aim of hurting the receiver and which manifests as distorted facts, manipulated content, defamation of peoples' faiths and beliefs, propaganda, and the spread of untrue news. The author further noted that individuals who develop and disseminate either part or all of false content could be taken as suffering from information disorder syndrome (IDS). The Council of Europe (2023) takes cognizance of the similarities in FNs, propaganda, and MI and thus considered the three concepts as a range of means in which information sharing results in damage or injury – intentionally or unintentionally – with respect to promoting a specific moral, point of view, and political motive. The Council further distinguished the abuse of information into misinformation (MI), disinformation (DI) and mal-information (MAL) as depicted in Figure 1.

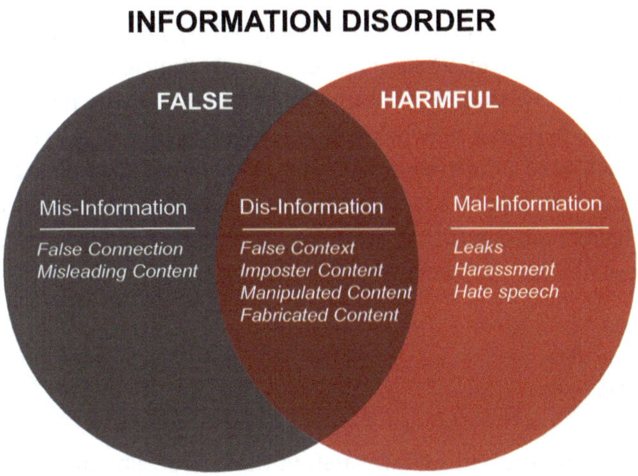

Figure 1: Intersection of three types of information disorder.
Source: Wardle and Derakshan (2017)

Types of Information Disorder

There are basically three different kinds of information disorder, namely: disinformation (DI), misinformation (MI), and mal-information (MAL). These are discussed in the sub-sections below.

Disinformation

DI is regarded as content or information that is deliberately created false and shared to cause injury or harm (Kandel 2020). DI is influenced by reasons such as financial needs (to earn money), local or international political leverage, and social or psychological intent (to cause disorderliness, distress, annoyance, worry, and complications) (Wardle 2020). When information or content that is deliberately false is created and transmitted among people, particularly among university students, it becomes MI and continues to be circulated or spread about.

Misinformation

MI refers to inaccurate or erroneous content that is shared by people without being aware that the content is untrue (O'Sullivan 2020). For instance, a part or a fragment of fake content could be shared by an unwitting individual who has the belief they are helping members in a network without knowing that such information is false. When such (dis) information is further disseminated, it usually results in MI. Elements such as psycho-social factors, socio-psychological factors (Kandel 2020), ethno-religious tension (Kurfi, Msughter, and Mohamed 2021), and lack of mindfulness (An et al. 2023) influence the spread of MI.

Mal-information

MAL is considered as genuine content or information that is rebranded or reframed to harm people (African Library & Information Associations & Institutions 2019; Wardle 2020). This includes information that is peculiar to an individual or an organization and shared by another to either cause harm or defame the reputation of the individual or the organization (Kandel 2020). Due to revolutions in the Internet and its technologies, actors, namely individuals, groups of individuals and organizations that engage in ID, now use reliable and validated content in new and misguiding ways under the pretext that such content is ineligible for AI checking.

There exist, however, artificial intelligence (AI) systems such as search engines, social websites, and organizations that are tightening up policies on fake accounts and fake content usage. These make MAL in many cases difficult to examine or investigate. It is important that undergraduates always ascertain the veracity of content before using or sharing any content, particularly on social media networks.

The Federal Government of Nigeria (FGN) between June 5 2021 and January 13 2022 banned Twitter from operating in Nigeria in a bid to stop the spread of FNs which incited hate speech that resulted into calumnies of personalities, associations, and organizations. The essence was to prevent the destruction of reputational image and, ultimately, disorderliness and violence.

Forms of Information Disorder

ID essentially manifests in seven different forms, namely: satire or parody, false connection, misleading content, false content, imposter content, fabricated content, and manipulated content. These forms are discussed using the illustration provided in Figure 2, which indicates a continuum of content with satire or parody at the starting point where no harm or injury is meant down to the other end of the spectrum, fabricated content, where there is potential harm to individuals, associations, organizations, and even nations.

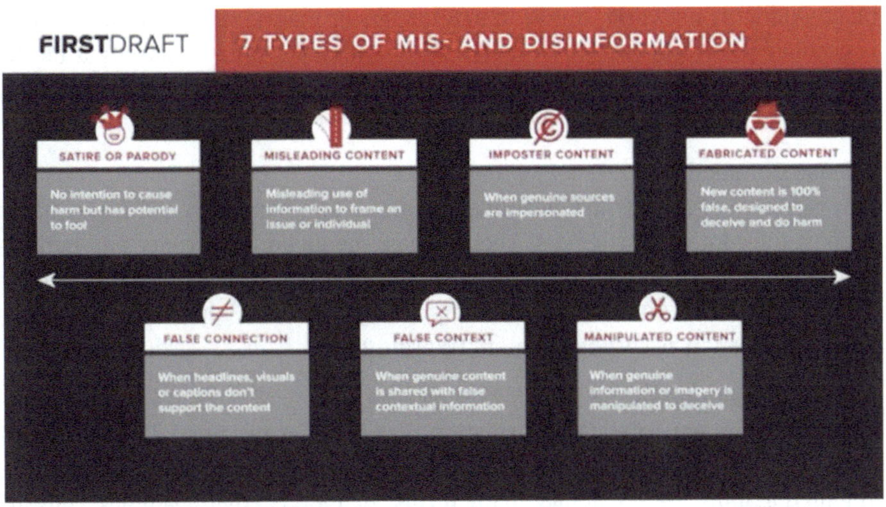

Figure 2: Seven types of mis- and dis-information.
Source: Wardle (2017).

Figure 2 as provided by Wardle (2017) is an illustrative chart that allows individuals to better understand the information war that people all over the world are presently experiencing and the interrelationship among the concepts.

1. Satire or Parody

Satire and parody are essentially creative works of art. However, satire is regarded as a genre of the visual, literary, and performing arts, which could be fiction and or non-fiction, in which vices, follies, shortcomings, and abuses are expressed to ridicule and expose perceived flaws of concerned governments, society, corporations, or individuals so as to bring about improvement (Elliot 2004). Satire is usually written by witty individuals in a clever and humorous manner to bring about productive criticism and topical issues in a society. Parody, on the other hand, is a literary or musical work designed to closely imitate the style of an author or a politician, an event, and movement for humorous effect or to ridicule. These may be found in art, culture, music, literature, television, film, theater, gaming and animation.

With the abundance of media and information, satire and parody are strategically used to bypass fact-checking tools in order to spread false content, rumors, clickbait, and polarizing content in less serious and deceitful manners. Individuals that first come into contact with satire or parody often recognize it as such, but as the information gets dispersed the original source is lost; thus, other receivers failed to appreciate the content as either a satire or a parody, especially on social media networks, unlike in newspapers or magazines where each section is peculiar to certain content or information. Although satire or parody is on the low side of the spectrum indicating no intent to harm people, it, however, has great potential to deceive individuals.

2. False Connection

This is a form of ID in which headlines, captions, and images do not align with or support the content. Participants such as mass media outlets, agents, and individuals involved in DI employ sensational languages to drive clicks or traffic to their websites, possibly with the intent of improving followers, popularity, and votes. This is referred to as clickbait content (Wardle 2020).

3. Misleading Content

Misleading content refers to information that is either intentionally or unintentionally false, inaccurate, or biased, and that leads to confusion or deception among the audience. It can take many forms such as clickbait headlines, conspiracy theories, fake news, manipulated images or videos, and propaganda (Faix and Fyn 2020). Misleading content can harm individuals, communities, and society at large by eroding trust, spreading disinformation, polarizing opinions, and undermining democracy. The harm of misleading content can be significant and long-lasting; for instance, it can lead to the spread of misinformation about public health, causing people to adopt risky behaviors or refuse lifesaving vaccines. It can also fuel prejudice and hate speech against marginalized groups, inciting violence and discrimination.

In governance, misleading content can manipulate public opinions and influence elections, causing lasting damage to democratic institutions and the rule of law. It can erode trust in traditional media sources, science, and expertise, leading to a fragmented information landscape and an erosion of the shared reality that underpins democracy, which can further polarize society and hinder progress on complex challenges such as climate change, inequality, and public health. Overall, the harm of misleading content is not only personal but also societal, and can affect the quality of public discourse, the functioning of institutions, and the well-being of individuals and communities.

4. False Content

False content, also known as fake content, is information that is intentionally fabricated or distorted to deceive and manipulate the audience. False content take many forms, such as fabricated news stories, photo shopped images, doctored videos, and manipulated audio recordings (Evanson and Sponsel 2019). False content can be created to have undue financial gain and political influence, and to spread disinformation and propaganda among many reasons. This can lead to widespread confusion, fear, and panic, as people struggle to distinguish fact from fiction. False content can also damage reputations, cause financial harm, lead to legal repercussions, and have serious consequences for public health and safety in particular, when false information is spread about vaccines or other medical treatments.

Further, false content has the capacity to undermine public trust in institutions and democratic processes, eroding the shared reality and social cohesion necessary for a functioning society. It may fuel extremism, polarize opinions, and incite violence against individuals or groups of individuals. Overall, the harm of false content is significant and multifaceted, capable of affecting individuals, communities, and society as a whole. It is crucial to be vigilant against false content and to seek out reliable sources of information to help prevent its spread.

5. Imposter Content

Imposter content refers to material that is falsely attributed to a person or entity, either by a machine or an individual. This can include but is not limited to forged signatures, fabricated documents, impersonated social media accounts, deep fakes, and manipulated videos or audio recordings that make it appear as though someone is saying or doing something they did not. This content can be used for a variety of malicious purposes, including identity theft, fraud, disinformation, and political propaganda. The effects of imposter content can be significant, particularly when it involves high-profile individuals and institutions. Imposter content can damage reputations, undermine public trust, and spread false information that can have serious consequences for individuals and societies.

Imposter content can also be used to intimidate or blackmail individuals, or to gain unauthorized access to sensitive information or resources. It can be difficult to detect and remove imposter content as it can be spread quickly and widely on social media and other online platforms. Imposter content can be sophisticated and convincing, making it hard for people to discern what is real and what is not. It is crucial for undergraduates to be vigilant against imposter content and to take steps to verify the authenticity of any material before sharing or acting on it.

6. Manipulated Content

Manipulated content refers to any form of digital material that has been altered or modified to deceive or mislead people. This can include edited images, videos, audio recordings, or text that has been manipulated to distort the truth and convey a false message. Manipulated content can be used for a variety of purposes, including political propaganda, advertising, satire, or entertainment. Manipulated content can be significant and far-reaching, particularly when it involves sensitive or controversial topics. Manipulated content can erode public trust in institutions, scientific research, or democratic processes, leading to a fragmented information landscape and a polarization of opinions which can fuel extremism, hate speech, or discrimination against marginalized groups, inciting violence and harassment.

In addition, manipulated content can have personal consequences for individuals, such as reputational damage, emotional distress, financial harm and lasting damage to livelihoods. It can also be used to perpetrate scams, phishing attacks, or other forms of cybercrime, exploiting people's trust and vulnerability. Overall, the harm of manipulated content is significant and multifaceted, affecting individuals, communities, and society at large. It is crucial for undergraduates to be aware of the risks of manipulated content and to take steps to verify the authenticity and accuracy of any material before sharing or acting on it.

7. Fabricated Content

Fabricated content refers to digital material that has been entirely concocted, meaning it never existed in the real world. This can include made-up news stories, photoshopped images, or deep fakes, which are manipulated videos or audio recordings that make it appear as though someone is saying or doing something they did not. Fabricated content can be created for various reasons, such as satire, entertainment, or disinformation campaigns. Fabricated content can undermine the credibility of traditional news sources, science, and expertise, leading to a loss of trust in individuals and institutions and a fragmentation of the information landscape. It can also fuel polarization and extremism, leading to the spread of false narratives and the erosion of social cohesion.

Moreover, fabricated content can be used to manipulate public opinion and influence political processes, leading to serious consequences for democracy and the rule of law. Fabricated content can be used to spread false information about candidates, parties, or policies, leading to voter confusion and mistrust. It can also be used to incite violence or harass individuals or groups, perpetuating discrimination and prejudice. Overall, fabricated content is a serious issue for societies because the content has the capacity to affect the quality of public discourse, the functioning of institutions, and the well-being of individuals and communities. It is crucial for undergraduates and other university community members to be vigilant against fabricated content and to take steps to verify the trustworthiness of any material encounter before using it.

Mechanism of Information Disorder

It is important to look at the fundamental processes that facilitate rumors and fabricated and misleading content that flourish among undergraduates worldwide. A better understanding and analysis of these processes would help university stakeholders (that is, university and library management, lecturers, researchers, library personnel, students, and other members) identify participants of ID and their motivators, evaluate its impact, and develop strategies to address the nuisances at hand. These factors are regarded as elements and phases of ID (Council of Europe 2023). This subsequent section describes the elements of the various forms of ID.

Elements of Information Disorder

The elements of ID are mainly agent, message, and interpreter.

1. Agent

The agent is essentially a participant such as an individual, a group of individuals, or a website including social media tools and organizations that engage in the design, production, and dissemination (transmission) of variants of ID. The motives behind involvement in the act could be to stimulate web traffic, generate income or revenue, influence citizens' opinions, and support political points of views (Belova and Georgieva 2018).

2. Message

The message is the content being created and distributed or disseminated. Questions that can assist undergraduates in identifying FNs, propaganda, MI, and MAL also assist them in avoiding being deceived, include: Is the message authentic or fabricated? What form does the message take? What are the features or characteristics of the message? Undergraduates need to be critical of messages or contents they come in contact with. Students should be able to determine the validity or authenticity of content they encounter either in newspapers, magazines, textbooks, journals, and particularly online content especially on social media networks like Facebook, Twitter, WhatsApp, and YouTube. Students should be able to assess if the information or content they encounter is educational, recreational, a joke, financial, or is just for the sake of fun. Being able to discern what comes along with the content – icons, small prints, and so on – would assist students to better evaluate the validity of the content and how to manage the information appropriately.

3. Interpreter
The interpreter is the individual who expounds or explains the content upon being received. The individual could be a student, working class, a researcher, old or young, male or female, literate or illiterate. How does the individual interpret the content? Does the person take the content with levity or seriousness? What action(s) is taken by the individual? Does the person just give a thumb up and move on, or add or remove widgets, icons, sounds, text, or visuals to the original information before sharing to other people and in different platforms? Undergraduates need to have good sense of using information and should be able to critically analyze, synthesize, and determine what content should and should not be shared on platforms or on the web.

Phases of Information Disorder

The three elements – agent, message, and interpreter – discussed above have relevance to the DI, MI, and MAL phases of ID. The phases are categorized into three, namely: creation, production, and distribution. Dissecting the phases of the cycle allows for a clear understanding of the actors and their motivations, while assessing the effects on society helps identify how the menace could be addressed among university students in the country.

1. Creation
This is the stage at which parody, or satire, false content, misleading content, and other types of ID are designed and created. The social web coupled with its variants, 3.0, 4.0, 5.0, and so on, along with sophisticated media devices (laptops and smart phones) allow text messaging, tweeting, collaboration, and so on and encourage all manners of content to be uploaded on the web with the slightest click of media tools. These concurrently have encouraged the creation, production, and dissemination of both validated and unscrutinized content – harmless and harmful – on the dynamic web. As such, individuals encounter good and bad information. Unwitting individuals usually get their hands burnt while trying to consume the content encountered, especially undergraduates. Some undergraduates have

lost their lives while some have been duped. However, some undergraduates have positively benefitted from the use of different web-based platforms and resources in terms of accessing educational content for learning and research, thereby improving their educational performance and other academic endeavors as well as personal enhancements (Adekunmisi 2021).

2. Production

Having created the content using smartphones, laptops, personal computers, and other devices, the content is turned into a media product which could be a print (for example, a poster or a textbook), a non-print like e-text, sound, video, or illustration (for instance, graphs, bar charts, and tables), a web-based form, or a combinations of these forms. The various information and communication technology (ICT) gadgets easily enable and assist in the design, production, dissemination, and storage of these media products, with little or no demand for fluency in ICT skills.

3. Distribution

At this stage, the designed content, now a potential media product usually in the electronic format, is distributed through a variety of channels such *as* Facebook, YouTube, WhatsApp, Instagram, websites, print media, and so on, depending on the format of the content.

Effect of Information Disorder

Information creates wealth and invariably gives power to individuals to learn and make right choices and decisions that can improve and better the lot and livelihood of individuals, especially undergraduates. ICT tools, particularly web-based sites and tools, also enable undisturbed access to content, whether useful, useless, harmful, or harmless. These tools have made life more comfortable, lively, impactful, and enriching to individuals, organizations, societies, and nations. Featuring some likely effects of FNs, propaganda, and other forms of ID would assist undergraduates and other members of university communities in understanding the complexities of the information ecosystem individuals currently live in. It becomes imperative for individuals, particularly undergraduates, to be aware and more knowledgeable about what constitute good and bad content and how to identify bad or harmful content so as to dissuade themselves and other people from getting involved in IDS.

Figure 2 provides an illustrative description of the extent of harm or damage ID variants can cause to unwitting undergraduates. The figure shows IDS (that is, the seven types of ID) on a continuum from low harm to high harm where satire or Parody, false connection, and misleading content have potential to cause no harm and rather just deceit. However, false context, imposter content, manipulated, and fabricated content have a high propensity to cause high or serious harm, with fabricated content having the capability to totally deceive and cause great harm (that is 100%) to unwise citizens. This harm can lead to violence and even loss of lives and properties.

This deduction has been supported by assertions and research documentation within and outside Nigeria. For instance, the Council of Europe (2023) asserted that FNS, propaganda, and MI reduce trust in democratic processes and weaken the democracies of countries. Belova and Georgieva (2018) asserted that ID spreads at a high rate and endangers the welfare of societies, democracy, and media in general. If ID is not controlled, it can lead to the proliferation of more conflicting and distorted content or information that is meant to deceive and disempower (African Library & Information Associations & Institutions 2019), thereby increasing confusion among people (Agim, Obiekwe, and Eneh 2020) and leading to severe effects such as debt and or death.

Importantly, young individuals, especially undergraduates, are becoming vulnerable to DI, MI, and MAL as a result of their poor media use habits (Perez-Escoda 2022; Farkas and Schou 2022; Faix and Fyn 2022). Researchers such as Sahoo and Gupta (2020) as well as de Oliveira et al. (2021) showed that undergraduates expend significant time playing online games, blogging, chatting, posting images of themselves, and engaging in other activities online at the expense of academic endeavors. Undergraduates depend heavily on online content for knowledge seeking and the reality of life in which they live and as such they are more likely to be prone to FNs (Leeder 2019; Evanson and Sponsel 2019; Sivasankari and Vadivu 2021). Unfortunately, parents and guardians lack the digital competence to control their children's activity online or educate them about the consequences they are likely to face (Council of Europe 2023). However, the syndrome presents opportunities for librarians and lecturers who teach research strategies on how to locate, analyze, evaluate, and use authentic and high quality sources to be more assertive of their roles.

Theoretical Framework

If the total elimination of IDS is unattainable, its reduction nonetheless impacts heavily on the job roles and expectations of librarians in higher education institutions. Satire or parody, false connection, misleading content, imposter content, fabricated content, manipulated and false content are all within the purview of the Framework for Information Literacy for Higher Education (HE) developed by the Association of College and Research Libraries (ACRL 2015), popularly referred to as the Framework. The ACRL Framework is an improvement over the Information Literacy Competency Standards for Higher Education that have been infused into academic curricula of Higher Education, in universities in particular (ACRL 2015). The swiftly changing Higher Education environment coupled with the dynamic and unpredictable information ecosystem occasioned the development of the new framework that envisioned the foundational ideas about the changing information landscape. The new Framework helped librarians in the university system contextualize and integrate information literacy as well as give an inner understanding of knowledge practices and dispositions an information literate undergraduate should possess (ACRL 2015; Durodolu and Ngoaketsi 2019).

The earlier standards of information literacy (for instance, Seven Pillars of Information Literacy developed by the Society of College, National and University Libraries (2011) and Media and Information Literacy Curriculum for Teachers developed by the United Nations Educational, Scientific and Cultural Organization (2011)) expected undergraduates to exploit and use technology to create or produce information for academic endeavours. The ACRL Framework in line with the conceptions of metaliteracies is highly relevant to this discourse because it expanded the worldwide acclaimed conceptions of information literacy and embraces the web-based (social) media, online communities, and open learning initiatives. The six frames of the Framework are: authority is constructed and contextual; information creation as a process; information has value; research as inquiry; scholarship as conversation; and finally searching as strategic exploration. All of these have the capabilities to reduce the various forms of ID among undergraduates.

The prescribed knowledge practices and dispositions of the ACRL Framework are highly flexible and inexhaustive in nature and, as such, it is applicable and highly relevant to academic and research endeavours in addition to being applicable to developing and improving metaliteracy skills of undergraduates. The wide range of skills, abilities, knowledge practices, and attributes can be employed to avoid ID in the present media age, depicting how libraries can fully deploy the framework when imparting research instructions to students. Librarians in the developed countries like United Kingdom, United States of America, and European countries have been proactive and have refocused library instruction programs to embrace news literacy and means to critically evaluate content students have encountered. This was done through a number of methods such as research guides and webinars (Batchelor 2017), teaching of information evaluation skills (Caulifield 2017), pop-up workshops (Wade and Hornick 2018), and use of assignments in courses (Neely-Sardon and Tignor 2018).

However, the Framework has been criticized by researchers like Bluemle (2018) and Faix and Fyn (2020) that it cannot alone adequately address the various forms of ID. Additional but complementary strategies such as metaliteracy that librarians could adopt to inculcate critical thinking and other twenty-first century skills in undergraduates with respect to information source evaluation and research techniques are advised.

Metaliteracy

Metaliteracy is considered the ability to evaluate information or content for its reliability, bias, and credibility and apply the content in the context of production and sharing of knowledge. Metaliteracy is applicable in the terrain of the Internet and social media (Mackey and Jacobson 2011). Mackey and Jacobson (2013) envisioned metaliteracy as a robust model for information literacy to advance critical thinking and reflection in open learning settings, social media avenues, and online communities. It is a new framework that takes cognizance of and addresses the challenging media era faced by all.

Metaliteracy is being used to help undergraduates become informed consumers and credible creators of content in online communities, including social media. This is because metaliteracy promotes the exploitation and management of existing and upcoming technologies. Metaliteracy simultaneously combines and embraces other literacies such as media literacy, digital literacy, and visual literacy (Jacobson and Mackey 2016) and encourages metacognitive reflection, that is, reflection on one's way of thinking and determining when and how to explore specific strategies for problem-solving (Metcalfe and Shimamura 1994; Schraw 1998). These skills, along with the comprehensive learning approach that encourages the creation and transmission of original and remixed content in digital and online communities, are equipped in learners with the metaliteracy model.

The metaliteracy model emphasizes the use of four learning domains (LDs), namely, metacognitive, affective, cognitive, and behavioural domains (Figure 3) that undergraduate learners specifically can be empowered with in the dynamic information environment. This empowers undergraduates to navigate the complex information landscape in a critical and ethical manner, making informed decisions about what information to trust, how to verify its accuracy and reliability, and how to use it effectively to achieve desired goals. These skills and attributes are imperatives in today's digital age, where the volume and diversity of information available is becoming overwhelming, and the potential for MI, DI, and propaganda is high. Metaliteracy supports undergraduates' curiosity to learn throughout their lifespan and appreciate that continuous learning is key to attaining knowledgeable and contributing citizens. Metaliteracy learners must endeavor to take ownership of their own learning perspectives and goals, be open to challenges as they adapt to new technologies, facilities, processes, and resources, reflect on self-growth, and analyze ideas and insights from multiple strategies. It is, however, important to look at each of the four domains earlier mentioned in this chapter, namely, the metacognitive, affective, cognitive and behavioral domains.

Metacognitive Domain of Learning

The metacognitive domain strengthens the individual undergraduate's ability to reflect (think) and regulate (modify, adjust) one's own learning pathway. This essentially entails an undergraduate analyzing, observing, and controlling their thinking and learning processes. For instance, a deeper understanding of how and why DI, MI, and MAL are perpetrated by actors, their ideas, knowledge, and bias about ID, and how to continue to think and learn about IDs and syndromes are critical to raising awareness of the severity of this menace. These skills are important for effective learning and evaluation of undergraduates' thought processes and biases. Although, the attributes may be assiduous to imbibe, mastery of the skills allows undergraduates to become critical learners and responsible citizens. Having a deeper reflection of one's thought processes about learning would assist undergraduates in identifying areas of strengths and weaknesses and always striving for improvements in those identified areas of weaknesses. The self-regulation cum self-directed activities usually pave the way to success for undergraduates that possess and demonstrate multiple literacy skills and attributes.

Cognitive Domain of Learning

The cognitive domain is the actual learning process and it embraces what individual undergraduate learner should know upon successful completion of learning activities. This may involve the level of dissection of information about ID, how ID ideas are organized, its applications, and the undergraduate's ability to evaluate and think critically about ID or information sources.

Behavioral Domain of Learning

The behavioral domain encapsulates what the individual undergraduate is able to do after a specific learning process, particularly the knowledge, skills, and attributes the undergraduate develops as outcomes of the training. The behavioral domain applies in formal learning situations and in everyday life.

Affective Domain of Learning

The affective domain relates to how the individual undergraduate feels about their emotions and attitudes after a specific learning activity. If undergraduates have had training on ID as well as its effects on victims and empathize with these affected individuals, they will not want to be involved in this bad practice because they have become more informed about it.

The affective domain, that is, emotions and attitudes interacting with other domains, can cause a feeling or an emotional stimulus that may motivate undergraduates to ask questions and learn more about ID or other particular concern or issues. Conversely, an attitudinal or emotional response may prevent undergraduates from looking at issues or perspectives objectively as a result of deeply held notions or viewpoints that may interfere with logic and reason. The four learning domains are interrelated and provide checks and balances to thinking processes, and it is thus important that undergraduates possess and demonstrate high abilities and good dispositions for all learning domains. This provides students with the opportunity to analyze how to think, what to do, what can be controlled, and how to respond, as well how to take up active roles such as being active participants, effective communicators, enlightened publishers, or information literate researchers as depicted in Figure 3.

Equipping undergraduates with these identified multi-literacies skills would promote critical thinking, creative and innovative thinking as well as the problem-solving skills of such students. As such, equipped learners are more able to think critically about wise creation and wise dissemination of content, and thus invariably such undergraduates would not want to participate in any form of infodemic. Such undergraduates would also want to be teachers, authors, or leaders and play active roles in discouraging colleagues, friends, family members and other members of the public from being perpetrators of FNs, propaganda, and other variants of ID and encouraging them to be responsible producers of content, translators, and other roles as depicted in Figure 3.

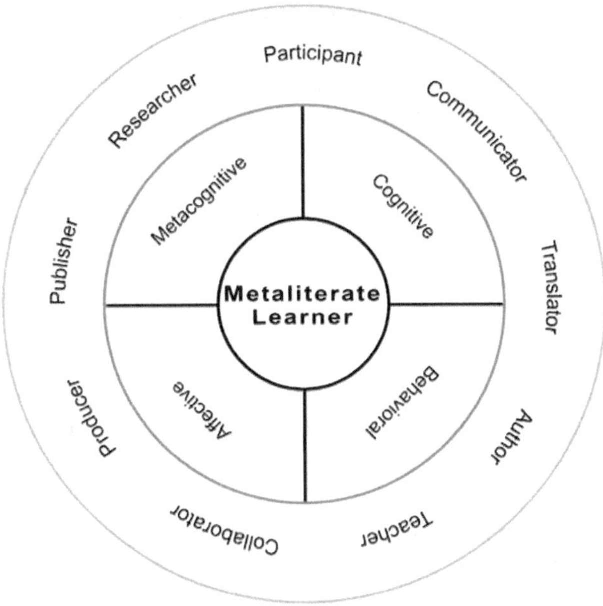

Figure 3: The Metaliterate Learner.
Source: Jacobson, Mackey and O'Brien (2018).

Some Roles of Librarians in Curbing Information Disorder among Undergraduates

Preventing and or controlling an infodemic among undergraduates requires many literacy skills which can be anchored by librarians in the university environment. The university is one of such places undergraduates can be trained to discern, comprehend, critique, condemn, evaluate, and produce credible content using imaginative attributes. Academic librarians have many roles in equipping students, particularly undergraduates, with these skills and attributes in a number of ways:

1. Teaching of media and information literacy skills: librarians can teach media and information literacy skills to undergraduates and other university community members to help them make informed decisions as users of media and information sources and become wise creators and producers of media and information messages. This can help reduce ID syndromes.
2. Teaching of information evaluation skills: librarians have a role to equip undergraduate students with critical information evaluation skills to responsi-

bly and ethically access, manage, and use web-based media and content credibly and responsibly.

3. Creation of research guides: librarians can create research guides that provide tips and resources for identifying and evaluating reliable sources of information.
4. Offer fact-checking websites and resources: librarians can provide fact-checking resources such as websites and tools that help verify the accuracy of information.
5. Curate collections: librarians can curate collections of reliable and credible sources to provide undergraduates with access to accurate and relevant information.
6. Promoting diversity and inclusion: librarians can promote diversity and inclusion in their collections by selecting materials that represent a variety of perspectives and experiences and by providing resources that support marginalized communities.
7. Develop collaborations with news organizations: librarians can partner with news organizations to promote accurate information and combat misinformation.
8. Host workshops, seminars, or webinars on information literacy: librarians can host information literacy workshops and seminars or webinars for students, particularly undergraduates, faculty, and other university community members to help them develop media and digital literacy skills as well as critical thinking skills. Librarians can also advocate for information literacy by participating in conferences, workshops, and other events.
9. Provide one-on-one reference services: librarians can provide one-on-one reference services to help undergraduates identify and evaluate information sources.
10. Monitoring social media and news sources: librarians can monitor social media and news sources to identify trends in misinformation and provide accurate information to undergraduates.
11. Advocate for information transparency: librarians can advocate for information transparency and open access to scientific research which can help combat the spread of misinformation and promote accurate, useful, and healthy information among undergraduates.

Recommendations

1. It is imperative for universities in the country to provide digital citizenship education. Individuals are expected to have begun their journey to becoming digital citizens from elementary education (to university level), however, it is also assumed that certain undergraduates may have missed such education along the way. It is not out of scope for universities to integrate digital citizenship education into curricula to develop and maintain high critical thinking skills in undergraduates.
2. Educators (library personnel and lecturers) do not have a clear understanding of why and how undergraduates access and consume MI, DI, and MAL. Educators must be well equipped in subjects that inculcate in students competencies that critically dissect, analyze, synthesize, and evaluate content encountered in all media forms.
3. Librarians should be competent in the use of ICT tools, particularly social media avenues which have become must-use sites, among digital natives and importantly the most prevalent tools for ID among undergraduates.
4. LIS professionals should be active change agents in the provision of instructions to undergraduates on how to best navigate and validate content they come in contact with.
5. Librarians and other educators should equip undergraduates with strategies to detect misinformation.
6. LIS professionals and lecturers should consider teaching methodologies that use problem-based learning approaches and be supportive of emerging technologies being intensified for undergraduates.
7. University management and curriculum developers need to adopt teaching methodologies that incorporate metacognitive reflection into academic curricula.

References

Adekunmisi, S. R. (2021). Media literacy, critical thinking and creative thinking as determinants of Web 2.0 competencies among undergraduates in universities in Ogun State, Nigeria. Unpublished Ph.D Dissertation, Department of School Library and Media Technology, University of Ibadan, Nigeria, pp. 1–193.

African Library & Information Associations & Institutions. (2019). *Weaponisation of Misinformation: the Role of Librarians in Curbing the Epidermic*. http://www.web.aflia.net/weaponisation-of-misinformation-on-the-role-of-librarians-in-curbing-the-epidemic. (22 January, 2024).

Agim, E. C., Obiekwe, O. L. and Eneh, A. E. (2020). The informative roles of librarians to their patrons in the COVID-19 Lockdown. *International Journal of Social Sciences & Humanities Review*, 10(2) June: 180–186.

Allcott, H. and Gentzkow, M. (2017). Social media and fake news in the 2016 election. *Journal of Economic Perspectives*, 31(2): 211–236. https://doi.org/10.1257/jep.31.2.211 (Accessed on 23 January 2024).

An, Y., Huang, Y., Danjuma, N. U., Apuke, O. D. and Tunca, E. A. (2023). Why do people spread fake news? Modelling the factors that influence social media users' fake news sharing behavior. *Information Development*. Online First. http://doi.org/10.1177/02666669231194357 (22 November, 2023).

Association of College and Research Libraries. (2015). *Framework for Information Literacy for Higher Education*. http://www.ala.org/acrl/standards/ilframework (22 December, 2023).

Batchelor, O. (2017). Getting out the truth: The role of libraries in the fight against fake news. *Reference Services Review*, 45(2): 143–148. https://doi.org/10.1108/RSR-03-2017-0006 (20 January, 2023).

Belova, G. and Georgieva, G. (2018). Fake news as a threat to national security. De Gruyter Open International Conference. *Knowledge-based Organization*, XXIV(1): 19–22.

Bluemle, S. R. (2018). Post-Facts: Information literacy and authority after the 2016 Election. *Portal: Libraries and Academy*, 18: 265–282.

Casero-Ripollès, A. (2021). The impact of COVID-19 on journalism: A set of transformations in five domains. *Comunicação e Sociedade*, 40: 53–69. Doi:10.17231/comsoc.40(2021).3283 (22 January, 2024).

Caulifield, M. (2017). *Web Literacy for Student Fact-checkers*. https://webliteracy.pressbooks.com/ (23 June, 2024).

Council of Europe. (2023). *Dealing with Propaganda, Misinformation and Fake News*. https://coe.int/en/web/camp . . . (03 February 2024).

De Oliveira, N. R., Pisa, P. S., Lopez, M. A., de Medeiros, D. S. V. and Mattos, D. M. F. (2021). Identifying fake news on social networks based on natural language processing: Trends and challenges. *Information*, 12: 38.

Durodolu, O. and Joseph M. Ngoaketsi (2019). Multilateral relationship between Information Literacy, Self-concept and Metacognitive ability. In N. Feza. *Metacognition in Learning*. London: Intech Open Limited ISBN 978-1-78985-114-4. https://www.intechopen.com/chapters/76617

Elliot, R. C. (2004). The nature of satire. Encyclopaedia Britannica. https://n.m.wikipedia.org.

Evanson, C. and Sponsel, J. (2019). From syndication to misinformation: How undergraduate students engage with and evaluate digital news. *Communications in Information Literacy*, 13(2): 228–250.

Faix, A. and Fyn, A. (2020). Framing fake news: Misinformation and the ACRL framework. *Portal: Libraries and Academy*, 20(3): 495–508.

Farkas, J. and Schou, J. (2022). *Post-Truth, Fake News and Democracy*. New York: Routledge.

Jacobson, T. E. and Mackey T. P. (Eds.) (2016). *Metaliteracy in Practice*. Chicago: Neal-Schuman/ALA Editions.

Jacobson, T. E. and Mackey, T. P. (2013). Proposing a metaliteracy model to redefine information literacy. *Communications in Information Literacy*, 7(2): 84–91.

Jacobson, T., Mackey, T. and O'Brien, K. (2018). *Metaliteracy in Practice: Metaliterate Learner Roles*. https://metaliteracy.org/ml-in-practice/metaliterate-learner-roles/ (22 January, 2024).

Kandel, N. (2020). Information disorder syndrome and its management. *Journal of Nepal Medical Association*, 58(224): 280–285.

Kurfi, M., Msughter, M. and Mohamed, I. (2021). Digital images on social media and proliferation of fake news on COVID-19 in Kano, Nigeria. *Galactica Media: Journal of Media Studies*, 3(1): 103–124.

Leeder, S. (2019). How college students evaluate and share "fake news" stories. *Library & Information Science Research*, 41(3): 100967. https://doi.org:10.1016lj.lisr.2019.100967(22 January, 2024).

Mackey, T. P. and Jacobson, T. E. (2011). Reframing information literacy as a metaliteracy. *College & Research Libraries*, 72(1): 62–78. https://doi:10.5860/crl-76r1. (22 January, 2024).

Metcalfe, J. and Shimamura, A. P. (1994). *Metacognition: Knowing about knowing*. Cambridge, MA: MIT Press.

Neely-Sardon, A. and Tignor, M. (2018). Focus on the facts: A news and information literacy instructional program. *The Reference Librarian*, 59(3): 108–121. https://doi.org/10.1080/02763877.2018.1468849 (01 March, 2024).

Ostenson, J. (2014). Reconsidering the checklist in teaching internet source evaluation. *Portal: Libraries and the Academy*, 14(1): 33–50. https://doi.org/10.1353/pla.2013.0045. (18 February, 2024).

O'Sullivan, S. (2020). *The Student's Guide to Information Disorder*. https://newscolllab.org/2020/03/ (11 March, 2024).

Peŕez-Escoda, A. (2022). Infodemic and fake news turning shift for media: Distrust among university students. *Information*, 13(11): 523–527. https://doi.org/10.3390/info13110523 (03 March, 2024).

Sahoo, S. R. and Gupta, B. (2020). Multiple features based approach for automatic fake news detection on social networks using deep learning. *Applied Soft Computing*, 100 March: 106983. https://doi.org/10.1016/j.asoc.2020106983 (01 March, 2024).

Schraw, G. (1998). Promoting general metacognitive awareness. *Instructional Science*, 26: 113–125. doi: 10.1023/A:1003044231033. (11 January, 2024).

Sivasankari, S. and Vadivu, G. (2021). Tracing the fake news propagation path using social network analysis. *Soft Computing*

Society of College, National and University Libraries. (2011). *Seven Pillars of Information Literacy: Core Model*. http://www.sconul.ac.uk/sites/default/files/documents/coremodel.pdf. (11 February, 2024).

United Nations Educational, Scientific and Cultural Organization. (2011). *Media and Information Literacy Curriculum for Teachers*. http://unesco.unesco.org/images/0019/001929/19297le.pdf (11 February, 2024).

Wade, S. and Hornick, J. (2018). Stop! Don't share that story! Designing a pop-up undergraduate workshop on fake news. *The Reference Librarian*, 59(4): 188–194. https://doi.org/10.1080/02763877.2018.1498430 (10 February, 2024).

Wardle, C. and Derakhshan, H. (2017). *Information Disorder: Towards an interdisciplinary framework for research and policy making*. Strasbourg: Council of Europe, 1–107.

Wardle, C. (2020 September, 22). *Understanding Information Disorder*. https://www.firstdraftnews.org/long-form-article/understanding-information-disorder/ (10 August, 2024).

Wardle, C. (2022 June, 14). *Update from First Draft Executive Director Claire Wardle*. https://www.firstdraftnews.org/first-draft-update-june2022 (15 August, 2024).

World Health Organization. (2021). *Infodemic Management: An Overview of Infodemic Management during COVID-19, January 2020 – July, 2022*. Geneva: World Health Organization. https://search.worldcat.org/title (10 March, 2024).

Husaini Musa

Information Disorder and the Need for Media and Information Literacy

Abstract: The chapter explores the escalating challenges posed by the surge in media and information, emphasizing the critical need for enhanced media literacy skills in the face of rampant information disorders. Its introduction underscores the relevance of media literacy in discerning between false and genuine information, emphasizing the cultivation of skills such as information access, analysis, and critical evaluation, with a focus on ethical behavior within the media and information environment paramount. The subsequent section delves into the intricacies of information disorders, categorizing them into disinformation, misinformation, and malinformation, with examples illustrating the deliberate spread of false information for various motives, highlighting the urgency of distinguishing between trustworthy and misleading sources. Media literacy assumes a central role in navigating the omnipresence of media in daily life, enabling individuals to think critically, engage with democratic societies, and express themselves effectively. The discussion on information literacy underscores its role in navigating diverse information systems, making ethical use of information, and utilizing ICT skills for effective information processing. The chapter culminates in an exploration of media and information literacy (MIL), positioning it at the nexus of universal human rights and freedom of expression. This abstract encapsulates the article's essence, emphasizing the dynamic nature of information landscapes and the imperative role of literacy skills in fostering an informed and discerning citizenry. The overarching goal is to equip learners with the tools necessary to navigate this complex environment, ensuring their active participation in society.

Keywords: information literacy, information disorder, disinformation, misinformation and, malinformation

Husaini Musa, College Library, Federal College of Education Technical Gusau, Zamfara State, Nigeria, e-mail: husainimzam@gmail.com

https://doi.org/10.1515/9783111176345-017

Introduction

As a result of the increase in media and information every day, anywhere, and anytime, the skills of media in this information explosion become relevant. The concept of information disorder will enable potential learners to differentiate false information and genuine information. Studying media and information literacy enables learners to have patience in the way of organization and dealing with information disorder, with various skills possible to acquire, which include some certain level of respect towards people that need and request the quality information. Learners need skills to access, analyze, and engage critically about the selection of information resources they utilize in order to make decisions about the everyday issues they face as regards to their studies. Acquiring skills in information and media allows an individual to gain the ability to conduct good research using various strategies, problems models, information types, and sources, and know how to use these skills to achieve their educational pursuit

This chapter will also engage learners in various topics related to selection of information/media and knowledge of information literacy that can allow access to quality choice of information. Information literacy and everyday life will cover the explanation about the awareness and understanding of information in day to day life, which includes media information and its limitation as well as how to access information and its value to society. Acquiring these skills will help individuals to behave ethically in media and the information environment. At the end of the chapter learners will be expected to be able to identify false information and the various sources of information for effective research in media and information literacy. The learners will also be able to select the most quality information in media and information centers as well as access information needs and strategies for information retrieval.

Information Disorder

According to Kandal (2022), information disorder comprises deceptive stories or fabricated information. It is the sharing, sending or creating of false information with or without the aim of causing trouble to individuals or society (IGI Global 2022). Information disorder can also be defined as distorting facts, manipulating information, or sharing information without knowing its consequences. In my view, information disorder is a false statement or a baseless story/report that spreads among people and often brings harms and chaos. The conglomeration of mis, dis, and malinformation creates what is called information disorder (Wardle and Derakhshan 2017).

Nowadays, social media are the media where the majority of the community are sharing and receiving information; elsewhere, some other media share genuine information, while others have no basis at all. Everyone knows that when the Covid-19 pandemic spread globally there was also a spread of all kinds of news and information all over the world, some truthful and some fake. The information disorder here is that some people are trying to change the belief and ideas of communities by spreading propaganda and fake news so that people can act vehemently.

Two studies (Wardle and Derakhshan 2017; Carlsson 2019) categorized three kinds of information disorder, shown below.

Disinformation: when information is false and spread with intention to cause harm or trouble to humanity this is called disinformation. Examples include when people spread fake information for the purpose of making money such as online advertisements that are false, or when politicians spread fake information against their opposition party with the intention to win elections.

Misinformation is fabricated and false information in which the sender does not know that it is fake or untruthful and does not have the intention of causing trouble to others. For example, during the spread of the Covid-19 pandemic around the world people fabricated information about the cure or protection against the Covid-19 pandemic, such as that whoever is taking garlic or ginger juice will not be exposed to the virus. Sometimes videos that have no relation with a particular event will be linked and shared with the intention of cautioning people.

Malinformation is information that is spread with the intention to cause trouble. This type of information can be seen on social media, where people send or share information of either related or not related to an event in order to cause trouble within the community. For example, in Sokoto State, Nigeria, in 2022, a video circulated on the killing of Deborah as a result of blasphemy to the prophet Muhammad (SAW). In such instances, because of select or trimmed video being related to the current situation, its source or indeed its factual basis is difficult to identify. However, the private information is usually spread to cause trouble to a particular society. In 2020 a video went viral showing a woman giving information to Nigerians on not to accept Covid-19 vaccine from America as, according to her statement that, America want Nigerians to be killed. Another example is when pictures or videos of Boko Haram members were shown terrorising or killing other people, with such pictures or videos attributed to another tribe simply to misrepresent them and cause trouble.

The above three types of information disorder describe fake news. Social media including WhatsApp, Facebook, and various online platforms provide huge amounts of information and knowledge, with some true but others fabricated. It is therefore very important for learners of today to differentiate information that is false from information that is true and that which is produced, created and disseminated by creators whose intention is to cause trouble.

Media and information Literacy

Media circulate all over the world. They are rapidly transformed into various platforms especially with the advent of information communication and technology (ICT), with different types of information disseminated through these media. People need to think of what to produce and share to avoid the spreading of fake news that can cause trouble in communities. Media literacy is very importance to human beings because it gives sincerity of what to consume and produce.

According to Global Media and Information Literacy (2017), the word media "is any channels of information delivery." Media is a plural of medium, meaning mode or carrier. It was originally called mass communication which meant newspapers (18 CE.) but, as a result of the advent of ICT (21 CE.), it now encompasses all channels of information delivery, including books, newspapers, social media, the Internet, television, radio and so on. It is intended to get to large audiences.

Media literacy is "the ability to access, analyse, evaluate, and communicate messages" (Tettey 2013). It is the process of acquiring skills through identification of different types of media and understanding the messages they are sending. There are various messages or information on social media, radio, advertising, television and more, but these media share information with aims of targeting the entire audience. The creation and dissemination of information by media houses was created with a reason, and understanding the reason of their information dissemination is the beginning of media literacy.

The individual needs to understand the cultural, economic, and political information of the organization that is producing and disseminating of information. These skills should be acquired to understand the content of the information and compare it with reality, by understanding the technological media, selection, and interpretation of information. The skills also provide reasonable access to information and allow people to choose and make a critical decision on what to produce or consume.

Literacy is the ability to read and write. Concerning information literacy, consumption and use of information require acquiring skills on various information systems, which include computer, audio-visual outlets, the news, digital media, and multi-media. When you organize these literacy by understanding the information they contain, information literacy is acquired. A person cannot become information literate unless they are versatile in various information sectors and how to decode and produce efficient results. "Information Literacy encompasses knowledge of one's information concerns and needs, and the ability to identify, locate, evaluate, organize and effectively create, use and communicate information to address issues or problems at hand; it is a prerequisite for participating effectively in the Information Society, and is part of the basic human right of life-long learning" (Information Literacy Meeting of Experts, 2003).

One can become an information literate person if they have the ability to organize information in meaningful ways. They also need to locate the various formats where information is stored and can be retrieved. Communities or society have the right to access information that are useful for their needs, and it is the right of government to provide various information media for the society, which include the provision of newspapers, radio, and television.

The Chartered Institute of Library and Information Professionals (CILIP) (2018) defined "Information literacy as the ability to think critically and make balanced judgments about any information we find and use. It empowers us as citizens to develop informed views and to engage fully with society."Finding true information is part of information literacy. Individual should always try to differentiate fake information and true information by critically identifying the channel in which the fake information is spreading and the appropriate channel for the truthful information. Another understanding of information literacy is that, "It is a practice that is enacted in a social setting. It is composed of a suite of activities and skills that reference structured and embodied knowledge and ways of knowing relevant to the context. Information literacy is a way of knowing" (Lloyd 2017).

According to Oso and Aknagba (2014), an information literate person is able to find out and organize any type of information they need. They are therefore able to:

1. Find out the level of information required
2. Use the needed information successfully and professionally
3. Appraise information and its sources critically
4. Integrate selected information into their knowledge base
5. Utilize information successfully to attain predetermined aims and objectives
6. Become aware of the political, economic, and social issues in the moral and efficient use of information.

The Society of College, National and University Libraries (SCONUL) is the membership organization for all academic and national libraries in the UK and Ireland which produced the concept of the seven information literacy landscapes. The SCUNUL believed that the information literate person should be able to acquire knowledge on the use of any information by the use of the following seven skills:

1. Identify: an information literate person should be able to identify various media in order to write or produce relevant information. These media include Google, YouTube, social media, websites, blogs, etc.
2. Scope: an information literate person should have a scope on the type of information they want to disseminate, which includes information on agriculture, technology, etc.
3. Plan: planning is very important. An information literate person should always prepare on how to get their information, which includes the various media they intend to check for the information, how they are going to access this information, etc.
4. Gather: gathering the various information resources for the purpose of writing or producing information, focusing on the selection of relevant media that will give the most up to date information.
5. Evaluate: the process of assessing the quality of information from any information media.
6. Manage: the process of managing information for better delivery. An information literate individual is able to organize information into many ways, which include online handling and its organization, websites, blogs, etc.
7. Present: an information literate person should be able to present quality information as not fake or fabricated information. In addition, they should always present information the way it was reported.

The concept of media and information literacy basically depends on the universal human rights and it is considered a basic condition for individuals, communities and entire nations to enjoy the right to freedom of expression and the right to access information. Article 19 of the Universal Declaration of Human Rights states that "everyone has the right to freedom of opinion and expression; this right includes the freedom to freely adhere to their convictions and the freedom to seek, receive and impart information and ideas by any means and regardless of state borders"(Ane, Daniela, and Angela 2019).

UNESCO believes that the concept of media and information literacy is essential in providing the people of all nations with the opportunity to understand their fundamental human rights and to effectively contribute in understanding the social development goals. Media and information literate people are able to access, receive, critically evaluate, create, use, and disseminate information and media content of all types.

According to Ane, Daniela, and Angela (2019), media and information literacy skills can be divided into five main areas:
1. Access
2. Analysis and comprehension

3. Critical evaluation
4. Use
5. Creation and engagement

Webber (2018), in her online video, described that "media and information literacy are both contextual. Information is different and media and information are different, depending on the person, the organization and nation. These differences matter. Media and information are important and complex and it is not just a mind they engage, they also engaged people on how to sense of spirit." In the previous topic media and information literacy were discussed, both of which are related because media carries information and information sometimes carries media. For example, video and pictures are both information as well as media. Media and Information Literacy (MIL) is a mixture of knowledge, feelings, talent, and practices required to use, investigate, assess, create, and communicate information and knowledge in creative, legal, and ethical ways that respect human rights.

The United Nations Educational, Scientific, and Cultural Organization (UNESCO) (2017), defines media and information literacy as the "set of competencies to search, critically evaluate, use and contribute information and media content wisely; knowledge of one's rights online; understanding how to combat online hate speech and cyber bullying; understanding of the ethical issues surrounding the access and use of information; and engage with media and ICTs to promote equality, free expression, intercultural/interreligious dialogue, peace, etc."

UNESCO came up with the below description of media and information literacy as contextual (Tettey 2013):

Information literacy	**Media literacy**
Define and articulate information needs	Understand the roles and functions of media in democratic societies
Locate and access information	Understand the conditions under which media can fulfill their functions
Assess information	Critically evaluate media content in the light of media functions
Organize information	Engage with media for self-expression and democratic participation
Make ethical use of information	Review skills (including ICTs) needed to produce user-generated content
Communicate information	
Use ICT skills for information processing	

Information literacy covers the ability to define and understand the information need. It locates, organizes and assesses information everywhere, while also making ethical use of information for effective communication and using ICT tools to process this information. Media literacy is all about understanding the position and importance of such information to the society; for example, the information should not be wrongly produced or produce fake news, and should not cause trouble to society.

Accessing information through media

A media and information literate person is able to acquire information everywhere and anywhere regardless of time and space. Information is therefore a simple thing to identify and access for a person with knowledge of media and information. Some information are physically and online – like library resources (textbooks, reference materials, and serial materials). Online resources are information materials stored on web resources or by using the likes of compact discs or a computer central processing unit (CPU). These resources are identified when the use of information is needed. Users of library and information centers need to know how to assess the relevant information they always require, but most people don't know how to access the relevant information whenever they have assignments or want to conduct research, and don't have the skills of assessing relevant information. There are two ways of assessing quality information:

a. Defining and outlining the need of information
b. Location of information and information retrieval.

a. Defining and outlining the need of information

Users of information need to define the types of information they need. An information literate person should always define their information by asking themselves questions may such as what type of information may I need? Why should I need that information and how should I get that information? Library users need various types of information but sometimes they don't know how to access reliable information. It is very important to library personnel to orient users on how to identify, use, and locate such information.

Defining the need for information involves identifying what specific information is required to address a particular question or problem, while the process of finding comprehensive and reliable information involves using various standard methods to obtain it. These methods include research initiation, research questions, writing techniques, and research strategy (information retrieval).

b. Location of information/information retrieval

Information can be located in many ways, such as libraries, the internet, and off-line resources.

a. Library: materials in a library can be located in its different sections, which include: the reserved section (rare textbooks, projects/theses, past question papers etc.) the reference section (reference material, which includes biographical works, dictionary, encyclopedias, indexes, maps, atlas, gazetteers, handbooks, directories etc.), the serial section (newspapers, magazines, pamphlets, leaflets), the circulation section (all textbooks of either science, arts, social science and technology), and catalogues entries (author, subject, and title entries)
b. Internet resources: social media (Facebook, YouTube, Gmail, Yahoo mail etc.), search engines (all search engines including Google scholar, Google, Yahoo, etc.), online public access catalogues, blogs, and websites of different organizations
c. Offline resources: offline resources are selected and downloaded resources stored in library computers and organized for easy retrieval of information by users.

Apart from the method of retrieving information, information can also be searched for using various methods which will give the user complete satisfaction in what they identified, as opposed to not finding relevant information which will lead to poor research. The approach to information findings may include the following:

1. Keywords
2. Using a thesaurus
3. Boolean searching
4. Truncation
5. Phrase search

1. Keywords:

Keywords are words that define the subject content of a word, signifying a subject in a statement. Information can be searched through using keywords which will give significant information for a particular research topic. Most of the peer review articles in journals provide keywords for adequate searching, with these keywords sometimes added to library software, for example online cataloguing such as Koha library software and Greenstone Digital Library.

Textbooks also provide keywords on the last pages. These words are highly important, with subjects in the book often arranged alphabetically through these words. An information literate person can thus identify their need if they are able to understand the uses of these keywords.

2. Using a Thesaurus

A thesaurus offers a standard list of terms or controlled vocabulary used for the finding and retrieving of information. Users of information may check a thesaurus for related subject matter in order to understand the word clearly. Searching words in a controlled vocabulary will give the best results and quality definition.

For example, if someone is attempting to understand words related to medicine, the best way of understanding these words could be found in a controlled vocabulary of Medical Subject Headings (Mesh), because it is the vocabulary used for indexing medical subjects and describes words literally.

Another example is that some academic libraries provide glossary terms for users. These terms are the most common words used in library and information science.

3. Boolean Searching

This is the use of the words "AND, OR, NOT" to search for information on the Internet. Using the above words or phrases will give better results and the most specifically relevant information. For example, if you want to search for information on the Internet or any database, you may use the following to get the most relevant information.

- search word1 AND word2 in order to find documents containing both word1 and word2;
- search word1 NOT word2 in order to find pages containing word1 and not word2;
- search word1 OR word2 in order to find pages containing either word1 or word2.
- AND operator directs search
- AND operator is used to connect various concepts in order to focus the search.
- OR expands the search area
- OR operator is used to connect synonymous terms. It may prove very useful in case a search returns less results than you require.
- NOT operator narrows the search and may be used to eliminate articles of no relevance.

4. Truncation

Truncation is very important it allow the researcher to use a wild card symbol for the last word. It involves adding an asterisk (*) after the initial root of a word and helps to see many results related to the original word.

Example:

- Test* = search for testy, testiness
- Protect* = search for protector, protecting
- Bore* = search for boredom, boring

5. Phrase Search

Using quotation marks around the phrases means telling the database or internet to give you or to search for the words together, not separately.

For example, "Sharia'ah in Zamfara" or "Media Literacy," "Hausa Fulani"

Conclusion

The evolving landscape of information and media necessitates the cultivation of robust literacy skills. This chapter aims to equip learners with the tools to navigate this dynamic environment, fostering a discerning and informed citizenry. Information disorder has been described in this chapter as false information, and there are various media that disseminate a variety of information to communities without consideration. An information literate person should use their skills of information literacy to identify the dependable information before consuming it. However, organizing and locating the quality information cannot be possible without the knowledge of information retrieval. Retrieving the quality information will also allow for quality research and study, with users of information able to differentiate fake information and real information.

References

Ane L., Daniela P., & Angela R. (2019). Collaboration in Designing a Pedagogical Approach in information literacy. https://doi:10.1007/978-3-030-34258-6_4

Carlsson, U. (2019). Understanding media and information literacy (MIL) in the digital age. A question of democracy. UNESCO

CILIP Information Literacy Group. (2018). What is information literacy? Retrieved from https://www.cilip.org.uk/page/informationliteracy?

Global Media and Information Literacy Week 2017-24–27 October, Jamaica Conference Centre Kingston

IGI Global Online Book Store (2022). What is information disorder? Retrieved from https://www.igi-global.com

Information Literacy Meeting of Experts. (2003). The Prague Declaration: towards an information literate society. Retrieved from http://www.unesco.org/fileadmin/MULTIMEDIA/HQ/CI/CI/pdf/PragueDeclaration.pdf

Kandal, N. (2022). Information disorder syndrome and its management. J. Nepal Medical Association, 58(224), 280–5. https://doi:10.31729/jnma.4968

Lloyd, A. (2017). Information literacy and literacies of information: a mid-range theory and model. *Journal of Information Literacy*, 11(), 91–105. doi:10.11645/11.1.2185

Oso, L., & Aknagba, O. (2014). Media Information Literacy and democracy in a multi-ethnic society. *Journal Komunikasi Malaysian Journal of Communication Jilid* 30(1): 170–186. https://doi:org/10:17576/jkmjc-2014-3001-09

Tettey, J. W. (2013). Media and information literacy, informed citizenship and democratic development in Africa: A handbook for media/information producers and users

UNESCO (2017). Media and information literacy, a critical approach to literacy in the digital world. Retrieved from https://www.unesco.org/en/articles/media-and-information-literacy-critical-approach-literacy-digital-world

Wardle, C., & Derakhshan, H. (2017). Information disorder: toward an interdisciplinary framework for research and policy making. Council of Europe report.

Weber, S. (2018). Media and information literacy: strength through diversity. http://youtu.be/hQn4OoSMF1M

Robert Stalone Buwule

Societal Implications of COVID-19 Misinformation in Uganda

Abstract: Fighting a global pandemic like COVID-19 can be compared to fighting a world war, however, the main difference here is that such a war is fought in medical terms. The World Health Organisation (WHO), through its partners at regional and national levels, actively engaged the media to carry out mass awareness campaigns geared towards preventing the further spread of the COVID-19 pandemic. While these mass awareness campaigns reached the uninfected population, the information was sometimes mixed with fear, speculation, and rumors, representing how the rapid spread of all kinds of information concerning a pandemic sometimes makes finding solution for the problem even more difficult. The purpose of this chapter is to analyze the societal implications of misinformation in the fight against the Covid-19 pandemic. It reviews the literature available on the COVID-19 pandemic misinformation and presents themes on the role of the mainstream mass media, social media and other alternative media, ethical dimensions of misinformation, initiatives geared towards countering COVID-19 misinformation, and the professional bodies actively involved in this fight against COVID-19 misinformation.

As a way forward of protecting an unsuspecting society from misinformation, it is recommended that there needs to be a delicate balance of the freedom of expression and censorship or swearing the oath of secrecy more so during a crisis time like the COVID-19 pandemic. Authorities need to effectively manage the online communication platform owing to their versatility and the speed at which they spread both reliable and false information and finally there is need to appreciate the exisitense of African or Ugandan indigenous knowledge and this should be demonstrated by protecting the intellectual property of innovations steming from this indigenious knowledge.

Keywords: COVID-19, misinformation, infodemic, societal implications

---

Robert Stalone Buwule, University Librarian, Mbarara University of Science and Technology, e-mail: rbuwule@must.ac.ug

https://doi.org/10.1515/9783111176345-018

Introduction

It was during the 1998 International Telecommunication Union (ITU) Plenipotentiary Conference in Minneapolis, United States, that it was proposed that Information Communication Technology (ICT) together with the Internet should be used for global communication and as such promote the achievement of the Millennium Development Goals (Mutula 2020). However, over two decades later and despite all the advantages that come with ICT and a fast internet, the world is experiencing a digital divide, poor internet governance, and failure of internet users to comply with the telecommunication standards. Therefore, the bombardment of misinformation during a crisis like the COVID-19 pandemic can exacerbate the problem instead of alleviating it.

Fighting a global pandemic like COVID-19 can be compared to fighting a world war. The main difference here is that such a war is fought in medical terms. The World Health Organisation (WHO) is normally at the helm of such a war and it marshals a number of arsenals such as spearheading scientists to develop new vaccines, testing the vaccines, conducting mass vaccinations, and spurring communities to practice prevention measures while treating and caring for the victims of the pandemic. The development and use of vaccines has faced a number of hurdles as a result of misinformation, however, this time it didn't only affect the general public but also healthcare personnel, with it reported that one of the leading reasons of vaccine resistance by Kampala urban refugee youths was the misinformation propagated by the healthcare teams (Logie et al. 2023)

Therefore the WHO strategies of developing vaccines are good but cannot be effective without the involvement of the media (Kihal et al. 2019). The WHO through its partners at regional and national levels have actively engaged the media to carry out mass awareness campaigns geared towards preventing the further spread of the COVID-19 pandemic from reaching the uninfected population but also at the same time provide the right information for treating infected people (Tran et al., 2020). The absence of this clear demarcation may breed misinformation or an infodemic and anxiety which might disrupt the good approaches of fighting the pandemic.

The term infodemic is a short form of "information epidemic" and was first coined by David J. Rothkopf following the outbreak of the Severe Acute Respiratory Syndrome (SARS) in 2003. Rothkopf (2003) defines the term infodemic as facts mixed with fear, speculation, and rumor, amplified and relayed swiftly all over the world by the modern information technologies. The World Health Organisation (2020b) picked it up from there, simplified it by calling it the overabundance of information, with some accurate and some not, thus making it hard for people to distinguish trustworthy and reliable sources of accurate information. It is basically false, malicious, stigmatising, and incorrect information spread quickly and widely (Islam et al. 2020; Mejora and Kalimeri 2020). The rapid spread of all kinds of information concerning an epidemic sometimes makes finding solutions for the problem even more difficult (World Health Organization 2018). Furthermore, the term infodemic outlines the dangers of such misinformation during outbreaks of epidemics like COVID-19 (Cinelli et al. 2020). Surprisingly, this fake information is spread on both mainstream and alternative or social media. Unfortunately, COVID-19 misinformation is curtailing the proper administration of treatment which indirectly fuels further the spread of the pandemic (Zlatí et al. 2020). The COVID-19 infodemic has provided various information on treatment such as high doses of vitamin C, raw garlic, black oil, and even, as suggested by some sources, the use of cocaine. According to the WHO (2020b), some of these approaches to the treatment of COVID-19 are dangerous as people primarily rely on them instead of the medically proven approaches.

In the United Kingdom for example, the rollout of COVID-19 vaccinations for schoolchildren sparked off a nasty wave of misinformation which brought anti-vax campaigners to protest right in front of the school gates using abusive and threatening messages. They falsely claimed that vaccines are untested and dangerous, while using leaflets and social media sites like Instagram to intimidate the teachers. The COVID-19 prevalence rates remain high in the UK with over 40,000 positive cases reported in one day, and addressing misinformation about the virus could possibly reduce the rate of hospitalization and death it causes (Smith-Spark 2021).

Since COVID-19 is a new disease, it requires a lot of research into new epidemiological areas. Sporadic cycles of new research information influence researchers' behavior and may alter the effectiveness of their findings. For example, epidemiologists design models to forecast the spread of the virus, but false information on the behavioral response of the population can inhibit epidemiologists from forming the appropriate public health interventions (Van Der Linden, 2022). The communication dynamics behind content consumption of the true information will not be guaranteed; therefore, proper treatment and model forecasts of the spread can work well if the population's preferences and attitudes are founded on accurate and fact based information (Cinelli et al. 2020).

As COVID-19 became a global pandemic, rumors about it appeared quickly in the virtual world. Much of these rumours were misleading and contradicted with the WHO's scientific facts, which was highly unethical as it led to many people becoming confused and terrified. Essentially, the media is supposed to be ethical and disseminate concrete, convincing, and scientifically proven information, all the more so if it is to do with epidemics and public health (World Health Organisation 2020a). This paper discusses the COVID-19 misinformation, the role of the mainstream and alternative media in this infodemic, and the information ethical implications of the COVID-19 infodemic.

In Africa research on misinformation has increased in recent years, though it is not clear whose role is it to control it. There is still insufficient amounts of empirical research addressing this problem on continent and thus a need for a sense of shared responsibility among individuals and institutions if Africa is to squarely address the spread of misinformation (Tully et al. 2022). For instance, in Uganda, there is widespread use of herbal medicine which is both approved and unapproved by the National Drug authority. Since the pronunciation of the COVID-19 pandemic, there was a profound psychological effect on people which caused a lot of fear, with many Ugandans as a result resorting to the use of the readily available, herbal remedies, some of which have toxicities. Much of this was propagated by the media which was recommending the use of natural remedies such as herbs and food supplements to fight COVID-19. Policymakers therefore need to emphasize on educational initiatives aimed at improving public awareness of using herbal medicines use disease prevention and treatment (Musoke et al. 2021).

Statement of problem

The Ugandan landscape of COVID-19 information experienced rapid and substantial changes during the pandemic which led to confusion, the spread of misinformation, and the public's mistrust of scientists and health workers (Fergus et al., 2021). Widespread dissemination of misinformation regarding COVID-19 from unreliable sources, curtailed efforts of the health workers of combating the pandemic in a timely manner. For example, Atuheirwe et al (2024) reported that misinformation increased COVID-19 vaccine hesitancy. A large proportion of the participants in his study (89.7%) encountered rumors regarding unverified adverse effects of the COVID-19 vaccine which significantly contributed to vaccine hesitancy. A number of Ugandans more so the urban youth and even those beyond in Sub Saharan Africa are knowledgeable but have various misconceptions about COVID-19 (Napakol et al., 2022). There is therefore need for empirical studies that can guide authorities to put in place preventive measures that can control the spread of such misinformation during a crisis like the COVID-19 pandemic and further still inculcate social mobilization interventions to influence the effects of the crisis misinformation in a community or local society.

Objectives of the study

This paper seeks to:
a) Assess the spreading patterns of COVID-19 misinformation Uganda.
b) Distinguish the reliable from the questionable sources of COVID-19 information in the Ugandan Society.
c) Analyze the preferences, attitudes, perception, influence, facilitation and promotion of COVID-19 infodemics social receivers in Uganda
d) Suggest ways of mitigating the negative effects of COVID-19 misinformation in Uganda.

COVID-19 Infodemic

Infodemics stem from infodemiology, a science portfolio of information science and epidemiology that addresses pressing concerns in public health and policy (Eysenbach 2009; Horvitz and Mulligan 2015). Infodemiology or infomediological studies also include meta-analysis on epidemiological discourses and chronological and historical research studies on pertinent key words on epidemics. A number of scholars have conducted historical studies on reciprocal relationships between infodemics and stigma in the communities affected by the epidemics (Michel et al. 2011; Pagel 2009; Pagel et al. 2007).

Unlike the previous infodemics, the COVID-19 infodemic was at a higher magnitude to the extent that the WHO Director General admitted that they were not only fighting the pandemic but also the infodemic (Hua and Shaw 2020). The COVID-19 infodemic has been a significant challenge to the extent that it engulfed the Communist government of China, the United States of America, and even the United Nations. It has been fuelled by individuals who post information on their social media sites like Twitter, Instagram, and blogs to mainly spread conspiracy theories on the origin of the coronavirus. One such narrative, picked by the Chinese government, was that the coronavirus is a bio-weapon created in an America lab and planted by the American military in Wuhan (Myers 2020). Since it was now owned by China, the mainstream media picked it up, leading America to retaliate, with the American Head of State later recorded in a press briefing referring to the coronavirus as a Chinese virus (Al Jazeera 2020). It was the COVID-19 infodemic that sparked the geo-political situation where the players involved wanted to change the balance of power, with some other G8 countries like Russia consequently getting sucked in. It should be noted that the United Nations (2019) advocates for provision of information with scientific facts so as to ease individuals' and society's making of informed choices.

The conspiracy theories didn't only involve the members of the G8, as leading philanthropical organizations like, the Rockefeller Foundation, and the Bill and Melinda Gates Foundations got involved (World Health Organisation 2020a). Theories claimed that these foundations had a role in the outbreak of the coronavirus for business interests. Such misinformation, such as in this situation, can cause the public to develop deep distrust for such benevolent societies (Emerald Insight 2019).

As the WHO was still grappling with the conspiracy theorists, unscrupulous businessmen also spotted an opportunity in COVID-19 and jumped on the bandwagon. According to Mejora and Kalimeri (2020), businessmen pushed a deluge of information, opinions, and prognoses on COVID-19 onto billions of internet users, in turn designing advertisements for their products and promoting them on social media since they are regulated. The adverts are fashioned to sell mainly public health products such as health insurance, food supplements, and herbal medicine, among others. Others posed as non-profit organizations and advanced their hidden agendas but blended them with the coronavirus, which was also done by politicians and civil right groups who gave donations to help vulnerable groups but used that as a platform to campaign and advance their agendas. The intention of helping the vulnerable is good but when this is used to provide misinformation on the coronavirus as a bioweapon, conspiracy theories, and other unverifiable claims then it all becomes tantamount to misinformation. The concerned authorities are therefore left with no option but to act in response to such opportunists.

Mainstream Media

The world had not yet experienced the magnitude of misinformation reported by the mainstream media like that of the COVID-19 pandemic. Notable and credible media houses even fell prey to COVID-19 misinformation, seen with the Cable News Network (CNN) which anticipated a likely lockdown in Lombardy, a town where COVID-19 broke out in northern Italy, and published the news before official communication from the Italian government (Cinelli et al. 2020). The release of such news from a mainstream media outlined sparked uncoordinated movements of people which was very dangerous when fighting the epidemic; expert epidemiologists assert that such sporadic movements fuel the spread of the virus outbreak.

Another example was in China where a map illustrating global air travel was broadcasted by popular media houses (World Health Organisation 2020a). This false map showed the travel patterns of Wuhan residents who had fled the city which in fact was not the case. There was also another piece of fake news that showed sulphur dioxide emissions over China as evidence of the mass cremation of COVID-19 victims. The biggest problem with this is that as soon as such infodemics are reported by the mainstream media, many believe the content and even cite it by other reputable media and organizations, therefore making it harder to discern the truth or challenge it in the eyes of the general public.

Predatory publishing is also increasingly becoming a menace, especially in the developing world. It is not only a hazard to scholarship and the intellectual discourse but also a major contributor to misinformation and disinformation amongst academicians and learners in developing countries (Otike, Bouaamri, and Barát 2022). Since predatory publishing defies the norms of the mainstream media and tradition publishing, it has indirectly become a catalyst for poor and misleading researched information, plagiarism, use and sharing of unethical sources of information, and promotion of unethical academic behaviour. This moves the students and academic faculty members who access this information towards great intellectual danger.

Social Media and Other Alternative Media

Misinformation has found an enabling environment in social and other alternative media. The COVID-19 infodemic has particularly been very popular on social media and its related communication technologies (World Health Organisation 2020c). Misinformation has thrived there because it travels faster, broader, and deeper more than accurate information. If one compares the infodemics of SARS and MERS to COVID-19, it appears the COVID-19 infodemic is the most exaggerated because of social media (Vaezi and Javanmard 2020). An essential characteristic of such an infodemic is the ability to use videos, pictures, interviews, and articles intermittently to substantiate false claims (World Health Organisation 2020a). The danger of this media is that as more of the infodemic is reported and shared on different social media platforms, the more people progressively start believing the misinformation. As the common adage goes, "if you keep repeating a lie, soon it becomes the truth." Misinformation can easily become viral when shared on the social media because the number of mobile phones and internet users has rapidly grown. As Mutula (2020) outlined, "Currently, there are over 4.4 billion internet users, 10 billion cell phone and over 3.5 billion social media users globally."

Adekoya and Fasae (2021) carried out a study on the Covid-19 misinformation spread by social media in Nigeria and found people's use of social media was for several reasons. While some used it for listening to announcements, knowing necessary measures to take when infected, and receiving updates, misinformation spread on social media was a disaster to the health sector. An infodemic is capable of misleading people and causing a loss of trust in government, health providers, and health regulatory authorities. A related study was done in Uganda which revealed that women and children in rural communities were the most vulnerable to the infordemic spread on social media and should be the people who should be given more attention in the Covid-19 misinformation national response (Kasozi et al. 2020).

Since social media has provided a rich ecosystem for COVID-19 misinformation, the WHO has joined with social media firms to promote the dissemination of accurate content, setting up structures that first check for facts before passing on the content. In they find content wanting, they make it harder for users to access it. When users search for COVID-19 on Google, the search engine directs them to WHO and other credible local health authorities. Elsewhere, Facebook run adverts encouraging users to seek information on COVID-19 from the WHO and other credible health local authorities (World Health Organisation 2020a). The Red Cross and meteorologists use a tool called Radian6 to monitor online risky communication on climate change (Roth and Brönnimann 2013); this allows them to monitor conversations on mainstream media, blogs, and Twitter, which helps them to analyze and gain insight on how they can best address the negative infodemic.

The COVID-19 has made the use of social media complex. According to Wiederhold (2020), social media is not only a source of entertainment, connection, and information, but also a reflection of what is on one's mind. So when the COVID-19 infodemics thrived on it, then it automatically triggers anxiety. Pieschel (2020) reported that in February 2020 when COVID-19 began spreading in the United States, online anxiety increased by 19% mainly because of the COVID-19 infodemic on the social media. If such an infodemic is not arrested, anxiety can increase and impact on mental health, largely because of the uncertainty about what is correct and false which keeps troubling the individual's already distressed nerves (Wiederhold 2020).

Additionally, social media platforms provide direct and quick access to an unprecedented amount of information like on virus outbreaks, most of which may be rumors and questionable information. A study was conducted to assess the patterns of misinformation from both reliable and questionable social media sources in Lombardy, the COVID-19 epicentre in Italy, to understand the difference in the spreading patterns of the virus (Cinelli et al. 2020). The study performed a comparative analysis on five social media platforms, Twitter, Instagram, Gab, YouTube, and Reddit, analyzing user preferences, attitudes, facilitation and promotion of infodemics, receivers' social perceptions, influence on policy, and ignoring of dissenting information. It was found out that misinformation easily proliferates on social media and spreads faster and wider than scientifically proven news on mainstream media (Vosoughi, Roy, and Arai 2018).

An infodemic can have a strong positive or negative influence on policy, seen with the WHO declaring the diffusion of the COVID-19 infodemic as a severe global threat (Ghebreyesus 2020). Normally the context of the infodemic muddles the attempts of policymakers to plan for issues like the undetected infected people, the number of required testing kits, early testing, and others (Zlatí et al. 2020). Infodemics can further influence public debate as it tries to adhere to popular world views and ignore dissenting views, thus creating polarization (Cinelli et al. 2020). For example false information can hinder policymakers from enforcing the benefit in early testing, causing both positive and negative people to shun going for early testing, amplifying the rate of spread of the epidemic.

Scientific and Societal Initiatives Geared to Counter COVID-19 Misinformation

Considering the damage that has been caused by the COVID-19 misinformation, the WHO has set up an information network for epidemics (EPI-WIN) platform for sharing information with key stakeholders (Hua and Shaw 2020). According to Zarocostas (2020), the WHO is combating the COVID-19 infodemic through the use of its EPI-WIN, using this platform to work with the different social media platforms to track the spread of specific falsified information and rumors and also ensure that content which is accurate and scientifically proven is displayed at the top of internet COVID-19 searches. On February 28, 2020, the WHO purposely joined the social media platform TikTok and shared information on the COVID 19 pandemic (World Health Organisation 2020c). This is all geared towards preventing misinformation, the absence of which gives birth to infodemics.

Wikipedia is another reliable source of factual content on COVID-19. Despite being a free to edit site, which may make it a breeding ground for misinformation, it has a group of editors who filter any kind of misinformation and also block contributors who proliferate it.

In fact, not all of social media is negative. There are many aspects of social media that can be used to convey the correct COVID-19 infodemic which can offer great assistance for those who are feeling anxious or isolated. The American Psychological Association recommends maintaining social networks by connecting with others through social media like via video chat, e-mail, or messenger apps (Wiederhold 2020). Further still, sharing useful information with friends and family is encouraged to deal with one's own anxiety (American Psychological Association 2020). The public is further advised to moderately consume media content to stay informed, but to avoid excess or incorrect information. Social media should only be used to unwind, connect, and tend to both the body and mind during the pandemic (Wiederhold 2020).

There are also other organizations that have taken the initiative to create spaces and platforms where researchers and other scientists can upload, deposit, and archive their research works on the COVID-19 pandemic. Figshare, one such institution, is a repository which allows researchers to openly share research data and any research output in the form of journal article preprints, conference presentations, data sets, code, and posters. Following the outbreak of this COVID-19 pandemic, Figshare (2020) created a platform for sharing scientific research works on COVID-19. Such initiatives aid the rapid dissemination of researched, secure, dependable, and ethical information on the pandemic.

Since COVID-19 misinformation has created a lot of fear and misconception which may or has affected policy formulation with regards to vaccination, mobility of people, and border closures (World Health Organisation 2020a), the Institute of Humanities in Africa, at the University of Cape Town, has come up with a novel idea as far as influencing social media is concerned. They have launched a blog called Corona Times (2020) where scholars from the humanities and social sciences can make humble and valuable contributions to help the masses understand this pandemic. Like Wikipedia, the editing is done collectively and this encourages social engagement, reflection, self-regulation, and academic freedom as one contributes information on COVID-19. If anyone perpetuates misinformation, there are submission guidelines to handle such contributors.

Similar to the above, the Research Data Alliance (RDA), an international community based organisation, has been asked to leverage the global information and data community by creating a RDA COVID-19 working group. This working group spearheads data sharing and reuse such that researchers can have reliable and dependable data to work with during the pandemic. The focus here is ensuring that the data comes from verifiable sources which can be fed into a system to maximize efficiency in epidemiological data on public health and scientific research. RDA strictly adheres to the fundamental principles of information ethics such as being Findable, Accurate, Intellectual property, and Re-usable (FAIR) data and information. RDA further embraces other information ethical principles including openness, consensus, balance, harmonization, community driven, non-profit, and technology neutral (Research Data Alliance 2020).

The world today is being significantly influenced by the Fourth Industrial Revolution (4IR). In this regard, attempts are being made to use Artificial Intelligence (AI) and Machine Learning (ML) to study misinformation (Bullock et al. 2020). It is interesting to note that AI and ML are not only being used to study the scale and spread of COVID-19 misinformation but also how it can be slowed down or controlled. This is done through staging epidemiological simulations and phenomenological modelling. When this was tested on leading social media platforms, it was observed that posts from less reliable sources had lower engagement compared to posts from reliable sources. For example, epidemiologists at the University of East Anglia in United Kingdom staged a similar simulation to scale the effect of misinformation on an infectious disease (World Health Organisation 2020a). They reported that reducing the harmful effect of an infodemic by just 10% greatly reduces people's risky behavior like not isolating themselves or sporadic movements between infected and un-infected areas. Another example is that of Brainard and Hunter (2019) who used an agent-based model to predict the spread of an infectious disease based on the behaviour of populations. They found out that reducing an infodemic by just 20% to 10% can drastically reduce the severity of virus outbreak.

Vaezi and Javanmard (2020) stressed the need for Risk Communication (RC) to counter misinformation like during COVID-19. They assert that RC is a two-way strategy, a dynamic art–science process which evolves as the pandemic develops. RC fills the information gap left between what scientists think people need to know and what people really want to know (Abdelwhab and Hafez 2011). It prepares the community to become ready for changes caused by the epidemics and is comprised of three elements: first, clearly and honestly talking of what is known and not known about the epidemic; second, listening to the community's fears and perceptions; and third, managing rumours and misinformation as quickly as possible (Vaezi and Javanmard 2020). Roth and Bronnimann (2013) also had earlier on observed that traditional communication can be adapted into the new information environment and be used to combat risky information in the soceity.

Though China is known for applying draconian measures, this time it had a systematic plan of handling the negative effects of the COVID-19 infodemic, combining strong governance, strict regulation, and spontaneous community/citizen participation (Hua and Shaw 2020). A QR code was designed by China's mobile network and big data system which screened people's online posts to establish the accuracy of information they were disseminating. The Supreme Court also directed the severe reprimanding anyone found sharing fake news and information that stirred up confusion and panic. Volunteers were recruited in local communities at the village level to enforce strict implementation of these measures and report the culprits. A website was also set up by Tencent to purposely to expose the rumours and fake news (Hua and Shaw 2020). In this way, the Chinese authorities directly got involved in reducing the negative effects of the COVID-19 infodemic.

Ethical Dimension of the COVID19 Infodemic

Ethics is derived from the Greek word ethos which means a conduct of character, moral upstanding, and socially accepted cultural values and standards (Donalson 1990). Ethics is a branch of philosophy and concerns itself with individual behavior and rational justification for any decision taken. In information science, it takes ethics for one to determine the right from the wrong, the good from the bad, the correct from the false, and making a commitment to always stand for what is right and good (Kaddu 2007). The COVID-19 pandemic came along with massive infodemics which demonstrated both ethical and unethical information practices, and one may wonder the ethical motivations of those who are creating and disseminating all these infodemics. The fundamental principles of information ethics rotate around privacy, freedom of access to information, confidentiality, anonymity, openness and transparency, accuracy, respect for intellectual property, and protection of indigenous knowledge, among other things (Capurro 2007; Kigongo-Bukenya 2007). Below is a discussion of the ethical implication of the COVID-19 infodemic in the light of some of the fundamental principles of information ethics.

Freedom of Access to Information

Free Access to Information and Freedom of Expression (FAIFE) is one of the core activities of the International Federation of Library Associations and Institutions (IFLA) (Lor 2007). Focusing first on unethical information behavior concerns the Chinese government as far as the COVID-19 infodemic is concerned, as it is known to employ secrecy, censorship, and propaganda. For instance, it was widely reported that Dr. Li Wenliang (RIP), who first raised the alarm on the outbreak, was investigated and harassed by Chinese security for spreading rumours (Jandrić 2020). Later Dong and Bouey (2020) observed that the purported rumours turned out to be true and, towards the end of 2019, after the government initially downplayed the epidemic's severity, public trust in its decision-making transparency and competency in information ethics dropped considerably

The world is now skeptical of all the information China officially releases relating to the pandemic, with the impression of inhibiting access to information greatly undermining the public's trust in the information that they provide (World Health Organisation 2020b). It is no secret today that the right of access to information and freedom of expression are basic human rights (United Nations 1995). When a government which is supposed to enable easy access to information hoards it and censors it, this poses a red flag as far as information access is concerned. It therefore takes magnanimous commitment and accountability on the part of governments and organisations to embrace this fundamental information ethic (Du Plessis 2007).

The COVID-19 infodemic first caused fear and racism against people who looked like Chinese and then later anyone who would cough and sneeze. After the outbreak of COVID-19, fewer customers were observed visiting businesses in United States belonging to the Chinese, Asians and even Asian-Americans. People who were ethnically profiled Chinese and Asians were attacked in subways, questioned, and refused services including accessing information. Sadly, the same xenophobic treatment was meted towards anyone who coughed or sneezed. This unethical behaviour curtails information access both from the information provider and receiver, with the COVID-19 infodemic creating that impression that all Chinese or those who coughed and sneezed were infected with the coronavirus, which was not necessarily the case.

On a positive note, the COVID-19 infodemic has shown how millions of people can easily access new forms of online media. The ever evolving communication technologies have made access to information fast or instant and thus decision-making also (Mheidly & Fares, 2020), with the only challenge remaining regulation and discernment of the true and false sources of information. COVID-19 has further stimulated the sharing of information between the north and south. Traditionally it has been the global north predominantly providing and the south consuming information; even when there was valuable local information in the south on the verge of destruction, it would take the north to inject resources to preserve that information or else it would perish(Jeppesen & Miklian, 2020). But with the COVID-19 infodemic, the north and south have been actively generating, sharing, and consuming each other's information on the pandemic. The unique aspect this time is that the information is shared not only through academic and scholarly papers, but also through posting photos, videos, songs, and blogs among others. The beauty of this era is that online technology is making reproducibility of information possible in so many different ways (Tullis & Kar, 2021). The information reproducibility and sharing is mainly facilitated by the posting of content without any restrictions and permission(Nosek et al., 2022).

Openness and Transparency

Free information and communication are at the heart of the fast spreading online content (Seeger, Sellnow, and Ulmer 2003). It is this very free and open information and communication that is currently being evidenced in the COVID-19 infodemic.

The faster and versatile communication technologies have created new opportunities for open information-gathering, coordination, and public communication.

Furthermore, many of the online communication channels that are open have been structured according to the information context-specific needs (Roth and Brönnimann 2013). During the COVID-19 infodemic, the structuring has enabled open debating of controversial issues and conflicts. When the online debates break up into smaller sub debates which are sometimes moderated with vigilante moderators, the discussions, though open, become more useful as the spread of false and inaccurate information is minimized.

Respect of Intellectual Property Rights

There is an emerging debate mostly in Sub-Saharan Africa about the preservation of Indigenous Knowledge (IK). Intercultural information ethics can never be fully developed without the promotion of pluralism. According to Ess, pluralism basically entails the respect and promotion of the different cultures and beliefs around the world. When COVID-19 broke out, being a new disease, scientists and researchers had to start from scratch to come up with a vaccine for the virus. Later it was found out that Chloroquine phosphate, a drug used to treat malaria, was shown to have "apparent efficacy and acceptable safety against COVID-19" (Cortegiani et al. 2020; Gao, Tian, and Xang 2020). Though it is recorded that this drug is made from sulfadoxine and pyrimethamine (Foster 1994), it already existed in Africa in the form of herbs (Willcox and Bodeker 2004) which are still used by some people to treat malaria, pneumonia, headache, and other illnesses. The problem is that no one has bothered to record and register this IK, even under communal innovations. Hopefully if this IK was patented, it could give African herbalists a stepping stone to even venture researching on the COVID-19 cure.

Over the years, there have been attempts to register IK as property on medicinal plants and foods, however, funding into this research has decreased by 40% (Gathegi 2007). The western pharmaceuticals companies who have ventured into this want to register these medicinal plants privately, yet Africans prefer registering them collectively. The COVID-19 infodemic has exposed how the western pharmaceutical companies patent drugs derived from African IK. On top of destroying IK intellectual property rights on communal innovations, this also limits access to such medical IK.

IFLA's committee on Copyright and other Legal Matters (CLM) promotes fairness in intellectual property. CLM was set up in 1997 to advise on a number of issues including the authenticity of electronic texts and posts (Lor 2007). As ICTs become more and more agile, CLM is expected to advise IFLA on the globalized use of ICTs, commodification of information, balancing of the powers of creators of false information, the inappropriate spread of information by intermediaries, and transmitters of content. As the COVID-19 infodemic demonstrated unprecedented levels of misinformation and fake news, CLM has a heavy duty on its shoulders to ensure that this behavior is arrested in its steps from IFLA to all libraries around the world.

Speaking of commodification of information and knowledge, one cannot help but think of the social pricing of information and services. Zielinski (2001) highlights the different use of information such as education, entertainment, business, and marketing. If information is viewed in such contexts, then it becomes a good or commodity which is valuable and can have a price. This view may, however, limit access to information for those who are less privileged and might not be able afford it (Ponelis 2007). The COVID-19 infodemic has, however, reformed the concept of information pricing, more so for the information at play which is desperately demanded for by both the haves and have nots. However, owing to the COVID-19 infodemic, the same information on COVID-19 that was originally only accessible from subscription journals was later opened up and made free to access.

Accuracy

A lot of the content on the COVID-19 pandemic has been generated in a variety of forms such as songs, poems, films, documentaries, press briefings, and other forms. This content has been produced in different languages and translators have taken it upon themselves to translate that content to English and other popular languages for mass dissemination. Some of the translations were accurate, which is very ethical, but some were misinterpreted with the meaning altered to conform to some of the conspiracy theories already doing the rounds. Since the world is multi-lingual, correct translation contributes to effective communication of a message that is socially acceptable (Woodbury 2007).

In a related study carried out by Lee (2015) it was revealed that infodemics can also present themselves through negative and unethical information behavior on Social Networking Systems (SNS) such as: misreporting, cyberbullying, stealing private information like photos and videos, and verbal abuse. The COVID-19 pandemic first exposed the Chinese communist government in this regard, who may be seen to have misjudged the potential of the virus outbreak and misreported about it (Emerald Insight 2020). This was, however, a miscalculation as the COVID-19 became worse, and the misreported figures could no longer be substantiated and later led to even the WHO having contradicting figures (World Health Organisation

2020a). The inaccurate and misreported information was very unethical as it caused speculation and distrust of the sources of such information.

One way of maintaining accuracy of information in the masses today is by monitoring social media data through tracking rumours in real time. However, the detecting, assessing, and responding to rumours, stigma, and conspiracy theories in real time can be a challenge (Islam et al. 2020). Misinformation associated with COVID-19 on social media can be managed by first confirming the source of the information before sharing it and trusting information from reliable sources (Adekoya and Fasae 2021). Administrators on social media platforms should always implore their members to only shared accurate and verified information. Health agencies and government stakeholders can then supplement this by tracking misinformation associated with the COVID-19 in real time and engage local communities to debunk misinformation.

Another strategy could be through policy formulation, mainly in institutions. Managers should put measures in place to police their own social media platforms more extensively. Social media companies should also be responsible for fixing problems emanating from social media misinformation (Himelfarb and Howard 2021).

Privacy

As already discussed earlier, the COVID-19 infodemic has predominantly thrived on social media, which has potentially made it a threat to people's privacy. For example the case cited earlier of the business people who were using the pandemic to advertise their health products indirectly invaded individuals' private spaces in an unethical, disturbing, and inconvenient manner. Members of the public using social media adverts are therefore advised to strictly follow strict privacy standards whenever they are collecting, reporting, and sharing commercial data and adverts (Mejora and Kalimeri 2020).

Recommendations for Deterring Covid-19 Misinformation

Based on the literature reviewed and experiences of the researcher, the study proposes the following recommendations to mitigate COVID-19 misinformation:

1. The Civic society can partner with government to conduct civic education on the proper use and management of social media and other related new technologies.
2. Establish set standards on the use and sharing of information on social media. The standards should ensure that evidence based information is shared and consumed in a healthy and ethical environment.
3. Identify social media champions that can influence and practically demonstrate the proper and responsible use of social media platforms.
4. Installation of smart applications that redirect information that appears to be false to authoritative and credible institutions like the WHO for verification before dissemination.
5. Support efforts of recording and registering all Covid-19 indigenous knowledge innovations.
6. Introduce misinformation literacy sessions in schools and communities

Conclusion

Through experience with previous epidemics like SARS, Ebola, and others, the WHO has come to appreciate the need to fight infodemics through directing substantial attention and resources to the media. It engages both the mainstream media like television, radio, print, and the social media. The approaches may vary from country to country but zeroing down to the Ugandan society, the Ministry of health has made an effort to standardize the scientific information on the epidemiological data, vaccines, clinical trials, and treatment disseminated across all media. If all players can stick to disseminating this standardized information, then the negative effects of infodemics can be reduced significantly

References

Abdelwhab, E. M., & Hafez, H. M. (2011). An overview of the epidemic of highly pathogenic H5N1 avian influenza virus in Egypt: Epidemiology and control challenges. *Epidemiol Infect*, *139*, 647–657.

Adekoya, C. O., & Fasae, J. K. (2021). Social media and the spread of COVID-19 infodemic. *Global Knowledge, Memory and Communication*. https://doi.org/10.1108/GKMC-11-2020-0165

Al Jazeera. (2020, March). Trump defends calling coronavirus the "Chinese virus." *Al Jazeera*.

American Psychological Association. (2020). *Five ways to view coverage of the coronavirus*.

Atuheirwe, M., Otim, R., Male, K. J., Ahimbisibwe, S., Sackey, J. D., & Sande, O. J. (2024).

Misinformation, knowledge and COVID-19 vaccine acceptance: A cross-sectional study among health care workers and the general population in Kampala, Uganda. *BMC Public Health*, *24*(1), 203. https://doi.org/10.1186/s12889-024-17678-9

Brainard, J., & Hunter, P. R. (2019). Misinformation making a disease outbreak worse: Outcomes compared for influenza, monkeypox, and norovirus. *Simulation*. https://doi.org/0037549719885021

Bullock, J., Pham, K. H., Luengo-Oroz, M., Luccioni, A. (Sasha), & Lam, C. S. N. (2020). Mapping the landscape of Artificial Inteligence application against COVID-19. *Arxiv*, 1–14.

Caftori, N. (2007). African women and the internet. In R. Capurro, J. J. Britz, T. J. D. Bothma, & C. Bester (Eds.), *Africa reader on information ethics* (pp. 259–266). University of Pretoria.

Capurro, R. (2007). Information ethics for and from Africa. In R. Capurro, J. J. Britz, T. J. D. Bothma, & C. Bester (Eds.), *Africa reader on information ethics* (pp. 3–14). University of Pretoria.

Cinelli, M., Quattroacchi, W., Galeazzi, A., Valensise, C. M., Brugnoli, E., Schmidt, A. L., & Scala, A. (2020). The COVID-19 social media infodemic. *Arxiv*, 1–18.

Coronatimes. (2020). *Understanding the world through the Covid-19 pandemic*. Web Document.

Cortegiani, A., Ingoglia, G., Ippolito, M., Giarratano, A., & Einav, S. (2020). A systematic review on the efficacy and safety of chloroquine for the treatment of COVID-19. *Journal of Critical Care*, 1–5.

Cunliffe-Jones, P. (2021) How Government Responses to Misinformation in Africa Restrict Freedom of Expression and Do Little to Tackle the Problem, *African Journalism Studies*, 42:2, 121–125, DOI: 10.1080/23743670.2021.1972532

Donalson, J. (1990). *Key issues in business ethics*. Academy Press.

Du Plessis, J. C. (2007). The spirit of open access to information as a key pillar in the African Renaissance. In R. Capurro, J. J. Britz, T. J. D. Bothma, & C. Bester (Eds.), *Africa reader on information ethics* (pp. 173–178). University of Pretoria.

Emerald Insight. (2019). *Conspiracy theories cost societies dear*. Web Document.

Emerald Insight. (2020). *China COVID-19 outbreak will weaken Xi*. Web Document.

Ess, C. (2007). Cybernetic pluralism in an emerging global information and computing ethics. In R. Capurro, J. J. Britz, T. J. D. Bothma, & C. Bester (Eds.), *Africa reader on information ethics* (pp. 91–120). University of Pretoria.

Eysenbach, G. (2009). Infodemiology and Infoveillance: Framework for an Emerging Set of Public Health Informatics Methods to Analyze Search, Communication and Publication Behavior on the Internet. *J Med Internet Res*, *11*(e11). https://doi.org/10.2196/jmir.1157.

Fergus, C. A., Storer, E., Arinaitwe, M., Kamurari, S., & Adriko, M. (2021). COVID-19

information dissemination in Uganda: Perspectives from sub-national health workers. *BMC Health Services Research*, *21*(1), 1061. https://doi.org/10.1186/s12913-021-07068-x

Figshare. (2020). *Discover COVID-19 research data on Figshare*. Web Document.

Foster, S. (1994). Economic prospects for a new antimalarial drug. *Transactions of The Royal Society of Tropical Medicine and Hygiene*, *88*(Supplement 1), 55–56.

Gao, J., Tian, Z., & Yang, X. (2020). Breakthrough: Chloroquine phosphate has shown apparent efficacy in treatment of COVID-19 associated pneumonia in clinical studies. *BioScience Trends*, P1–2. https://doi.org/10.5582/bst.2020.01047

Gathegi, J. N. (2007). Intellectual property, traditional resources rights and natural law: A clash of cultures. In R. Capurro, J. J. Britz, T. J. D. Bothma, & C. Bester (Eds.), *Africa reader on information ethics* (pp. 179–186). University of Pretoria.

Ghebreyesus, T. A. (2020). *WHO director-general's opening remarks at the media briefing on covid-19 – 13 March 2020*. Web Document.

Himelfarb, S., & Howard, P. (2021, October 7). What's stunning about the misinformation trend – and how to fix it. *HKT*. https://edition.cnn.com/2021/10/07/opinions/facebook-misinformation-and-how-to-fix-it-himelfarb-howard/index.html?utm_source=feedburner&utm_medium=feed&utm_campaign=Feed%3A+rss%2Fcnn_topstories+%28RSS%3A+CNN+-+Top+Stories%29

Hongladarom, S. (2007). Information divide, information flow and global justice. In R. Capurro, J. J. Britz, T. J. D. Bothma, & C. Bester (Eds.), *Africa reader on information ethics* (pp. 71–76). University of Pretoria.

Horvitz, E., & Mulligan, D. (2015). Data, privacy, and the greater good. *Science*, *349*, 253–255. https://doi.org/10.1126/science.aac4520.

Hua, J., & Shaw, R. (2020). Corona Virus (COVID-19) "Infodemic" and Emerging Issues through a Data Lens: The Case of China. *Int. J. Environ. Res. Public Health*, *17*(2309), 1–12. https://doi.org/10.3390/ijerph17072309

Islam, S., Sarkar, T., Khan, S. H., Kamal, A.-H. M., Hasan, S. M. M., Kabir, A., Yeasmin, D., Islam, M. A., Chowdhury, K. I. A., Anwar, K. S., Chughtai, A. A., & Seale, H. (2020). COVID-19–Related Infodemic and Its Impact on Public Health: A Global Social Media Analysis. *American Journal of Tropical Medicine and Hygiene*, *103*(4), 1621–1629. https://doi.org/10.4269/ajtmh.20-0812

Jandrić, P. (2020). Postdigital Research in the Time of Covid-19. *Postdigital Science and Education*, *2*, 233–238.

Jeppesen, S., & Miklian, J. (2020). Introduction: Research in the Time of Covid-19. *Forum for Development Studies*, *47*(2), 207–217. https://doi.org/10.1080/08039410.2020.1780714

Logie, C. H., Okumu, M., Isha B., Alyssa M., Musoke, D. K, Hakiza, R., Amaya P., Baral, S. &Kyambadde P. (2023) Multi-method findings on COVID-19 vaccine acceptability among urban refugee adolescents and youth in Kampala, Uganda, *Global Public Health*, 18:1, DOI: 10.1080/17441692.2023.2185800

Kaddu, S. B. (2007). Information ethics: a student's perspective. In R. Capurro, J. J. Britz, T. J. D. Bothma, & C. Bester (Eds.), *Africa reader on information ethics* (p. 321). University of Pretoria.

Kao, A. C. (2020). Oscaring Parasite. *AMA Journal of Ethics*, *22*(3), 179–182. https://doi.org/10.1001/amajethics.2020.179

Kasozi, K. I., MacLeod, E., Ssempijja, F., Mahero, M. W., Matama, K., Musoke, G. H., Bardosh, K., Ssebuufu, R., Wakoko-Studstil, F., Echoru, I., Ayikobua, E. T., Mujinya, R., Nambuya, G., Onohuean, H., Zirintunda, G., Ekou, J., & Welburn, S. C. (2020). Misconceptions on COVID-19 Risk Among Ugandan Men: Results From a Rapid Exploratory Survey, April 2020. *Front. Public Health*, *8*(416). https://doi.org/10.3389/fpubh.2020.00416

Kigongo-Bukenya, I. M. N. (2007). Towards professionalism and commitment in Africa: The case for theory and practice of information ethics in Uganda. In R. Capurro, J. J. Britz, T. J. D. Bothma, & C. Bester (Eds.), *Africa reader on information ethics* (pp. 289–296). University of Pretoria.

Kihal, F. E., Abouelkheir, I., Ruchik, M., & Elmouki, I. (2019). Role of Media and Effects of Infodemics and Escapes in the Spatial Spread of Epidemics: A Stochastic Multi-Region Model with Optimal Control Approach. Mathematics,. *Mathematics*, *304*.

Lor, P. J. (2007). International advocacy for information ethics: the role of IFLA. In R. Capurro, J. J. Britz, T. J. D. Bothma, & C. Bester (Eds.), *Africa reader on information ethics* (pp. 187–200). University of Pretoria.

Mejora, Y., & Kalimeri, K. (2020). Advertisers jump on the Corona virus bandwagon: Politics, news and business. *ACM*.

Mheidly, N., & Fares, J. (2020). Leveraging media and health communication strategies to overcome the COVID-19 infodemic. *Journal of Public Health Policy, 41*(4), 410–420. https://doi.org/10.1057/s41271-020-00247-w

Michel, J. B., Shen, Y. K., Aiden, A. P., Veres, A., Gray, M. K., & Pickett, J. P. (2011). Quantitative Analysis of Culture Using Millions of Digitized Books. *Science, 331*, 176–182. https://doi.org/10.1126/science.1199644

Musoke,P., Nantaayi, B.,Ndawula, R. K.,Wannyana, B.,Ssewante, N.,Wekha, G.,Olum, R.,Nakyagaba, R., Nassozi, D. R.,Nabukeera, J.,Kanyike, A. M.,Ojilong, D.,Madut G.,Akech, Kajjimu, F., Kiwumulo, J., Agira, D.,Okot J&Bongomin F (2021) Fear of COVID-19 and the Media Influence on Herbal Medication Use in Uganda: A Cross-Sectional Study, *Risk Management and Healthcare Policy*, 14:, 3965–3975, DOI: 10.2147/RMHP.S332325

Mutula, S. M. (2020). *Infodemic during the COVID-19 Pandemic: Ethical Implications*. College of Law and Management. https://clms.ukzn.ac.za/infodemic-during-the-covid-19-pandemic-ethical-implications/

Myers, S. L. (2020, March). China Spins Tale That the U.S. Army Started the Coronavirus Epidemic. *New York Times*.

Napakol, A., Kitego, E., & Dralega, C. A. (2022). (Mis)Information, Sources and Credibility: Targeting Youth in Public Health Response to COVID-19 in Uganda. In C. A. Dralega & A. Napakol (Eds.), *COVID-19 and the Media in Sub-Saharan Africa: Media Viability, Framing and Health Communication* (pp. 77–98). Emerald Publishing Limited. https://doi.org/10.1108/978-1-80382-271-620221006

Nosek, B. A., Hardwicke, T. E., Moshontz, H., Allard, A., Corker, K. S., Dreber, A., Fidler, F., Hilgard, J., Kline Struhl, M., Nuijten, M. B., Rohrer, J. M., Romero, F., Scheel, A. M., Scherer, L. D., Schönbrodt, F. D., & Vazire, S. (2022). Replicability, Robustness, and Reproducibility in Psychological Science. *Annual Review of Psychology, 73*(1), 719–748. https://doi.org/10.1146/annurev-psych-020821-114157

Otike, F., Bouaamri, A. &Barát, A. H (2022) Predatory Publishing: A Catalyst of Misinformation and Disinformation Amongst Academicians and Learners in Developing Countries, *The Serials Librarian*, 83:1, 81–98, DOI: 10.1080/0361526X.2022.2078924

Pagel, M. (2009). Human language as a culturally transmitted replicator. *Nat Rev Genet, 10*, 405–415. https://doi.org/10.1038/nrg2560

Pagel, M., Atkinson, Q. D., & Meade, A. (2007). Frequency of word-use predicts rates of lexical evolution throughout Indo-European history. *Nature, 449*, 717–720. https://doi.org/10.1038/nature06176.

Peischel, W. (2020). *Mental health professionals are preparing for an epidemic of anxiety around the coronavirus*.

Ponelis, S. R. (2007). Implications of social justice for the pricing of information goods. In R. Capurro, J. J. Britz, T. J. D. Bothma, & C. Bester (Eds.), *Africa reader on information ethics* (pp. 215–220). University of Pretoria.

Research Data Alliance. (2020). *RDA Working on*. Web Document.

Roth, F., & Brönnimann, G. (2013). *Risk analysis using the internet for public risk communication*.

Rothkopf, D. J. (2003, May). When the Buzz Bites Back. *The Washington Post*.

Seeger, M. W., Sellnow, T. L., & Ulmer, R. R. (2003). *Communication and Organizational Crisis*. Praeger.

Smith-Spark, L. (2021, October 12). Schools are new battleground in war of disinformation over Covid-19 vaccines. *CNN*. https://edition.cnn.com/2021/10/12/europe/schools-vaccine-disinformation-covid-19-cmd-intl/index.html?utm_source=feedburner&utm_medium=feed&utm_campaign=Feed%3A+rss%2Fcnn_topstories+%28RSS%3A+CNN+-+Top+Stories%29

Tran, B. X., Dang, A. K., Thai, P. K., Le, H. T., Le, X. T. T., Do, T. T. T., Nguyen, T. H., Pham, H. Q., Phan, H. T., Vu, G. T., Phung, D. T., Nghiem, S. H., Nguyen, T. H., Tran, T. D., Do, K. N., Truong, D. V.,

Vu, G. V., Latkin, C. A., Ho, R. C. M., & Ho, C. S. H. (2020). Coverage of Health Information by Different Sources in Communities: Implication for COVID-19 Epidemic Response. *International Journal of Environmental Research and Public Health, 17*(10), 3577. https://doi.org/10.3390/ijerph1710357

Tullis, J. A., & Kar, B. (2021). Where Is the Provenance? Ethical Replicability and Reproducibility in GIScience and Its Critical Applications. *Annals of the American Association of Geographers, 111*(5), 1318–1328. https://doi.org/10.1080/24694452.2020.1806029

Tully, M., Madrid-Morales, D., Wasserman, H., Gondwe, G.&Ireri, K (2022) Who is Responsible for Stopping the Spread of Misinformation? Examining Audience Perceptions of Responsibilities and Responses in Six Sub-Saharan African Countries, *Digital Journalism*, 10:5, 679–697, DOI: 10.1080/21670811.2021.1965491

United Nations. (1995). *UN Doc. E/CN.4/1995/32.*

United Nations. (2019). *The future is now science for achieving sustainable development.*

Vaezi, A., & Javanmard, S. H. (2020). Infodemic and Risk Communication in the Era of CoV-19. *Advanced Biomedical Research.*

Van Der Linden, S. (2022). Misinformation: Susceptibility, spread, and interventions to immunize the public. *Nature Medicine, 28*, 460–467.

Vosoughi, S., Roy, D., & Arai, S. (2018). The spread of true and false news online. *Science, 359*(6380), 1146–1151.

Wiederhold, B. K. (2020). Using Social Media to Our Advantage: Alleviating Anxiety During a Pandemic. *Cyberpsychology, Behavior and Social Networking Volume, 23*(4), 1–2. https://doi.org/10.1089/cyber.2020.29180.bkw

Willcox, M. L., & Bodeker, G. (2004). Traditional herbal medicines for malaria. *BMJ, 329*, 1156.

Woodbury, M. (2007). Information integrity in Africa: exploring information corruption issues. In R. Capurro, J. J. Britz, T. J. D. Bothma, & C. Bester (Eds.), *Africa reader on information ethics* (pp. 229–237). University of Pretoria.

World Health Organisation. (2020a). *Misinformation undermines the fight against Corona Virus*. Emerald Insight.

World Health Organisation. (2020b). *Novel corona virus (2019-nCov): situationreport—13.*

World Health Organisation. (2020c). *Strategic Preparedness and Response Plan.*

World Health Organization. (2018). *Managing Epidemics: Key Facts about Major Deadly Diseases.*

Zarocostas, J. (2020). How to fight an infodemic. *The Lancet, 395*(10225), 676. https://doi.org/https://doi.org/10.1016/S0140-6736(20)30461-X

Zielinski, C. (2001). The changing role of information in development. *IIS/IDF Conference.*

Zlatí, V., Barjaˇsí, I., Kadoví, A., Štefančí, H., & Gabrielli, A. (2020). Bi-stability of SUDR+K model of epidemics and test kits applied to COVID-19. *Arxiv*, 1–6.

Racheal Opeyemi Odunlade, Olatunde Barber,
Oluwayimika Esther Adegbite and Bibire Nurat Badmus

Misinformation, Disinformation, Malinformation and Propaganda: The Ills of Social Media

Abstract: The advent of social media has transformed the landscape of information dissemination, presenting both opportunities and challenges. Social media platforms have become breeding grounds for misinformation, disinformation, mal-information, and propaganda, which can have significant consequences for individuals and societies. This chapter examines the concepts of misinformation, disinformation, mal-information, and propaganda within the context of social media, exploring their definitions, types, and implications for society. It delves into the erosion of trust in traditional media and the role of social media in shaping public opinion, highlighting the ease with which false information can be shared and amplified. The chapter discusses the various forms of misinformation and disinformation, including unintentionally-spread misinformation, intentionally-spread misinformation, and unverified information, as well as the types of propaganda such as agitative, integrative, white, grey, and black. It also addresses the impact of these phenomena on public health, social trust, and political outcomes. Finally, the chapter considers the challenges of identifying and combating misinformation and propaganda on social media and offers recommendations for individuals and society to mitigate their effects.

Keyword: social media, misinformation, disinformation, mal-information, propaganda, fake news, information dissemination, information manipulation

Racheal Opeyemi Odunlade, Olatunde Barber, Oluwayimika Esther Adegbite, Bibire Nurat Badmus, University of Lagos Library, Akoka Yaba, Lagos. Nigeria

https://doi.org/10.1515/9783111176345-019

Introduction

The advent of social media has ushered in both good and bad, especially in the area of information dissemination. Many have capitalized on the weapon of social media to perpetuate mischief, especially against individuals or organizations considered as enemies. In some climes, they do not even need to be in dispute or disagreement with their victims before using these weapons to destroy the reputation. The concepts of misinformation, disinformation, mal-information, and propaganda have enjoyed wide proliferation with the help of social media handles. Often times, it is impossible to differentiate between correct information and falsehood as a result of the level of manipulation that is achievable on social media. These variables have become a significant menace in society in this era because social media platforms have created an environment where false information spreads rapidly and widely. This development is essentially disturbing because of the attendant repercussions that could ensue from the activities of misinformation, disinformation, mal-information, and propaganda.

Without mincing words, all of these variables are hinged on fake news and are used to manipulate public opinions, undermine democratic institutions, and sow division and conflict among individuals and organizations. The consequences of these on society can be severe, with various authors explaining the concepts and effects of misinformation, disinformation, and malinformation (MDM). In February 2022, the Canadian Centre for Cyber Security described "MDM as three main forms of informational activity that can cause minor or major harm."

The erosion of trust in the media parallels the declining confidence in other institutions, and is a noteworthy phenomenon in itself. Within Western societies, there exists a deep-seated belief among significant segments of the population that the current structures of governance, along with the information systems responsible for ensuring their transparency and accountability, are no longer functioning effectively. Social media platforms have a unique influence in this context, as users who come across news articles through social media tend to assign equal importance to the social identity of the person sharing the news and the reputation of the original source when assessing the credibility of the information. Even in other parts of the world, this phenomenon undermines the potential moderating influence of mainstream journalistic media, which traditionally played a central role in providing balanced perspectives. Thus, the spread of misinformation, disinformation, mal-information, and propaganda on social media have serious consequences. It can lead to misunderstandings, confusion, and panic, as well as the spread of hate speech and violence. In some cases, it can even influence political outcomes, as has been seen in recent elections in many parts of the world.

In recent times, the global communities of professionals in the field of media and individuals who consume media content have developed a strong interest in and concern about various forms of media genres, which encompass terms like "fake news," "misinformation," "disinformation," "media manipulation," "coordinated inauthentic behaviour," and "propaganda." The emergence of these phenomena has captivated and alarmed people worldwide, who fear these unreliable and misleading messages might undermine the effective functioning of the nation's political spaces. As a result, governments, citizens, and scholars have been driven to comprehend and confront this menace. The recognition of the importance of disseminating content through social media networks has resulted in a growing dependence on indicators of public attention trends, which has been a source of frustration for numerous journalists.

One of the main reasons why these mischiefs thrive on social media is the ease with which false information can be shared and amplified through social media handles. Unlike traditional media, social media platforms have very few barriers to entry, and anyone can create and share content without vetting or fact-checking. This can lead to people making decisions based on false information, such as refusing to heed warnings or engaging in harmful behavior. It can also undermine trust in individuals and institutions, as people may no longer believe in the accuracy and impartiality of the media, the government, or other sources of authority. This has led to proliferation of fake news/stories, conspiracy theories, misconceptions, and misleading information.

The most pertinent questions which require answers are: why do people deliberately create false information? Why spread incorrect or malicious information on social media? How can one identify false or malicious information on social media platforms? What are the effects of these variables on human existence? What can we do as individuals and society to curb these menaces? This chapter takes a cursory look at these concepts with a view to providing meanings to them, discusses the implications on humanity, and suggests ways to identify them even in the midst of the information overload characterizing the times we live in.

Conceptual Overview of the Variables of the Study

This chapter is anchored on the following five concepts: social media, misinformation, disinformation, mal-information, and propaganda.

Social Media

Social media has revolutionized the way we communicate, interact, connect with others, and share information. It has become an integral part of our lives, connecting us with people across the world, providing us with a platform to express our thoughts and opinions, and influencing how our decisions and choices are made.

Social media happens to be one of the easiest channels used for spreading and consumption of information. Social media can be defined as websites which give room for profile creation and the visibility of relationships between users. They are interactive websites and web-based applications that facilitate social networking among virtual communities engaging in information sharing, relationships, group conversations, profiles, and sharing of ideas using laptops, computers, and mobile phones. Social media platforms like Facebook, Twitter, Instagram, YouTube, WhatsApp, Snapchat, LinkedIn, Telegram, blogs, and so on allow people to share and update opinions, activities, photos, statuses, personal messages, memories, stories, recommendations, news, articles, text, audios, and videos with one another in real time.

Social media is a valuable tool for cooperation, communication, and knowledge sharing on an open access platform (Lamidi, Gana, and Alafiatayo 2014). It plays an important role in society as it has been embedded in people's lives with the innovations in information technology. Social, ethical, environmental, and political issues have gained increased visibility through social media, which has shifted the balance of power as people can express their concerns on social issues without actually having to engage actively with others. It is typically used for social interaction and to share, create, disseminate, and retrieve news and information for decision making, thereby revolutionizing the way people communicate and interact.

While all of the aforementioned are true and factual about the social media, it can also be used to manipulate almost anything to suit the handler's purpose, ranging from images to figures and words. So many destinies have been destroyed on social media as a result of fabrications that stem out of jealousy, hatred, and the deliberate act of maligning the victims. This is not peculiar to gender, race or status; people as well as organizations are not left out, with government parastatals and religious bodies also sharing in this evil.

Information is live, information is power, and information enables man to be informed and to take decisions. However, the source or sources of information available to an individual go a long way to determining whether one is rightly informed or not. As much as information can be used to enhance peace within society, it can likewise be used to disrupt the peace of a society. The intention of the disseminator and the receiver also determines the interpretation of such information. Whether fake or correct, social media plays a significant role in information dissemination and consumption. Information becomes abusive when the provider uses it with the ulterior motive of deceiving or maligning the victim. At that point, a piece of information can either become propaganda, mal-information, disinformation, or misinformation.

Misinformation

Misinformation refers to false or inaccurate information that is spread unintentionally. False information that is not intended to cause harm or false information without an ulterior motive is referred to as misinformation, which can happen when people share information without verifying its authenticity. In this age of social media, misinformation can spread quickly and easily. For example, false news about a natural disaster or a political event can spread rapidly, causing panic and confusion. Simply put, misinformation is incorrect or misleading information by accident or otherwise.

Misinformation on social media is a growing concern and many similar terms such as disinformation, fake news, rumor, and so on may be seen as the same as misinformation. What differentiates them from misinformation, however, is intention of deception. Generally, misinformation as an umbrella term includes all false or inaccurate information that is spread on social media.

Types of Misinformation

Unintentionally-Spread Misinformation: this is a type of misinformation that is not intentionally spread to manipulate the recipients. Regular users may contribute to the propagation due to trust of information sources, such as their friends, family, colleagues, or influential users in the social network. Instead of wanting to deceive, they usually try to inform their social network of a certain issue or situation.

Intentionally-Spread Misinformation: some misinformation is intentionally spread to deceive recipients, so as to trigger the intensive discussion about the information. "There are usually writers and coordinated groups of spreaders behind the popularity, who have a clear goal and agenda to compile and promote the misinformation" (Wu et al. 2019). Typical examples of intentionally-spread misinformation include conspiracies, rumors, and fake news etc.

Unverified Information: unverified information is another form of misinformation, though it can sometimes be true and accurate. Before a piece of information is verified, it can be described as unverified. Verified information which can be found to be false or inaccurate also belongs to the class of misinformation. One thing that is common between unverified information and other types of misinformation is that they are capable of causing fear, hatred, and bewilderment.

Rumor: a rumor is unverified information that can be true or not true. Therefore, it is a type of misinformation.

Satire/Parody: news satire is a form of parody that imitates genuine news sources and derives its comedy from its deadpan, sarcastic delivery. This is because not every reader will see the irony, and satire like this might be mistaken as reality. It is characterized by irony, exaggeration, and humor. While satire is not meant to be manipulative, it can have the same impact as false news if not read correctly.

Clickbait: this is sensationalized material on the internet meant to attract readers' interest by appealing to their emotions or interest. Clickbait includes hyperlinks intended to entice the viewer to click through to another website. Usually, the target of clickbait is to get readers engrossed with the material in order to make money. It is generally devoid of facts or other useful information and may also disseminate false information; just like satire, it is hyperbolic in nature and readers may be affected in turn and share it with their social connections, thereby spreading incorrect information to a large number of people.

Misleading titles: misleading titles, like clickbait, may also be used to attract attention by appealing to people's emotions and curiosity. Even if the piece is verified and factually accurate, the content might convey the wrong impression if the headline is deceptive. As a result, even if they don't match the characteristics of "clickbait," such headlines may be referred to as clickbait. Unfortunately most social media users share stories based on the title without really reading the text. Misinformation on social media can have serious aftermath effects on public health and social trust, and therefore the role of social media in misinformation needs to be addressed.

Disinformation

Basically, disinformation has to do with intentional dissemination of false and malicious information in order to cause harm or misdirection. It is a deliberate and intentional lie which results in people being actively misguided by malicious actors. There is no definition that can sufficiently describe the term disinformation, as there are multiple and different contexts in which it may arise. Disinformation includes malicious content such as hoaxes, spear phishing, and propaganda, spreading fear and suspicion among the population. Most often, it is well-funded and supported by automated technology, which has a harmful impact on society. The spreaders of disinformation prey on the gullibility of their target audience in an effort to recruit them as multipliers and amplifiers by taking advantage of their readiness to share knowledge for various reasons and to excite them into acting as their messengers. Disinformation affects people's perceptions of what is essential and clouds the truth in order to reduce the influence of rationality on people's decisions. Most times, the perpetrators of disinformation are governments, political organizations, and individuals with a specific agenda. They can be difficult to detect, and when they are successful they can significantly impact public opinion.

Freelon and Wells (2020) define the term "disinformation" as encompassing various forms of false, inaccurate, or misleading information that is intentionally crafted, presented, and promoted with the explicit purpose of causing harm to the public or for personal gain. They analyzed this definition as incorporating three essential criteria: (1) the presence of deception; (2) the potential for causing harm; and (3) an intent to inflict harm. Within this definition, disinformation messages are likened to ammunition in information warfare, serving as non-lethal weapons intended to subjugate adversaries rather than engage in rational discourse.

Malinformation

Malinformation is a term used to describe information that is true but is shared with the intention of causing harm or damage. This can include sharing personal information about individuals without their consent or sharing private information that is intended to embarrass or shame someone. It is information based on reality but utilized to harm a person, institution, or society. It can be used as a tool of revenge or to harm someone's reputation.

The concepts of disinformation and malinformation share some characteristics. First, both concepts are endemic to society. Second, the spread of disinformation and malinformation is made possible through social media. Finally, studies reveal that some of the factors that influence the spread of disinformation and malinformation on the social media platform are users' social media habit, political beliefs, and lack of critical reasoning (Madrid 2023).

There are different kinds of disinformation and mal-information. These include:

1. Fabricated Content: information that has been spread but is entirely fake;
2. Manipulated Content: actual data or images that have been altered;
3. Imposter Content: content that impersonates reliable sources, such as by adopting an established agency's branding;
4. Misleading content: this type of information is used to frame topics or people in particular ways by cropping images, choosing statements or data at random, or presenting an opinion as fact;
5. False Context: repurposing truthful material outside of its original context or factually accurate information mixed with false contextual information, such as when an article's headline fails to accurately describe its substance;
6. False Connections: when the headlines, pictures, or captions contradict the information in the text. This occurs when pictures or captions are utilized, especially on websites or social media platforms, to convey an idea that isn't supported by the written word.
7. Sponsored Content: advertising or PR disguised as editorial content;
8. Propaganda: information sharing intended to sway public opinion. The systematic use of symbols (words, gestures, banners, monuments, music, dress, emblems, hairstyles, designs on coins, and postage stamps, etc.) to influence the attitudes, beliefs, or behaviors of others is known as symbolism. Managing attitudes, values, and knowledge through content;
9. Error: a mistake made by agencies in their reporting.

Propaganda

Despite the numerous benefits of social media in the present century, some people indulge in the negative use of social media to create tension in the society. One of the negative uses of social media is the dissemination of fake news through propaganda. Propaganda in Latin means "to propagate" or "to sow"; it is the dissemination of biased and potentially inaccurate information with the intent of harming a person, group, or movement. It is the use of misleading information to promote a particular point of view or agenda, especially in politics, advertising, and media, to shape public opinion and influence decision-making. Social media has made it easier for propaganda to spread and has made it difficult to separate fact from fiction. Historically, in 1622, the word propaganda was established by the Vatican, "the *Sacra Congregatio de Propaganda Fide*," which means sacred congregation for propagating the faith of the Roman Catholic Church and was used to spread the faith to the New World while at the same time antagonizing Protestantism. Subsequently, the usage of propaganda lost its objectivity (O'Donnel and Jowett 2005). Therefore, propaganda messages suggest something negative and dishonest. Gossip, falsehood, prejudice, misleading, deception, manipulation, etc. are common words used synonymously for propaganda. Thus, to grab readers' attention and affect them, propaganda, like other sorts of misinformation, typically depends on emotional appeals, lies, or incomplete truth. It occasionally exploits partial quotations, misleading facts, or other deceptive approaches. Most of the time propaganda is linked to political issues although it comes from other issues too.

Forms of Propaganda

Five forms of propaganda have been identified (O'Donnel and Jowett 2005):

Agitative propaganda: this form of propaganda affects the audience to a certain extent with the aim of bringing about notable change.

Integrative propaganda: the goal of integrative propaganda is to make the listener submissive and accept the story unchallenged. Most times, the audience is made to believe an ideology that is not real without them realizing it.

White propaganda: this form of propaganda originates from verified sources with factual information. The aim is to establish credibility with the audience in the hope that this will be advantageous in the future.

Black Propaganda: here, the source of propaganda is hidden or attributed to a bogus authority; black propaganda spreads lies, fabrications, and deceptions, and can be dubbed with significant lies which encompasses all forms of creative deception.

Grey Propaganda: grey propaganda is a mixture of white and black propaganda. The fact and source about the information may or may not be established. Governments are not the only parties who use grey propaganda, with companies twisting annual statistics reports also engaging in this practice.

Role of Social Media in Propagating Misinformation, Disinformation, Malinformation, and Propaganda

In today's information age, social media plays an increasingly important role in individuals' daily lives, careers, businesses, communication etc. It has transformed communication of information such that people now prefer to use it as a medium of communication. Before the advent of social media, conventional media such as television, radio, newspaper, magazines etc. were the reliable formal ways that individuals sorted and monitored information, meaning fake news, misinformation, and other forms of online deception were not issues.

But today the Internet has amalgamated society at the global level, which is attributed as one community, and has exposed information at an unprecedented volume and velocity for public consumption. It has become easy for people to consume and disseminate information freely through social media at a low cost, which has in turn led to the spread of wrong and ill-motivated information due to the ability to reach a large and diverse audience quickly and easily. Social media has virtually promoted all kinds of inaccurate and false information because of its accessibility, visibility, and ease of use.

Historically, social media development can be grouped into three rough-hewn temporal divisions or ages: the age of electronic communications (late 1960s to early 1990s); the age of virtual community (early 1990s to early 2000s); and the age of social media (early 2000s to date). Lately, there has been a growing concern regarding the perceived rise of "fake news," particularly in discussions surrounding media and politics in the United States and Europe. The majority of research on the impact of contemporary instances of "fake news" has primarily focused on the Global North, with only a limited number of case studies examining the situation in other parts of the world. However, UNESCO asserts that disinformation and misinformation constitute an "emerging global problem" that affects societies in general, including the field of journalism.

Within the context of African journalism studies, groundbreaking research in South Africa has shed light on the utilization of social media platforms such as Twitter as tools for disseminating political misinformation. Additionally, there has been a rise in the presence of fraudulent news websites online that mimic established satirical media critiques. Similar trends can be observed in other regions of the Global South, as seen in examples from India. In African countries such as South Africa and Kenya, misinformation has also taken the form of extreme speech, inciting violence or propagating messages of racism, misogyny, and xenophobia, often transmitted through popular mobile apps like WhatsApp.

According to Wasserman and Madrid-Morales (2019), instances of misinformation campaigns have been employed to manipulate political agendas, prompting governments to implement measures to counteract them.

One notable example is the "white monopoly capital" campaign orchestrated by supporters of former South African president Jacob Zuma, with assistance from the British public relations firm Bell Pottinger. This campaign involved the utilization of fake Twitter accounts and bots to disseminate false information. In Nigeria, the Minister for Information and Culture, Alhaji Lai Mohammed, expressed repeated concerns about the impact of "fake news" and pledged to combat misinformation during the 2019 presidential elections (Vanguard, 2017). This might be because more often than not, fake news thrives mostly for political reasons in other to cause commotion (Odunlade, Ojo and Agbo, 2021). In other instances, governments, like those of Cameroon, Ethiopia, and Togo, resorted to internet shutdowns as a response to online criticism.

There appears to be a significant increase in the amount of focus in both journalism scholarship and public discussions on the topics of truth, accuracy, and deceit. Although the United States is often the primary subject of these discussions, numerous other countries, such as Myanmar, Nigeria, and India, have also been grappling with the consequences of various forms of misinformation and disinformation, commonly referred to as "fake news." The widespread presence of false or misleading information on a global scale, or at least the perception that it has become more prevalent, combined with the exploitation of the term "fake news" by populist politicians, has been identified as factors contributing to a decline in public trust towards news media. However, it is important to note that these factors are not the sole causes of this phenomenon.

Nonetheless, the Internet has also presented an alternative avenue for reliable information. In 2018, when the Kenyan government silenced privately-owned television stations to suppress the reporting of opposition leader Raila Odinga's unofficial "inauguration" event, online platforms emerged as a safe haven for narratives that were not endorsed by the state. Many had to result in the use of satires and jokes on social media for political criticism when mainstream media were captured by the state or elites. In African contexts where public trust in the media has been eroded, these informal networks of news and information, characterized by registers that imitate, criticize, or undermine mainstream news, have frequently played a positive role in expanding the public sphere.

This has resulted in the tendency of users to favor information obtained on social media over information obtained in the mainstream media, due to government censorship of mainstream media. The decline in trust in formal media will ultimately lead audiences to shift attention to other channels for the satisfaction of their information needs. However, the strength and popularity of a shared piece of information may be misconstrued for its authenticity. The significance of this drive could be particularly applicable in African nations where the government exerts substantial influence or ownership over the media, resulting in a diminished level of confidence in mainstream media.

The widespread concern surrounding the alarming rise of "fake news" and its potential consequences has prompted actors in the field to engage in self-reflection. They have expressed apprehension regarding the influence of misinformation and disinformation on the media's essential role in governance and, consequently have contemplated various approaches to combat the dissemination of false information. These strategies often involve a renewed emphasis on upholding professional values and adhering to established standards, however, the issue of trust goes beyond the mere existence of "fake news," as the presence of misinformation and disinformation in African contexts is influenced by broader structural elements such as the manipulation, suppression, and ownership of news media.

Aside from the aforementioned, the growing number of ways through which social media contributes to information manipulation keeps increasing by the day. In a recent study, the United Nations High Commissioner for Refugees (2022) identified a number of emerging ways that are being employed by people to spread false information on social media platforms. These include:

– **Sockpuppet:** A Sockpuppet adopts an online fictitious or penname to hide his/her identity. This individual, in disguise, deceives and proliferates misleading content such as praise, support, or defense of an individual or organization. They manipulate public opinions with ease and because a Sockpuppet presents themselves as an independent third party, it is easy to evade restrictions or sanctions from a website. Though many online communities do not encourage the activities of Sockpuppet, they wangle their way onto most websites.

– **Sealioning:** this is a type of online trailing that allows perpetrators to pursue their victims with persistent requests for evidence or repeated questions. In this case, the attacker pretends to be sincere and civil while sending incessant invitations which have the intent to dupe the victim.

– **Astroturfing:** this has to do with sponsoring of a message or messages to make it appear as though it comes from grassroots participants. The message(s) could be religious, political, advertising, and so on. Astroturfing aims at providing organizations with credibility by withholding information about their motives or financial connections.

– **Catfishing:** this is a scam where a person creates a fake identity aimed at attacking a particular victim on social media. It is mostly used for love scams on dating websites to compromise a victim or as a form of ensnaring, sometimes done for financial gain.

Influence of Misinformation, Disinformation, Malinformation and Propaganda on Society

The influence of misinformation, disinformation, mal-information, and propaganda on society cannot be over-emphasized. This influence can be positive or negative but in this study our focus is on the negative influence because the dangers posed by fabricated content are real and tangible. Some of these are highlighted as follows:

1. It promotes negative health behaviour. Studies have established that misinformation and its associates have the capability of subjecting people to mental distress due to misleading and false health related content, especially during pandemics, health emergencies, and humanitarian crises.
2. It engenders negative decisions on the part of the citizenry. Wrong information leads to information pollution that affects people's ability to make informed decisions. Anytime people are fed with wrong information, especially in the form of propaganda or hate speech about an individual, organization, or government, doubts and social divisions are created. Public communication is in turn disrupted, which could pose a threat to unity amongst the people.
3. It is capable of sweeping up sentiments that threaten freedom of thought, right to privacy, and democratic participation, consequently destroying democratic processes in societies.
4. It aids the authenticity and reliability of fallacy. Social media enhances the authenticity and reliability of misinformation and propaganda in the sense that when more than one social media platform publish the same inaccurate information, people tend to believe and act on the information. The influence of this on society can be delusion, misjudgment, or wrong decisions.
5. It polarizes facts and figures. Social media has increased polarization of misinformation and propaganda in the society by bringing and allowing like-minded individuals to find and reinforce each other's negative views, creating filter bubbles and reverberating unnecessary hollows that further establish the existing fake and erroneous information.

6. Social media enhances the quick dissemination of information among large audiences within a short period of time. This feature allows the spread of wrong information and falsehood from various actors, such as individuals, insurgent organizations, governments, and so on, to reach out to large audiences quickly to manipulate public opinion and shape the discourse around the issues.

Steps to Take to Combat the Menace of Information Manipulation ON Social Media

To counteract the propagation of manipulated contents on social media, digital and media literacy and the evaluation of the information and its source must be encouraged.

Digital and media literacy is the capacity to engage critically, effectively, and responsibly with all forms of media. In order to do this, one must learn how to use digital networks and media tools, access media, analyze and evaluate media critically using a few key concepts, create media for one's own interests, work, and school projects, and interact with media to express oneself and engage with communities both online and offline.

Evaluation of the information and its source has to do with evaluating the information landscape critically and taking time to review the sources and the messages, which is an important criterion in identifying the intention of an author of particular information. When viewing content in any form, it is essential to ask the following questions:
- Is there an emotional reaction?
- Does it make a strong argument on a contentious subject?
- Is the assertion extraordinary?
- Is there clickbait in it?
- Does it contain current information that is relevant to the situation?
- Does it make use of sparse, accurate information that is misinterpreted or exaggerated?
- Has it been widespread on untrusted or dubious platforms?

These and many more questions should be asked when obtaining information through social media, especially information that has not been subjected to verification or review and is available on an open site.

Conclusion

While social media has transformed the way we communicate and access information, it has also become a fertile ground for the spread of misinformation, disinformation, malinformation, and propaganda. These ills pose serious threats to individuals, societies, and democratic processes and have thrived on social media in the recent time. They are crafted to appeal to a core audience to generate participation, with the quick traction of the content leading to the possibility of turning viral. Social media platforms have helped in expanding the reach of individuals spreading false information in a unique way, influencing the approach and opinion of the people.

By understanding the different forms of misleading information and taking collective action, we can mitigate the negative impacts and foster a healthier, more informed digital ecosystem. It is essential to recognize the importance of responsible use, critical thinking, and concerted efforts to address the challenges posed by social media's dark side.

Addressing the challenges posed by social media requires a comprehensive strategy. Social media companies bear the responsibility of implementing policies that prioritize information accuracy, ensuring transparent content moderation practices, and restricting the spread of misinformation. Fact-checking organizations are instrumental in debunking falsehoods and increasing public awareness, while promoting media literacy education is essential to empower individuals in critically navigating the digital realm. Additionally, governments can contribute by enacting legislation that fosters transparency, accountability, and responsible conduct within social media platforms.

Bibliographies and References

Boyd, D. M. & Ellison, N. B. (2008). Social Network Sites: Definition, History, and Scholarship. Journal of Computer-Mediated Communication. 13, p. 210–230.

Brianna Steph (2020). The Impact of Fake News on the Society. INOSR HUMANITIES AND SOCIAL SCIENCES 6(1) p. 21–26. http://www.inosr.net/inosr-humanities-and-social-sciences/

Cherilyn Ireton and Julie Posetti (2018) Journalism, 'Fake News' & Disinformation: Handbook for Journalism Education and Training. France: United Nations Educational, Scientific and Cultural Organization.

Deen Freelon & Chris Wells (2020). Disinformation as Political Communication, *Political Communication*, p. 1&2, DOI: 10.1080/10584609.2020.1723755

Diga, M., & Kelleher, T. (2009). Social media use, perceptions of decision-making power, and public relations roles. Public Relations Review, 35, 440–442.

Durodolu, O.O; Chisita, C. T; Dube, T.V. (2021). Flattening the Curve of Fake News in the Epoch of Infodemic: An Epistemic Challenge. In R.J. Blankenship (Ed.). Deep Fakes, Fake News, and Misinformation in Online Teaching and Learning Technologies (pp. 143–161). IGI Global. DOI: 10.4018/978-1-7998-6474-5.ch007

Herman Wasserman & Dani Madrid-Morales (2019). An Exploratory Study of "Fake News" and Media Trust in Kenya, Nigeria and South Africa, African Journalism Studies, p. 4, DOI: 10.1080/23743670.2019.1627230

Kietzmann, J. H., Hermkens, K., McCarthy, I. P., & Silvestre, B. S. (2011). Social media? Get serious! Understanding the functional building blocks of social media. Business Horizons, 54(3), 241–251.

Lamidi, D.A., Gana, M and Alafiatayo, B.O. (2014). Use of social media in information dissemination. Paper Presented at Workshop/seminar organized by the Association of Government Libraries (AGOL) Section of the Nigeria Libraries Association (NLA). Held at Nasarawa State University, Keffi on the 20th November, 2014.

Wu, L; Morstatter, F; Kathleen M. Carley, K.M and Liu, H (2019). Misinformation in Social Media: Definition, Manipulation, and Detection https://dl.acm.org/doi/abs/10.1145/3373464.3373475

Lipschultz, J. H. (2020). Social media communication: Concepts, practices, data, law and ethics Routledge.

Madrid Pamela (2023). USC study reveals the key reason why fake news spreads on social media. University of Southern California. https://news.usc.edu/204782/usc-study-reveals-the-key-reason-why-fake-news-spreads-on-social.

Marshall, J. (2017). Disinformation Society, Communication and Cosmopolitan Democracy. Cosmopolitan Civil Societies: An Interdisciplinary Journal. 9(2), 1–24. http://dx.doi.org/10.5130/ccs.v9i2.5477

Misinformation in Social Media: Definition, Manipulation, and Detection https://www.kdd.org/exploration_files/8._CR.10.Misinformation_in_social_media_-_Final.pdf

Mohammed, Lai (2017, 10/25). Fake news, misinformation biggest obstacles to 2019 election – Lai Mohammed. Vanguard. P. 1. https://www.vanguardngr.com/2017/10/fake-news-misinformation-biggest-obstacles-2019-election/

O'Donnel, V. and Jowett, G. S. (2005). Propaganda and persuation. 4th ed.

Odunlade, R.O; Ojo, J.O. and Agbo, N.A. (2021). Effects of fake news and propaganda on management of information on Covid-19 pandemic in Nigeria. International Journal of Knowledge Content Development & Technology.11 (June 2021). http://www.ijkcdt.net/xml/29480/29480.pdf

Oestreicher-Singer, G., & Zalmanson, L. (2013). Content or community?A digital business strategy for content providers in the social age. MIS Quarterly, 37(2), 591–616

Olan, F., Jayawickrama, U., Arakpogun, E.O. *et al.* Fake news on Social Media: the Impact on Society. *Inf Syst Front* (2022). https://doi.org/10.1007/s10796-022-10242-z

United Nations High Commissioner for Refugees (2022). Factsheet 4: Types of Misinformation and Disinformation. https://www.unhcr.org/innovation/wp-content/uploads/2022/02/Factsheet-4.pdf

Tinyiko Vivian Dube and Oluwole Olumide Durodolu

Analysis of information disorder caused by COVID-19 and how to practice information hygiene in academic libraries: a librarian perspective

Abstract: The coronavirus (COVID-19) pandemic has drastically changed academic libraries' regular operations, leading libraries to adapt to a new normal of providing library and information services to remote clients. Since meeting students' information needs is a primary function of academic libraries, library management had to devise means to figure out how library and information services should be provided amidst the COVID-19 pandemic. Librarians had to learn to practice social distancing, sanitization of hands, and wearing masks at all times while on the library premises. To maintain social distancing as one of the significant regulations for COVID-19 and ensure that remote clients' information needs were met, librarians with underlying medical conditions had to work from home, while those whose key responsibilities could not be done from home had to report to work. The work interrogates the challenge of the world's health security that is approaching criticality, examines information disorder and its related challenges, and outlines ways to address the infodemic. Furthermore, this research critically investigated the casualties prompted by misinformation that was peddled through social media during the pandemic. The article finds that most people turn to social media, which is bedevilled with lots of false information on COVID-19, for news that may not conform with the guidelines of the World Health Organization (WHO). Recommendations are advanced in the hope that it will bolster the effect of these twin misfortunes and help the affected organizations and bodies in charge of disseminating information to the public follow conformity.

Keywords: academic libraries, COVID-19, information disorder, information hygiene

Tinyiko Vivian Dube, Department of Library and Information Services, University of South Africa, South Africa, e-mail: tinyikov@hotmail.com
Oluwole Olumide Durodolu, Department of Information Science, University of South Africa, Pretoria, South Africa, e-mail: woledurodolu@gmail.com

https://doi.org/10.1515/9783111176345-020

1 Introduction

The COVID-19 pandemic took the whole world by surprise and affected many nations severely. The preliminary growth of the virus was exacerbated by inaction and an absence of reliable evidence on how best to control the worsening situation (Del Rio and Malani 2020). As soon as thousands of people started getting infected all around the world, governments and health institutions began working to respond to the global pandemic that subsequently claimed many lives across the globe. The primary role of academic libraries is to conserve the prevailing knowledge by ensuring that their clients' needs are met and relevant information resources are accessed in a preferred format (Dube 2016; Ugwu and Orsu 2017). These library services, however, were interrupted by academic libraries' closure during the COVID-19 pandemic to adhere to the hard-national lockdown.

The growing reliance on social media for reporting, together with increasingly escalating media inaccuracies, has threatened the frontiers of knowledge and trustworthy information. This predicament has resulted in disorderliness in the information ambience because of the inability to standardize activities on social media allied to the problem of interfering with the fundamental right of free speech, with such circumstances leading to the new concept in the health emergency of COVID-19 known as an infodemic, an epidemic magnified by the inability to access accurate information in relation to the virus (Durodolu and Ibenne 2020).

With the improvement of the current social technology, the world has witnessed a phenomenal use of social media, which has led to larger-scale pollution of information at a broad scale, which has brought about a significant watering down of the quality of information. This is worrisome with the advent of a global health emergency like COVID-19, which called for accurate information to manage the pandemic. While information disorder is not an entirely new predicament, the complication and scale of information pollution in a digitally connected world brings to light the unprecedented challenge (Wardle and Derakhshan 2017). This research attempts to argue the urgent need to pursue practical solutions to bring about information hygiene to stem the tide of information pollution. This effort should be driven by experts in information use and dissemination, like librarians, media experts, and information literate personnel.

This study discusses the librarian's perspectives and responses to the challenges of information disorder caused by the COVID-19 pandemic. Fake news spread recklessly, at equal proportions to the virus, which compromised the ability to access life-saving information during the pandemic, with this factor thus constituting a grave danger to human lives. Even though libraries closed, and some changed their usual modes of operation, librarians were still able to assist in combating the spread of fake news by conducting online information literacy and sharing information on social media. Misinformation persists and thrives mainly because of digital illiteracy, whereby people cannot evaluate and understand sources of information, thus leading them to believe everything available on social media. In this way, fake news about COVID-19 generated fear that disrupted the WHO's efforts to disseminate helpful information scientifically proven to curb the Coronavirus spread.

As said, the advent of COVID-19 has led to the proliferation of false and misleading information, often compromising the ability to access truthful health information and thus putting people's lives at risk and generating undue anxiety. In a post-truth epoch where deceptive information has been heightened, citizens must acquire a new paradigm shift in dealing with the issue. Library and educational institutions have highlighted various information literacy strategies for a more information-literate society where citizens can evaluate information and identify fake news (Durodolu, Ibenne, and Dube 2021; Durodolu, Nduka, and Tinyiko 2022).

2 Problem Statement

It was a significant concern that erroneous information of different types may impact public knowledge and lead to an epistemic challenge that would undermine citizens' ability to access accurate information in dealing with the dreaded COVID-19 pandemic, especially in the post-truth era where facts based on evidence are less influential in determining public opinion and views than appeals to emotion and personal belief (Mackey 2019). The manifestation of the pandemic in the post-truth age advances many questions about the role of facts in dealing with the crisis caused by the pandemic's potential long-term repercussions for the post-truth age (Schulman 2020). Indeed, as COVID-19 continues to affect the whole world, there is the possibility that ambiguity resulting from a lack of knowledge will continue to make it problematic for even experts to answer all questions concerning the extent of the crisis and how to manage and cope with it.

In a situation where the world is still searching for accurate knowledge to deal with the pandemic, misinformation, charlatanism, fake news, and distortion of facts associated with the post-truth age have become the order of the day. The dilemma that arises from the need to fill the void left by the limited information accessible during the crisis has also impacted on people's viewpoints and the information they receive (Schulman 2020). At the initial stage of the COVID-19 outbreak, there was an overwhelming amount of information, particularly on social media platforms, as a meeting point for citizens' engagement relating to the sources and repercussions of the virus, both locally and internationally. Sadly, a higher proportion of information relating to the pandemic, its symptoms, transmission methods, and response mechanisms have mainly been undependable. Because of this, social media spectators have been serenaded with misinformation, disinformation, misconceptions, and myths through propaganda and fake news. Nevertheless, part of this can be traced to social media platforms' technical landscape that categorizes users through echo chambers and algorithms (Ogenga 2020). Another predicament is the contradictions emanating from "reliable institutions" about facts relating to the pandemic and how they have promoted the infodemic significantly due to people's desire to find the solution to an unknown virus and comfortable facts that do not cause cognitive dissonance.

3 Objective

This chapter is based on in-depth analysis of the subject matter of information disorder in the context of COVID-19 and how to practice information hygiene.

4 Methodology

This research is conducted by using an interpretive research paradigm, a structure and practice within social science research anchored on theoretical and methodological techniques of appreciating social reality (Given 2012). The research used qualitative content analysis which is well-defined as an exploration method in a subjective interpretation through the methodical classification process of identifying themes or patterns to collect data (Hsieh and Shannon 2005). Extant literature like journals, conference papers, dependable online resources, and policy documents were gathered and evaluated on the subject matter of information disorder in the context of COVID-19, information hygiene, and librarian perspective. The study identified a gap in the study to give birth to new knowledge, with relevant keywords identified and used as a search strategy in different academic databases like EBSCOhost and Web of Science.

5 Literature Review

Since COVID-19 had been confirmed a global pandemic, higher learning institutions in South Africa and academic libraries had to close due to the national lockdown. The closure of institutions of higher learning posed challenges in the provision of library and information services, which caused unequal treatment amongst librarians because the regulations instructed people with underlying medical illnesses and those above the age of 60 to stay and work from home while others worked in shifts. Realizing that the spread of COVID-19 was increasing at a fast pace and students had to be provided with a service, the library management responded to the COVID-19 alert levels by encouraging librarians except for those working at the front desk to work from home.

The new normal of library operations was motivated by the information technology in academic libraries that brought benefits to the library, librarians, and students, but has unfortunately also presented challenges. These challenges were brought about by changes in academic libraries' operational aspects and the work done by library staff. The traditional role of academic libraries is to provide library and information services to walk-in clients. Considering how COVID-19 changed how academic libraries operated posed questions such as whether reference desks will exist in the new normal. During the COVID-19 era, students were encouraged and trained to be self-sufficient; they can now download online information resources such as e-books, e-journals, e-reserves, and e-theses/e-dissertations without library staff's assistance.

Currently, there is a rise in open content whereby librarians are required to educate faculties about the importance of creating and adopting open educational information resources and low-cost or free prescribed books to supplement and support learning and research (Durodolu, Ibenne, and Dube 2021). For example, the closure of laboratories and restrictions on social interaction and travel created problems for faculties to complete academic year curricula. However, in this regard, academic libraries were in a good position to ensure that teaching, learning, and research can be carried out online, as they use platforms like Open Researcher and Contributor ID (ORCID) to enable students and researchers to connect to gateways like ArXiv with institutional repositories to create a wide-ranging research platform (Cox 2020). Based on this, academic libraries must reconsider their online services now that they represent the main pathway for students to access information resources.

The new normal of academic library operation affected numerous aspects whereby librarians had to wear masks, work behind plexiglass shields, often sanitize hands, work in shifts, and work from home. The drastic implementation of librarians to work from home posed challenges, as librarians were obliged to learn new technologies such as Zoom, Microsoft Teams, Google Docs, and WebEx to host and engage in meetings and training. This might seem like a good initiative by the academic library management, but COVID-19 aggravated the division between essential workers who had to work onsite and those who were granted the privilege of working from home. Furthermore, to work from home, librarians needed to have a laptop or PC and reliable internet connection or have produced a medical certificate declaring that they were at high risk of being infected by COVID-19.

Hygienic behaviour has been identified as one way to combat COVID-19 but maintaining information hygiene will equip citizens with life-saving information to prevent and control the disease. Information hygiene entails purposeful and ethical use and consumption of available information as well as responsible dissemination and suspicious appraisal of information that is consumed and disseminated into public space. This is all the important against the backdrop of highly toxic information that has flooded the internet, with experts calling for a new attitude in information use by acquiring information hygiene practices (Ilyin 2013).

Various educational and professional librarianship organizations have designed many information literacy frameworks to help identify fake news. For instance, the International Federation of Library Associations and Institutions (2017) acknowledged that critical thinking is a significant skill in media and information literacy, with the mission of libraries to educate and advocate its importance. The inference here is that information literacy knowledge must be acquired along with other educational skills to enable citizens to cope with misinformation. According to the assessment by the United Nations Educational, Scientific and Cultural Organization (UNESCO 2020), many people use social media to access information, especially youths. Consequently, social media can be utilized to train people in a social context to be more media information literate. Social media can also be used to convey persuasive information to raise consciousness about the significance of media and information literacy at various levels of society (UNESCO 2020). This is especially important as information disorder can twist experts' opinions and reliable sources to generate anxiety and sow cynicism and mistrust (McDonald 2020).

In the confusion that accompanied the spread of COVID-19, scientific institutions were also affected. As a result of the general public's expectations, research articles were hurriedly published with little or no peer review, with much of this research thus presenting contradictory findings and debatable expertise which, however, was viewed by the public and decision-makers as factual rather theories and assessments requiring validation and proof. However, when this research was eventually revealed as erroneous, it resulted in uncertainty (Schulman 2020).

As a result, collective reliance on such institutions has diminished to the extent that academia and the media are perceived as outlets for the establishment to control the societal narrative. Because of this confidence crisis, the information age has become a breeding space for propagating numerous competing narratives, distrust, misunderstanding, sensationalism, exaggeration, polarization, and generally a "post-truth" atmosphere. Undoubtedly, social media plays an enormous role in the "post-truth" perception (Loftus 2020).

Misinformation is when people are fed with information that is not accurate or false; it is unfounded information predicated on ignorance. Nyhan and Reifler (2010) posit that misinformation occurs when people's beliefs about factual matters are not supported by clear evidence and expert opinion. It is, however, challenging to determine the accuracy and authenticity of information in the age of information overload on the internet. In the midst of an overwhelming amount of information, it is often hard to distinguish which information is accurate and which is not (Warner-Søderholm et al. 2018).

However, assessing the credibility of information on social media is increasingly challenging due to structural changes in the information environment (Lazer et al. 2017). Kim and Dennis (2019) highlight that since the start of this century the quantity and dissemination of misinformation have grown exponentially. In turn, Brennen et al. (2020) raised a concern that most of what we learn about the COVID-19 is difficult to separate into information and misinformation, true and false, reliable and unreliable. Presently, the internet is flooded with all sorts of fake information on the COVID-19 pandemic. Since the source and authenticity of such information may not be verified before it is sent out, flooding various social media, gullible and unsuspecting individuals may see such news as real and quickly forward it to their loved ones, unknowingly increasing the chain and coverage of such fake news.

Health Sciences Librarians (HSLs) play an essential role in fighting fake news within this context since they are knowledgeable about health issues. These professionals can play a leading role in educating society through information literacy programs on how to appraise facts and cross-check health information's validity. HSLs stimulate public discourse to work out modalities and advance mechanisms to avert and counter the spread of fake and adulterated health information. The only defense should be training, using various information literacy programs to alert the public on recognizing fake news (Naeem and Bhatti 2020).

6 Responsibility of Librarians in Promoting Information Hygiene

Accurate information provides people with the capacity to facilitate learning and make the right and intellectual decisions to better their livelihood. On the other hand, inaccurate information, especially posted or circulated on social media to disempower and deceive people, becomes harmful when used to influence or sway public opinions.

Munsanje (n.d., n.p.), a notable Zambian librarian and former President of the Zambian Library Association, stated that "Librarians have a key role to play in fighting and correcting fake news." It is the responsibility of librarians to educate people on how to look for legitimate information. In addition, librarians need to educate people about the dangers or consequences of spreading and reading fake news. Munsanje (n.d.) further highlighted that librarians' aims should be to take pre-emptive measures because some of the damage caused by fake news is irreparable and irreversible, such as guiding users on how to recognize reliable sources of information. Most notably, librarians must alert people on how to use the Internet effectively. It is vital to take note that information technologies and social media platforms enable people to write, post, or disseminate information that is not regulated, or peer reviewed, and that such information might be fake and harmful to recipients. This calls for librarians to be proactive by being in the front line, reaching people using various mediums and making them aware of the dangers of fake news.

Librarians should conduct information literacy training to enable people to evaluate information before believing or circulating it to avoid more damage that might be irreversible. Librarians should conscientiously ensure that fake news is not disseminated and train society on detecting it. Information literacy is the solution that librarians should consider to combat fake news.

7 Managing Information to Combat Information Disorder

Combating information disorder in a democratic society seems problematic as there is a growing concern about addressing this problem without undermining the democratic system and the benefit of open access to online information (West 2017). In order to maintain the democratic system and open access to online information, the government, in collaboration with libraries and other relevant stakeholders, should work together to devise means to solve the problem of information disorder, while information technology companies and social media platforms should invest in tools that detect fake news and information disorder, thus preventing the dissemination of such information. In addition to this, librarians should conduct awareness for people to be skeptical of what they read or watch on the internet and social media.

Media platforms such as Google, Facebook, Twitter, Google Chat, WhatsApp, and Instagram facilitate and simplify access to various sources of information and expose users to the threat of information disorder (Goguadze 2019). In turn, young people are prone to the dangers of radicalization, hate speech, and propaganda of information disorder. Therefore, the emphasis should be placed on awareness to distinguish between legitimate information and propaganda or fake news.

8 Setting Agenda for Fighting Fake News

Previously, information was ethically managed and carefully controlled by media professionals, journalists, information experts, and librarians. However, the open and universal landscape of the digital world has made it challenging for these traditional caretakers to regulate and validate information. In view of this, misinformation and distortion of facts has made information management difficult, with potential cataclysmic and precarious consequences for social disorder (De Paora and Heravi 2020). Therefore the knowledge of information literacy remains the panacea for sanitizing the polluted information environment (Durodolu, 2018; Morudu 2019).

Library and Information Science (LIS) authors have conducted studies to find the solution for fighting fake news (Sullivan 2019). However, there are proposed solutions to ensure libraries continue to maintain their values and the services they intend to provide. The ALA president indicated that fake news could provide an awakening for libraries and serve as a reminder that one of their services should be to combat the spread of fake news (Dollinger 2017). This is supported by Fallis (2015) in his discussion about fake news that libraries and other information services are at the vanguard of efforts to ensure that people have access to quality information. Subsequently, most of the studies conducted over the past decades, along with the recent call of action due to the emergence of the COVID-19 pandemic where people lived in fear because of the fake news circulating on social media, suggest that information literacy is the primary weapon to fight the spread of fake news. Cooke (2017) emphasized that the bulk of disinformation on the internet could be combated with basic evaluation skills whereby the WHO could impart those skills to the librarians so they are able to address and handle the spread of fake news through library awareness. Alvarez (2016) supports this, affirming that because of their distinctive positions as associates, educators, and community champions, librarians have the opportunity to teach information and media literacy.

9 Promoting Information Literacy in the Student Community: Librarians' Obligation

Libraries are influential, authoritative, and prevailing pedagogical resources whereby librarians are considered collaborators, teachers, and instructional designers. Librarians teach students and the community how to think about their information needs, reflect and reproduce their experiences, and find relevant and up-to-date information. Moran (2019) believes that information literacy is an indispensable learning outcome to combat spreading fake news which, in general and regarding COVID-19, generates fear, stress, and anxiety. Since COVID-19 is new, the WHO and health workers are trying to understand how the virus spread and how to combat the spread of the virus. At the same time, librarians should teach the general public about the dangers of spreading fake news and believing everything posted on social media. Librarians can achieve this by promoting information literacy when using online platforms, in particular as it was the only suitable means to adhere to the responses to the COVID-19 pandemic, such as maintaining social distancing. The International Federation of Library Associations and Institutions (IFLA) (2017) supports online information literacy and further alluded that libraries have always played a vital role in disseminating information using informal online platforms such as social media to promote lifelong learning. Therefore, libraries have taken a logical step in developing information literacy for those who are technologically disadvantaged, lack technological skills, are reluctant to use information technology tools, and lack understanding of why they should use information technology.

10 Conclusion

As COVID-19 is still a problem affecting the world, citizens are warned to authenticate the information they read on the internet and social media before circulating. This is also the era for librarians to be on the frontline and consider information literacy to increase and alert citizens about misinformation. Most importantly, citizens should also evaluate information from the Internet and social media before sharing to be certain whether the information is legitimate or not. Citizens should be aware that not all information found on the Internet is legitimate and that they should not believe anything they read on the internet or social media without validating the source of information.

11 Recommendations

- Community support groups, civil society, libraries, and mass media should all be empowered by the government to help disseminate authentic news across all communities. Online and offline/physical visits to local communities should be organized by these bodies to reach out to those without access to online resources. Flyers and banners showing information about COVID-19 in local languages should be circulated in local communities by these groups.
- The government should set up a task force to immediately work on ways to remove all hoaxes and rumors from all online platforms, which will help in solving the problems that arise from information disorder. As long as those messages remain, there is a risk of the public getting access to them and believing in their content. All scientifically proven and well-researched information about COVID-19 should be encouraged and widely spread. The task force should work closely with social media and other online providers to aid in the achievement of this task.
- Laws guiding the kind of information that can be put to the public during this pandemic should be made and strictly adhered to. Perpetrators of any form of misinformation that goes against the law should be brought to justice.
- Healthcare providers should be equipped with recent research findings on COVID-19 containing valid information to help them leverage cordial and more productive patient-provider communication, which could yield better health outcomes.

References

Alvarez, B. (2016). Public libraries in the age of fake news. *Public Libraries* Vol. 55 (6),pp. 24–27.

American Library Association (ALA). (2006, July 14). Core Values of Librarianship. http://www.ala.org/advocacy/infreedom/corevalues (Accessed March 29, 2021).

Brennen, J. S., Simon, F., Howard, P. N., & Nielsen, R. N. (2020). *Types, Sources, and Claims of COVID-19 Misinformation*. Oxford: Reuters Institute for the Study of Journalism.
Cooke, N. A. (2017). Posttruth, truthiness, and alternative facts: Information behavior and critical information consumption for a new age. *Library Quarterly* 87(3), 211–221.
Cox, C. (2020, June 5). *Changed, changed utterly*. https://www.insidehighered.com/views/2020/06/05/academic-libraries-will-change-significant-ways-result-pandemic-opinion (Accessed March 29, 2021).
De Paora, S. & Heravi, B. (2020). Information literacy and fake news: How the field of librarianship can help combat the epidemic of fake news. *The Journal of Academic Librarianship*, (46(5), 102218. https://doi.org/10.1016/j.acalib.2020.102218
Del Rio, C. & Malani P. N. 2020, February 28. COVID-19-new insights on a rapidly changing epidemic. https://jamanetwork.com/journals/jama/fullarticle/2762510 (Accessed May 9, 2021).
Dollinger, A. (2017, March 21). Can librarians save us from fake news? https://www.vice.com/en_us/article/can-librarians-save-us-from-fake-news (Accessed March 29, 2021).
Dube, T.V. (2016). *Information Technology skills and competencies of staff members in the Information Resource Distribution Directorate of the University of South Africa*. Masters. University of South Africa, Pretoria.
Durodolu, O.O (2018). Secondary School Teachers' Perceptions of Information Literacy Skills. Mousaion: South African Journal of Information Studies. Vol.36 (1). https://doi.org/10.25159/0027-2639/3873
Durodolu, O. O., Ibenne, S.K., & Dube, T.V. (2021). New mental costume: leveraging ACRL's value of information frame in dealing with infodemics in a post-truth era. *College & Research Libraries News* 82(4), 175–176.
Durodolu, O. O., Nduka, S. C. and Tinyiko, T.V (2022). Media Literacy and Fake News: Evaluating the Roles and Responsibilities of Radio Stations in Combating Fake News in the COVID-19 Era. International Journal of Information Sciences Management (IJIMS). Vol. 6 (1), pp. 32–49
Durodolu, O.O. & Ibenne, S.K. (2020). The fake news infodemic vs information literacy, *Library Hi Tech News*, 37(7), 13–14. https://doi.org/10.1108/LHTN-03-2020-0020
Fallis, D. (2015). What is disinformation? *Library Trends* Vol. 63 (3), pp. 401–426.
Given, L.M. (2012). Interpretive Research: The SAGE Encyclopedia of Qualitative Research Methods. DOI: https://dx.doi.org/10.4135/9781412963909.n235
Goguadze, N. (2019, November 29). Media education is essential to combat the information disorder. https://pace.coe.int/en/news/7710 (Accessed May 8, 2021).
Hsieh, H. and Shannon, S. E. (2005). Three Approaches to Qualitative Content Analysis. https://doi.org/10.1177/1049732305276687
Ilyin, Y. (2013, October 2). Information hygiene at work. https://www.kaspersky.com/blog/information-hygiene-at-work/14899/ (Accessed March 29, 2021).
International Federation of Library Association. (2017, August 18). IFLA statement on digital literacy. https://www.ifla.org/publications/node/11586 (Accessed March 29, 2021).
International Federation of Library Association. (2017, March 13). How to spot fake news. https://www.ifla.org/publications/node/11174 (Accessed October 19, 2020).
Kim, A., & Dennis, A. R. (2019). Says who? The effects of presentation format and source rating on fake news in social media. Mis quarterly, 43(3), 1025–1039.
Lazer, D., Baum, M., Grinberg, N., Friedland, L., Joseph, K., Hobbs, W., & Mattsson, C. (2017, May 2). *Combatting fake news: An agenda for research and action*. Final report from conference held February 17–18, 2017. https://shorensteincenter.org/combating-fake-newsagenda-for-research/ (Accessed March 29, 2020).

Loftus, S. (2020, April 17). COVID-19: Post-Truth Age – Or Facts Making a Comeback?. https://www.theglobalist.com/coronavirus-pandemic-covid19-media-social-media-misinformation-post-truth-populism/ (Accessed September 16, 2020).

Mackey, T. P. (2019). *Metaliterate Learning for the Post-Truth World*. Neal-Schuman Publishers, ISBN 978-0838917763

McDonald, J. (2020, January 24). Social Media Posts Spread Bogus Coronavirus Conspiracy Theory. https://www.factcheck.org/2020/01/social-media-posts-spread-bogus-coronavirus-conspiracy-theory/ (Accessed September 8, 2020).

Moran, C. (2019, February 8). Using librarians to foster information literacy. https://www.ccdaily.com/2019/02/using-librarians-foster-information-literacy/ (Accessed March 29, 2021).

Morudu, S. A. (2019). *The relationship between the use of Information and Communication Technologies (ICTs) and document delivery service at an Open Distance Learning institution*. Masters. University of South Africa, Pretoria.

Munsanje, M. W. ([n.d]). Weaponisation of misinformation – the role of librarians in curbing the epidemic. https://web.aflia.net/weaponisation-of-misinformation-the-role-of-librarians-in-curing-the-epidemic/ (Accessed May 8, 2021).

Naeem, S, B. & Bhatti, R. (2020). The Covid-19 'infodemic': a new front for information professionals. *Health Information and Libraries Journal*, 37(3), 233–239. https://doi.org/10.1111/hir.12311

Nyhan, B., & Reifler, J. (2010). "When Corrections Fail: The Persistence of Political Misperceptions." *Political Behavior*, 32 (2), 303–330.

Ogenga, F. (2020). Myths, Misconceptions and Politics of COVID 19 in Africa on Social Media Platforms. https://www.frontiersin.org/research-topics/13822/myths-misconceptions-and-politics-of-covid-19-in-africa-on-social-media-platforms#overview (Accessed September 8, 2020).

Schulman, R. (2020, June 15). COVID-19 and the Post-Truth Age: The Role of Facts in Public Policy | Summary of an Online International Conference. https://www.inss.org.il/publication/coronavirus-and-post-truth-seminar/ (Accessed September 8, 2021).

Sullivan, M. C. (2019). Why libraries can't fight fake news. *Journal of Library and Information science*, 51(4), 1146–1156.

Ugwu, C. I & Orsu, E. N. (2017). Challenges of utilisation of online information resources by undergraduate students: implications for information services. *Library Philosophy and Practice*, 1–30.

UNESCO (2020), Media and Information Literacy: Critical-thinking, Creativity, Literacy, Intercultural, Citizenship, Knowledge and Sustainability (MIL CLICKS). https://en.unesco.org/MILCLICKS (Accessed September 19, 2020)

Wardle, C & Derakhshan, H. (2017, September 27). Information Disorder: Toward an interdisciplinary framework for research and policy making. Council of Europe: Strasbourg Cedex. https://rm.coe.int/information-disorder-report-version-august-2018/16808c9c77 (Accessed January 30, 2020).

Warner-Søderholm, G., Bertsch, A., Sawe, E., Lee, D., Wolfe, T., Meyer, J., & Fatilua, U. N. (2018). Who trusts social media? Computers in Human Behavior, 81, 303–315. Doi: https://doi.org/10.1016/j.chb.2017.12.026

West, D. M. (2017, December 18). How to combat fake news and disinformation. https://www.brookings.edu/research/how-to-combat-fake-news-and-disinformation/ (Accessed May 8, 2021).

World Health Organisation (2020). First WHO Infodemiology Conference 30 June – 16 July 2020. https://www.who.int/news-room/events/detail/2020/06/30/default-calendar/1st-who-infodemiology-conference (Accessed September 18, 2020).

MA Shoroma

Contribution of Academic Libraries in Addressing Misinformation Pre and Post Covid-19 Through User Education Programmes

Abstract: In the rapidly evolving landscape of information dissemination, academic libraries play a pivotal role in combating misinformation, particularly in the context of the pre and post-COVID-19 pandemic period. This is shown in this chapter through a mixed-methods approach, combining a comprehensive review of existing literature with insights gathered from librarians to provide a nuanced understanding of the multifaceted challenges posed by misinformation. This study investigates the contribution of academic libraries in mitigating the impact of misinformation through user education programmes, encompassing library instruction and information literacy initiatives.

The findings of the study reveal the roles of librarians, educators, and policymakers in fostering information literacy skills and combating misinformation, as well as contribute to the ongoing dialogue on the vital role they play in empowering users to discern credible information from misinformation in the context of the period of the Covid-19 pandemic and beyond.

Keywords: academic libraries, misinformation, fake news, user education programmes

Introduction

Assisting students and faculty with research and teaching information literacy skills are major responsibilities of librarians. For this reason, academic libraries today continue to act as hubs of information, which are not only sources of relevant and accurate information but are also helpful in curbing misinformation. Academic libraries, in the process of providing information in academia and beyond, for teaching, learning, and research, have encountered their fair share of challenges in providing information whilst managing the changes of collection,

MA Shoroma, University of Limpopo Library, South Africa, e-mail: abigail.shoroma@ul.ac.za, Orcid.org/0000-0002-6946-8096

https://doi.org/10.1515/9783111176345-021

curriculum, evolution of technology, and the booming internet world. Consequently, academic libraries have had to embrace new frontiers and expand their services in providing information that is relevant and accurate. In turn, the development of new technologies has also resulted in rapid and fundamental changes in the way libraries conduct business (Bailin and Grafstein 2005, 317). In light of these changes, librarians have reconfigured the way their time and resources are utilised in order to maximise access and repurpose their mission in providing relevant, accurate, and current information towards literary skills within academic institutions. Information literacy skills involve evaluating information sources in terms of reliability, validity, currency, and bias, as well as knowing when information is necessary and how to use and cite it properly. Information literacy is not only important in academia, but also has practical applications in everyday life, such as making informed decisions and forming opinions based on accurate and relevant information.

With the growth of online resources and media, the scope of information literacy has expanded to include various platforms, including print, broadcast, online, and social media. The role of librarians in teaching information literacy has also expanded significantly, as this skill is now essential across disciplines and platforms. Librarians have also had to come up with ways to reform their services, repackage, and reassemble the resources for their users to ensure the relevance of library and information services in the increasingly complex information environment, as prevailing expectations posit that literacy interventions protect users against any harmful effects of misinformation (Jones-Jang, Mortensen, and Liu 2021). Librarians have also had to adapt to the multimodal training methods for user education programmes as well as commence the offering of online teaching, learning, and assessment through the likes of Blackboard Collaborate interface. Consequently, librarians faced various challenges during the pandemic, including the need to reimagine and reassemble library resources to maximise accessibility.

Statement of the Research problem

Misinformation has become increasingly prevalent in today's society and could negatively impact the quality of research. Throughout the Covid-19 lockdown period and beyond, librarians had to work hard with limited resources to reimagine and reassemble the library resources to maximise accessibility. During this period, librarians encountered various challenges, and were required to show resilience by adapting to multimodal teaching (Mehta and Wang 2020, 354). The purpose of this study is to provide guidance for librarians on how to educate patrons about identifying and avoiding misinformation in their research. Librarians, as said, encountered difficulties during the Covid-19 period in providing adequate distant services to researchers, academics, and students for teaching, learning and research. With members of academic institutions being heavy users of social media, the likelihood of their exposure to misinformation, which circulates easily, is now a matter of concern to all who seek and generate accurate knowledge. Students and faculties must make evaluative judgments about the believability and trustworthiness of materials found on the internet, especially on social media platforms, and decide whether to share such information as accurate information or not. Although initial research has studied academic libraries' ability to identify and address misinformation, their behaviours around identifying, evaluating, and sharing misinformation through user education programs is yet to be elaborately investigated (De Paor, and Heravi 2020). Moreover, the ability of students in accurately evaluating their own skills at identifying misinformation requires more research. Previous research has also documented students' reliance on social media as an information source but is yet to specifically investigate any effect this behavior may have on their ability to accurately evaluate misinformation or fake news (LaGarde and Hudgins 2018). There is therefore an urgent need to increase the interrogation of the contributions of academic libraries in addressing misinformation through user education programs. In order to address these gaps, this study explores the following research questions:

Research Questions

RQ1. How are academic libraries addressing misinformation through their institutions' user education programs?

RQ2. What are the information-seeking behaviors relating to the ability of students in accurately identifying misinformation.

RQ3. What information-seeking behaviors relate to students' ability to accurately recognise and evaluate misinformation?

These research questions were investigated through a data analysis of a collection of literature on addressing misinformation through user education programs.

Methodology

This study employed both the qualitative and quantitative research methods. The researcher utilised a literature review and content analysis to examine user education programs offered by academic libraries. The documents were categorised based on the analysis, which was conducted using a systematic literature review that involved searching for relevant articles using specific keywords. The focus of the study was on academic libraries and their user education programs, also known as information literacy and library instruction. Additionally, the experiences of reference librarians were also considered during this study. To collect data and address the research questions, the researchers conducted a systematic literature review that involved examining scientific publications and shared experiences of librarians.

Literature Review

This section deals with literature findings on this study. A literature review involves the careful selection of available documents related to a particular topic and this section, therefore, presents pertinent literature pertaining to this study in historical and contemporary terms. It features a body of texts, which are reviews of critical points of current knowledge, including substantive findings and contributions to the topic under investigation. Kitchenham (2004) attests that a systematic literature review is a method of identifying, evaluating, and interpreting available research relevant to a specific research question, topic area, or phenomenon of interest and a comprehensive search approach with limited bias. Consulted literature for this study provide information, ideas, data, and evidence from specific perspectives in order to achieve certain aims as well as express various views on the nature of the topic under investigation. The effectiveness of these documents, in relation to the proposed research went through evaluation.

While conducting the literature for this study, a set of selection criteria was used to identify primary studies that offer direct evidence relevant to the research question. It is important to establish these criteria at the outset in order to minimise bias, as noted by Kitchenham (2004, 9). For this particular study, the primary aim was to identify recent research that examines or discusses library practices related to misinformation. As such, only documents that fulfilled this purpose were selected, without any bias related to language, country, type of library, or author perspective. In essence, literature for this study distinguishes what has been done from what needs to be done; synthesising and gaining a new perspective; identifying the relationship between ideas and practice; discovering important variables to relevant topics; relating ideas and theory to application; and identifying methodology and techniques that have been used (Akinola, Ebewo, and Abodunrin 2017). The review in this chapter is expected to bring the reader up to date with current findings on the intersections of academic libraries in relation to misinformation, social media, and user education programs and to form the basis for understanding the prevalence of disinformation, misinformation, and fake news in communities. This is important, as it would be a major basis for this study.

Defining Misinformation

Misinformation has become a pervasive problem in our society, and libraries are stepping up to combat it. Many libraries provide resources and tools to help their patrons develop the critical thinking skills necessary to distinguish between accurate and misleading information, including offering fact-checking services, hosting workshops on media literacy, and providing access to reliable sources of information (Cooke 2017). Additionally, libraries collaborate with community partners to promote media literacy and combat misinformation beyond the library's walls. Furthermore, libraries also work to provide guidelines for evaluating sources and encouraging patrons to be mindful of their own biases.

As trusted institutions which provide access to information, libraries play a crucial role in promoting accurate and reliable sources of information and in equipping individuals with the skills needed to navigate an increasingly complex media landscape (Das and Tripathi 2023). Libraries take a proactive approach to combating misinformation and promoting media literacy in their communities. By doing so, they help to ensure that individuals have the tools and knowledge needed to make informed decisions about the information they consume and share.

Misinformation, colloquially regarded as "fake news," is related to types of incorrect, false, or misleading information. Disinformation is false information deliberately spread to deceive readers or hearers (Lazer et al. 2018). What distinguishes disinformation from misinformation is whether the error and the likelihood of others' deception are purposeful, intended, and spread for economic gain or to purposely mislead, and therefore we are dealing with a "creation" of disinformation (European Commission 2018). On the other hand, if there was no intention to deceive but rather a misinterpretation inherent in the receivers, or if the search took place in untrustworthy sources of information and then the unverified data was shared, incomplete, or inconsistent with the course of events, we are dealing with misinformation (Szpitalak and Polczyk 2019). Implicitly, misinformation has to do with error of omission or commission while disinformation relates with intention to deceive. Misinformation may be unintentionally inaccurate, but disinformation, on the other hand, is always deliberately false and misleading (Jack 2017). Furthermore, popular media has widely used misinformation coated with politicises and contested meanings that presents partisanship along political lines (Caplan, Hanson, and Donovan 2018).

Contemporary literature characterises misinformation as deliberately and demonstrably false news articles, capable of misleading readers. It encompasses problematic content employing misleading, deceptive, or incorrect information presented as authentic news, fabricated articles disseminating falsehoods while maintaining an appearance of credibility, and fabricated information imitating news media content in form but deviating from editorial norms and processes. Additionally, misinformation includes sensational, partisan narratives designed to boost readership and generate revenue (Allcott and Gentzkow 2017; Caplan, Hanson, and Donovan 2018; Howard et al. 2017; Knight Foundation and John 2018; Lazer et al. 2018; Marwick and Lewis 2017). Snopes.com, the well-known fact-checking website, refers to misinformation as fabricated stories intended to fool readers and generate advertising revenue (Mikkelson 2016). For the purpose of this study, fake news is agreed to be news or information that are prepared and disseminated with the sole purpose to deceive readers or viewers into thinking those are real information (Tandoc, Lim, and Ling 2018).

Misinformation Education Guide for librarians

This study seeks to provide guidance for librarians on ways to educate patrons on how to identify and avoid misinformation in their research. This is due to the prevalence of misinformation in today's society, which could impact negatively on the quality of research. To combat misinformation, librarians can play crucial roles by developing and implementing effective strategies that educate patrons about the importance of conducting research with integrity and using reliable sources (Bangani 2021).

The first step in educating patrons about misinformation is to define what it is misinformation and how it differs from accurate information. Misinformation is defined as false or inaccurate information that is spread unintentionally. To effectively educate patrons, it is important to provide practical examples of misinformation. Additionally, it is important to explain the consequences of relying on unsound sources of information and thereby spreading misinformation. Another approach to user education by librarians towards combating misinformation is through information literacy. A few strategies that academic libraries should adopt to address misinformation using information literacy are as follows:

1. Collaborate with Faculty
Librarians should collaborate with faculty to integrate information literacy instruction into course curricula. Embedding information literacy will help ensure that students are equipped with the skills they need to evaluate information critically and recognise misinformation. Librarians in collaboration with faculties can develop assignments that require students to evaluate sources and identify potential biases.

2. Online Learning Materials
Academic libraries have created online learning materials to help students and faculty develop their information literacy skills. This includes instructional videos, tutorials, and interactive quizzes that cover topics such as evaluating sources and recognizing misinformation.

3. Workshops and Training Sessions
Information literacy curriculum is developed in collaboration with research departments to ensure that throughout the research process, students adhere to good research ethics and academic integrity. Academic libraries provide information literacy initiatives encompassing programs such as library orientation, fundamental library and research skills, an introduction to online catalogs, utilization of electronic databases, ethical information access and use, incorporating anti-plagiarism tools, and reference management software and tools. Many libraries have integrated information literacy into course curricula and research programs, aiming to bolster graduate success and contribute to the efforts against misinformation (Committee of Higher Education Libraries of South Africa. (2021).

4. Access to Trusted Sources
Although academic libraries provide access to high quality, peer-reviewed sources, students need to be taught how to use and evaluate these sources. Librarians curate collections of trusted sources and provide guidance on how to find use and evaluate them. This includes, amongst others, evaluating accuracy, relevancy, authority, objectivity, and many other criteria.

5. Promote Critical Thinking
Academic libraries promote critical thinking by encouraging students and faculty to question information and seek out multiple perspectives. This includes using multiple resources when writing assignments to get varied arguments and avoid biasness in resource usage. Librarians also provide guidance on how to analyse sources and recognise potential biases.

Students' information-seeking behaviours in identifying misinformation

Misinformation has become a prevalent issue in today's digital age, and students are not immune to its effects. As such, understanding students' information-seeking behaviors when it comes to identifying and combating misinformation is crucial for helping them to navigate the online landscape. Providing students with the tools and strategies necessary to identify accurate information from misinformation will not only benefit students in their academic endeavours but also help them become informed and responsible digital citizens.

Educational institutions, educators, and researchers hold a crucial role in arming students with the skills and knowledge required to discern reliable sources of information and cultivate a critical mindset while interacting with online content. Therefore, it is imperative to undertake studies and research that explore the information-seeking behaviors of students concerning the identification of misinformation. Through such investigations, we can gain insights into their approaches to information consumption and discernment, pinpoint gaps in their knowledge or skills, and formulate intervention strategies to enhance their effectiveness as consumers and producers of information. Ultimately, by empowering students to actively engage with information and question its validity, we can assist them in evolving into informed and responsible citizens who are better prepared to navigate the intricate and ever-evolving digital landscape.

Required evaluation skills in order to recognise misinformation by students

Developing the ability to recognise and evaluate information sources is a crucial skill for students to navigate the informational landscape. This is achievable through critical thinking skills, providing fact-checking training, and the offering of media literacy education with continuous support through user education training. To address this imperative, universities and colleges should prioritise providing resources that instruct students on how to discern and assess the accuracy of information (Zuo 2022). These resources ought to be designed to educate students on distinguishing between credible and unreliable sources, evaluating the quality of presented evidence, and critically analysing information from diverse perspectives.

Moreover, colleges and universities can enhance these efforts by collaborating with media literacy organisations, offering workshops, or integrating courses on media and digital literacies. This collaborative approach equips students with the tools to assess the credibility of information, enabling them to make informed decisions based on accurate data and ultimately fostering responsible and engaged citizenship (Jacobson and Mackey 2013). It is pivotal for students to cultivate a keen sense of awareness and skepticism towards information sources, allowing them to effectively discern fact from fiction. Lastly, the integration of media literacy into the curriculum contributes to the development of critical thinking skills essential for navigating a rapidly changing and complex information environment.

Limitations of librarians in Fighting Misinformation

Librarians plays a crucial role in promoting information literacy and critical thinking skills among their patrons. However, their efforts are limited in fighting misinformation due to the vast amount of information available on the internet and the inability to fully monitor and control online content. Librarians also face challenges in addressing the entrenched beliefs and biases of some individuals that contribute to the spread of misinformation (Herrero-Diz and López-Rufino 2021). Furthermore, they may face difficulty in reaching broader audiences outside of their institution and the immediate community they serve.

To overcome these limitations, librarians can collaborate with other organisations and individuals to promote reliable sources of information and educate the public on how to identify misinformation (Schroeder 2021). This can involve collaborating with media literacy and education organisations, participating in community outreach programs, and providing resources for digital literacy training to reach a wider audience. It is also essential for librarians to stay up-to-date with the latest developments in technology and media trends, enabling them to respond appropriately and provide relevant information to their patrons.

Recent evidence has also shown that the field of librarianship can help counteract the phenomenon of misinformation (De Paor and Heravi 2020). Librarians can engage in proactive approaches by providing guidance and support to their patrons on how to critically evaluate information sources, fact-check claims, and use reliable resources to obtain accurate information. Revez and Corujo (2021) posit that while the majority of studies have highlighted practices in academic libraries, with a primary focus on information literacy instruction, the ongoing discourse centres on strategies that aim to reinforce an authority-based approach to source evaluation. This contrasts with the challenge of discerning emotional-based reactions to misinformation in a post-truth world.

Keisling (2018, n.p.) attests that "adapting to remote services, librarians have also had to adjust their user education programmes to accommodate online learning." This has included offering online training and workshops, as well as using virtual platforms like Blackboard Collaborate for teaching, learning, and assessment. This shift to online learning has required librarians to develop new skills and approaches to engage and support their users in a digital environment.

Despite the challenges, many librarians have risen to the occasion and found innovative solutions to support their users during the pandemic. Some librarians have developed new online resources and tools to support remote learning and research. Others have partnered with faculty to integrate library resources and services into online courses and assignments (Mehta and Wang 2020).

Misinformation as a global challenge

The issue of misinformation has become increasingly prevalent in recent years, fuelled by the proliferation of social media platforms and the ease with which information could be shared. As such, combating misinformation is a critical challenge that requires the collective efforts of individuals, organisations, and governments across the globe to ensure the integrity and accuracy of information. To address this challenge, strict government laws and legislation are necessary to prevent the spread of misinformation, social media rumours, disinformation, and misinformation (Pate, Gambo, and Ibrahim 2019).

Wang et al. (2019) attest that the proliferation of misinformation has become a pressing concern for individuals and institutions around the world. Detecting the elusive effects of misinformation thus requires a multifaceted approach that involves utilising advanced algorithms to monitor online activity, examining social media trends, analysing patterns of information spreading on social networks, and developing novel approaches for detecting misinformation.

Findings

The librarianship profession has endured and prospered over the years, successfully navigating through various pandemics, including the challenges posed by Covid-19. In academic settings, the conditions have not been excessively harsh; however, the identified issues are complex, with multifaceted forces at play, diverse constituencies to accommodate, various curricula to support, and competing values to reconcile. Despite these challenges, librarians continue to thrive in their profession (Otike et al. 2021).

Leeder (2019) infers that the study of misinformation contributes to the field of Library and Information Science by identifying specific information-seeking behaviors that correlate with students' ability to accurately identify and evaluate misinformation stories but has not identified specific behaviors associated with effective evaluations thereof.

Misinformation serves as a powerful form of disinformation, posing as a legitimate journalistic article and thereby undermining the credibility of journalism. This deceptive tactic exploits the timely nature of content and enables the dissemination of articles on sensitive topics, including politics, through platform newsfeeds (Tandoc, Lim, and Ling 2018). Recognising misinformation is challenging, primarily due to the difficulty in distinguishing intentional deception from the genuine presentation of a controversial perspective (Wang et al. 2019)

Limitations of the Study

Whereas this study's finding is essential, there are a few limitations to it. One limitation is the inability to ascertain the degree to which participants read news stories. This limitation aligns with previous research, which suggests that a significant proportion of the public's interaction with news on social media is restricted to reading the article headlines and not delving into the actual content of the story (Gabielkov 2016). Bastick (2021, n.p.) posits that "the behavioural effects of disinformation are not obvious through current means of content analysis or fact-checking, and such misinformation is consequently difficult to detect." In turn, although the behavioural impacts of misinformation could be empirically observed, refined, and anticipated by malicious content creators, the manipulation processes and their comprehensive consequences may not be fully grasped by the misinformation creators themselves (Lim 2020). This highlights a limitation for librarians, as they may struggle to detect these behavioral effects.

Conclusion

This study sought to establish the contribution of academic libraries in combating misinformation through user education programs, provision of quality and credible information to users, support for research and collection development, creating awareness, and using various tools to evaluate information. Despite the enormous threats and high risks of behavior modification through misinformation, the least librarians and libraries generally can do is to make their clients aware of this pandemic (misinformation).

Librarians can also help patrons identify reliable sources of information by providing them with access to credible databases and online resources. Librarians should provide guidance on how to critically evaluate sources of information, which has been achieved by offering training sessions, workshops, and creating online resources that teach patrons how to evaluate information for its currency, relevance, accuracy, authority and purpose.

Sullivan (2019a, 106) posits that librarians have admirably taken on the challenge, positioning themselves on the frontline to combat misinformation. The solutions presented were primarily derived from the traditional arsenal of information literacy tools. Sullivan (2019b, 1153) additionally asserts that individual responses have served to reaffirm the fundamental values of librarianship and underscore traditional services as effective means to counter misinformation.

This study, like most on this subject, confirmed the attempt, strategies, and efforts by academic libraries and librarians to combat misinformation or misinformation. Through information literacy initiatives such as collaborating with faculty, creating online learning materials, offering workshops and training sessions, providing access to trusted sources, and promoting critical thinking, academic libraries are trying to address misinformation pre and post Covid-19.

References

Akinola, O.C., Ebewo, P. and Abodunrin, O.J., 2017. Christian worship as dramaturgical model: a study of the use of drama in Nigerian churches. Imbizo, 8(1), pp.42–56.

Allcott, H., & Gentzkow, M. 2017. Social media and fake news in the 2016 election. Journal of Economic Perspectives, 31, 211–236.

Bailin, A. and Grafstein, A., 2005. The evolution of academic libraries: the networked environment. The Journal of academic Librarianship, 31(4), pp.317–323.

Bangani, S., 2021. The fake news wave: Academic libraries' battle against misinformation during COVID-19. The Journal of Academic Librarianship, 47(5), p.102390.

Bastick, Z., 2021. Would you notice if fake news changed your behaviour? An experiment on the unconscious effects of disinformation. Computers in human behaviour, 116, p.106633.

Borchers, C. 2017. Fake news' has now lost all meaning. The Washington Post. February 9. Retrieved from https://www.washingtonpost.com/news/the-fix/wp/2017/02/09/fake-news-has-now-lost-all-meaning.

Caplan, R., Hanson, L., & Donovan, J. 2018. Dead reckoning: Navigating content moderation after fake news. Data & Society Research Institute. Retrieved from https://datasociety.net/pubs/oh/DataAndSociety_Dead_Reckoning_2018.pdf.

Committee of Higher Education Libraries of South Africa. 2021. State of South African Academic Libraries Report: Embracing New Frontiers. [Unpublished report]. Pretoria: CHELSA. Available from: https://chelsa.ac.za/wpcontent/uploads/2021/08/StateOfSouthAfricanLibrariesReport.pdf

Cooke, N.A., 2017. Posttruth, truthiness, and alternative facts: Information behavior and critical information consumption for a new age. The library quarterly, 87(3), pp.211–221.

Das, A.K. and Tripathi, M., 2023. Strategies for checking misinformation: An approach from the Global South. IFLA journal, 49(1), pp.3–17.

De Paor, S. and Heravi, B., 2020. Information literacy and fake news: How the field of Librarianship can help combat the epidemic of fake news. The Journal of Academic Librarianship, 46(5), p.102218.

European Commission. 2018. COMUNICAÇÃO DA COMISSÃO AO PARLAMENTO EUROPEU, AO CONSELHO, AO COMITÉ ECONÓMICO E SOCIAL EUROPEU E AO COMITÉ DAS REGIÕES: Combater a desinformação em linha: uma estratégia europeia. Retrieved from: https://secure.ipex.eu/IPEXL-WEB/dossier/document.do?code=COM&year=2018&number=236&appLng=P

Gabielkov, M., Ramachandran, A., Chaintreau, A. and Legout, A., 2016, June. Social clicks: What and who gets read on Twitter?. In Proceedings of the 2016 ACM SIGMETRICS international conference on measurement and modeling of computer science (pp. 179–192).

Herrero-Diz, P. and López-Rufino, C., 2021. Libraries fight disinformation: An analysis of online practices to help users' generations in spotting fake news. Societies, 11(4), p.133.

Howard, P. N., Kollanyi, B., Bradshaw, S., & Neudert, L. M. 2017. Social media, news and political information during the US election: Was polarizing content concentrated in swing states? Oxford Internet Institute. Retrieved June 26, 2019, from: https://comprop.oii.ox.ac.uk/research/working-papers/social-media-news-and-political-informationduring-the-us-election-was-polarizing-content-concentrated-in-swing-states/.

Jack, C., 2017. Lexicon of lies: Terms for problematic information.

Jacobson, T.E. and Mackey, T.P., 2013. Proposing a metaliteracy model to redefine information literacy. Communications in information literacy, 7(2), pp.84–91.

Jones-Jang, S.M., Mortensen, T. and Liu, J., 2021. Does media literacy help identification of fake news? Information literacy helps, but other literacies don't. American behavioral scientist, 65(2), pp.371–388.

Keisling, B., 2018. Blended learning: Scaling library services and instruction to support changing educational landscapes. Library Management.

Kitchenham, B., 2004. Procedures for performing systematic reviews. Keele, UK, Keele University, 33(2004), pp.1–26.

Knight Foundation, J.L. and John, S, 2018. American Views: Trust, Media, and Democracy.

Lazer, D.M., Baum, M.A., Benkler, Y., Berinsky, A.J., Greenhill, K.M., Menczer, F., Metzger, M.J., Nyhan, B., Pennycook, G., Rothschild, D. and Schudson, M., 2018. The science of fake news. Science, 359(6380), pp.1094–1096.

LaGarde, J. and Hudgins, D., 2018. Fact vs. fiction: Teaching critical thinking skills in the age of fake news. International Society for Technology in Education.

Leeder, C., 2019. How college students evaluate and share "fake news" stories. Library & Information Science Research, 41(3), p.100967.

Lim, S., 2020. Academic library guides for tackling fake news: A content analysis. The Journal of Academic Librarianship, 46(5), p.102195.

Mackey, T. P., & Jacobson, T. E. 2011. Reframing information literacy as a metaliteracy. College & Research Libraries, 72(1), 62–78.

Marwick, A.E. and Lewis, R., 2017. Media manipulation and disinformation online.
Mehta, D and Wang. 2020. COVID-19 and digital library service – a case of study of a university library. Digital Library Perspectives. 36(4): 351–363.
Mikkleson, D., 2016. We Have a Bad News Problem, Not a Fake News Problem: Fake news may be a bad thing, but not all bad news reporting is "fake," and that distinction should be kept clear. Retrieved 2023 from: https://www.snopes.com/news/2016/11/17/we-have-a-bad-news-problem-not-a-fake-news-problem/.
Otike, F.W., Bouaamri, A., Barát, A.H. and Kiszl, P., 2021. Emerging Roles of Libraries and Librarians During and Post COVID-19 Pandemic: Challenges and Opportunities. Handbook of Research on Information and Records Management in the Fourth Industrial Revolution, pp.1–16.
Pate, A, U., Gambo, D. and Ibrahim, M, A. 2019. "The Impact of Fake News and the Emerging Post-Truth Political Era on Nigerian Polity: A Review of Literature," Studies in Media and Communication, 7(1), p. 21.
Revez, J. and Corujo, L., 2021. Librarians against fake news: A systematic literature review of library practices (Jan. 2018–Sept. 2020). The journal of academic Librarianship, 47(2), p.102304.
Schroeder, B, S. 2021. "Incoming Undergraduate Students Struggle to Accurately Evaluate Legitimacy of Online News," Evidence Based Library and Information Practice, 16(1),p. 95–97.
Sullivan, M.C., 2019a. Libraries and fake news: What's the problem? What's the plan? Communications in Information Literacy, 13(1), pp.91–113.
Sullivan, M.C., 2019b. Why Librarians can't fight fake news. Journal of Librarianship and information Science, 51(4), pp.1146–1156.
Szpitalak, M., & Polczyk, R. 2019. How to induce resistance to the misinformation effect? Characteristics of positive feedback in the reinforced self-affirmation procedure. Psychology, Crime and Law, 25(7), 771–791. https://doi.org/10.1080/1068316X.2019.1574791
Tandoc Jr, E.C., Lim, Z.W. and Ling, R., 2018. Defining "fake news" A typology of scholarly definitions. Digital journalism, 6(2), pp.137–153.
Wang, Y., McKee, M., Torbica, A. and Stuckler, D., 2019. Systematic literature review on the spread of health-related misinformation on social media. Social science & medicine, 240, p.112552.
Zuo, F. 2022. "College English Teaching Evaluation Model Using Natural Language Processing Technology and Neural Networks," Mobile Information Systems, 2022, p. 1–9.

Oluwole Olumide Durodolu and Onyinye Alexander Osedo

Information Pollution, Threat to Knowledge and Application of Information Hygiene as a Panacea for Combating Information Disorder

Abstract: This study explored the modern predicament of information pollution as a threat to knowledge and the application of information hygiene as a panacea for combating information disorder. The theory of information laundering was used to rationalize the emergence of information and news, fabricated or otherwise, from unsubstantiated sources into a conventional platform. The study was a descriptive survey in which qualitative and quantitative research methods were used, with the quantitative research method used through systematic empirical investigation in which data were obtained from reliable existing data and analyzed to reveal information relevant to this study. The study evaluated the correlation between the growing use of the internet and the world population, assessed the extent of the global use of social media in 2021 based on the available public records, and appraised the extent of reliability and accuracy of the information on the internet. The study affirms the fact that the use of the Internet has improved the quality of education and access to knowledge remarkably and, in the same vein, opened the floodgate to an abundance of information, knowledge, and educational resources, increasing prospects and opportunities for learning beyond the traditional classroom environment by using online resources to prepare lessons and help students to improve their spectrum of learning through and with the use of social media. Information pollution is a risk to knowledge, and consequently practicing good information hygiene can mitigate against misinformation; thus, sanitizing the procedure around information manners can help in preventing deceptive and false information.

Keyword: information pollution, threat to knowledge, information hygiene, misinformation, fake news, information disorder

Oluwole Olumide Durodolu, Department of Information Science, University of South Africa, Pretoria, South Africa, e-mail: woledurodolu@gmail.com, https://orcid.org/0000-0003-2734-8165
Onyinye Alexander Osedo, Micheal Okpara University of Agriculture Library, Umudike, Abia State, Nigeria, e-mail: osedo.onyinye@mouau.edu.ng, https://orcid.org/0000-0003-0344-5710

https://doi.org/10.1515/9783111176345-022

Introduction and Background to the Study

The advent of social media has signalled a turning point in information management, where information is disseminated to the public without going through peer review that is customary for traditional media or academic publication. We are in the era where individuals with unidentified and unverifiable skills propagate potentially unsafe information content via text, audio, or video, usually with the ambition to deceive, mislead, and ambush innocent information seekers and control public opinion. Based on this premise, many have called for social media reform through government policy and self-regulatory mechanisms as a result of the danger it poses to the quality of knowledge in the public domain. Despite this, it should be noted that social media can facilitate a method of overcoming knowledge management obstacles, expedite the acceleration of problem-solving practices by providing essential information, find erudite and knowledgeable people across the world, harness mutual wisdom, and build collaboration (Ward et al. 2016).

In recent times, the explosive development of the Internet and related information and communication technologies, combined with society's deeper incursion into the electronic world and the new media environment, has changed the way information is generated, curated, disseminated, consumed, and assimilated, bringing into attention the glitches of information surplus and the developments in society which allow for the spread of fake news and misinformation. Whenever fake news or misinformation is discussed, two recent events readily come to mind: the 2016 US presidential election (Rose-Wiles 2018; Sullivan 2019) and the outbreak of the Coronavirus in Wuhan, South China in November, 2019 later code-named COVID-19 by the Worl Health Organisation (WHO) (Giwa and Desai 2020). Nowadays society is highly exposed to information at a devastating proportion and volume, which has impacted so much on how information is interpreted, shared, and utilized by its members. Gallagher (2016) is of the view that as the online environment becomes more complex and harder to navigate, doubt and uncertainty continues to grow, with people less likely to value reliable facts. These developments can be said to have brought about the pollution of the information landscape.

The phenomenon of fake news is not new; it has always been present, albeit taking alternative forms. Bhaskaran, Mishra, and Nair (2017) stated that as far back as "1835, the *New York Sun* published a 6-part series claiming there was life on the Moon." Nine years later, in 1844, some newspapers in Philadelphia published false reports about Irishmen stealing bibles from public schools, leading to riots (Soll 2017). Elsewhere, Bhaskaran, Mishra, and Nair (2017, n.p.) assert that "Joseph Pulitzer's the *New York World* and William Hearst's the *New York Herald* may be considered as the first proponents of 'Fake News' publishing highly dramatised and sensationalist versions of events in their newspapers to increase revenue." The only difference in today's practice of fake news is the rise the internet and different social media stages that allow it to spread at much faster rates. Nelson and Taneja (2018, in Egelhofer and Lecheler 2019) stated that the so-called "fake news" crisis had been one of the most deliberated topics in both public and scientific discourses since the 2016 U.S. presidential campaign. Egelhofer and Lecheler (2019) asserted that the term fake news was originally applied to political satire, but now seems to be applied to all things "inaccurate," and is even applied in contexts that are completely unrelated to mediated communication.

All information is now viewed with scepticism because, according to De Paor and Heravi (2020), as the digital world continues to expand with its attendant large quantities of free data, people have responded by sifting information to a form that is not only more comprehensible but also applicable and consistent with their worldview. In turn, Batchelor (2017) asserts that where an environment is full of erroneous information, the value of critical thinking skills is more pronounced than ever. In essence, what fake news represents, nevertheless, is something greater than the term itself: a fundamental shift in political and public attitudes towards information and news, as well as how facts and information may be obtained in an electronic world. So much feasting of information powered by cyberspace has produced a "post-truth" society in which people feed on information that reaffirms their pre-existing views and thoughts as opposed to endeavouring to undertake the difficult task of finding the reality. There is no doubt that fake news, also known as misinformation, disinformation, false information, misleading information, and inaccurate information, is influencing how people react to societal issues, which in turn is having an adverse effect on society. In the opinion of Lewandowsky, Ecker, and Cook (2019), the post-truth world arose due to super public inclinations such as a weakening in social capital, growing economic disparities, increased divergence, declining confidence in science, and an increasingly fractionated media landscape.

The pollution of the information arena with false or fake news, disinformation or misinformation, inaccurate and/or distorted information poses a significant threat to the acquisition of knowledge. Fake news leads to lesser levels of confidence in media and less correct proof of real news; consequently, regular conversations on fake news affect the way individuals trust news media and the standards with which information is evaluated, which in turn may also prompt the spreading of false information by individuals without background checks and care.

Thorson (2016) detailed that misinformation which is hard to correct may have a resounding effect over time if it goes uncorrected and may follow philosophical lines (Weeks 2015). According to Lewandowsky, Ecker, and Cook (2019), terms such as "post-truth" and "fake news," which were largely unknown until 2016, have now exploded into media and public discourse, especially among the elitist class. Elsewhere, Allcott and Gentzkow (2016, n.p.) term fake news is seen as "news articles that are deliberately, and verifiably incorrect information, given out to mislead receivers." The upsurge in fake news has thus unsurprisingly expanded the method of misinformation and led to a new line of misinformation research (Van Duyn and Collier 2018). Lewandowsky, Ecker, and Cook (2019) state that in today's information world, the power of control lies with those who are most vocal and influential on social media, from celebrities and big corporations to botnet puppeteers who can mobilize millions of tweet bots or sock puppets – that is, fake online personalities through which a small group of operatives can create an illusion of a widespread opinion (Bu, Xia, and Wang, 2013; Lewandowsky 2011).

In this information world, professionals are ridiculed as deceitful or discriminatory each time their informed facts threaten the control of the well-financed and influential or the preconceptions of the ignorant. Unsurprisingly, the prevalence of misinformation, disinformation, fake news, and false news led to the term "post-truth" being announced as the world of the year in 2016 by the Oxford Dictionary (Flood 2016). This led on from the World Economic Forum – a not-for-profit institution "committed to improving the state of the world" – earlier warning that, due to the speed of growth of misinformation, the spread of misinformation would be one of the ten most significant problems facing the world in 2013 (WEF 2013).

The information landscape, and society at large, has undoubtedly been dangerously polluted by fake news merchants threatening knowledge acquisition. It is important to observe that fake news, false information, misinformation, disinformation, inaccurate information, or whichever term that describes deliberate misleading information, causes fear and anxiety that can result in distressing effects on the mental health of people and have an adverse psychological and social impact. It is equally necessary to note that providing accurate and timely information can result alleviate such impacts, with the clearance of elusiveness and relief of apprehension. Kuklinski et al. (2000, in Lewandowsky, Ecker, and Cook 2019) believe the maxim that an effective democracy depends on a well-informed public. In the same vein, if people are persistently misinformed, it is likely that societal conclusions will be suboptimal. Correspondingly, if an individual is misinformed, that person's decisions may not be in their best interest and can have adverse consequences (Lewandowsky, Ecker, and Cook 2019).

Durodolu and Ocholla (2017) asserted that knowledge of information literacy is important in evaluating information, likening the knowledge of information literacy to a compass; just as this instrument is useful to find the right direction for travel, knowledge of information literacy is a prerequisite to navigate the overwhelming world of information. Since information pollution is a threat to knowledge, practicing good information hygiene can reverse the negative trend of misinformation and sanitizing the process around information conducts can assist in thwarting deceptive and false information content, some of which is harmful to knowledge acquisition.

This chapter generally aims at identifying ways of applying information hygiene as a panacea for combating information disorder. It will specifically discuss the havoc fake news has wreaked in the information hemisphere and society at large, how fake news gains popularity, the motive behind spreading fake news, how to identify fake news, the advanced means of establishing trust in traditional information merchants such as conventional media houses, and how information disorder can be controlled. The motives behind conspiracy theories is also included.

Literature Review

There is growing literature on fake news, misinformation, disinformation, inaccurate information, false news, deliberate falsehood, and related areas. Mustafaraj and Metaxas (2017), in their definition, see fake news as online falsehoods configured and disseminated in a way that allows them to give the impression of authenticity and legitimacy to the readers. Jeffries et al. (2017) define fake news as a term that includes falsehoods as well as stories with no iota of truth that are misrepresented. Equally, fake news, defined as news that intentionally presents misinformation with the intent to deceive the audience (Bakir and McStay 2018; Horne and Adali 2017; Kumar and Shah 2018), also known as disinformation (Hannah, McCarthy, and Kietzmann 2015), has severe consequences for societies. Accordingly, Van Duyn and Collier (2018) stated that fake news had expanded the method of misinformation, and a new line of misinformation research has emerged. Therefore, any information which is intentionally or unintentionally created and shared to mislead or misinform people about a given situation is fake news. The concept of fake news is said to have originated from the field of psychology but has, in recent times, gained widespread attention and become a dominant theme among scholars in different disciplines, including political communication, health communication, and cognitive psychology (Waisbord 2018). Oftentimes, the term is used synonymously with misinformation, disinformation, and false information.

Dewey (2016) stated that fake news is not new, has been existing in one form or another for eras, and has assumed prevalent dimensions now more than ever.

Indeed, the creation and impact of fake news have risen to new highs, driven principally by the growing digitization of information and the explosion of social media (Baccarella et al. 2018; Berthon and Pitt 2018), consequently driving the interest of scholars in learning more about it within the past few years, with a large number of the existing fake news studies conceptual and qualitative (Jin et al. 2016; Mustafaraj and Metaxas 2017; Shu et al. 2017). Today, the terms "post-truth" and "post-facts," which were literarily not known a few years ago, have become commonplace in the media arena with thousands of topical allusions (Lewandowsky, Ecker, and Cook 2019). Misinformation or fake news misleads, with an inconceivable adverse effect on individuals and society. There are, nevertheless, other more deceptive and debatably more threatening elements of misinformation.

The eruption of the Coronavirus in Wuhan, South China, in November 2019, later code-named COVID-19 by the World Health Organisation (WHO) (Giwa and Desai 2020), provides a perfect example of the devastating effect fake news and misinformation, especially when laced with conspiracy theory, can have on the mental health of people, as well as the negative psychological and social impacts due to the distress and apprehension that went along the pandemic. From the inception of the outbreak, much fake news, misinformation, and disinformation replete with myths and fallacies flooded both traditional and new media, even when it was unclear what the specific cause(s) of the virus was. Prominent among such fake news and misinformation was that the virus does not and cannot survive in hot weather, that taking a high dose of chloroquine medication can protect against its infection, and that consuming a large amount of ginger and garlic can prevent the virus (WHO 2020). Other misinformation suggested that Covid-19 is a bio-weapon allegedly created by China, ostensibly to decimate the United States and assume the position of the world's leading power; that Covid-19 is God's punishment for the sins of the world; that it was created as a population control mechanism; and that it was caused by the fifth generation (5G) network. This situation of an excessive amount of misinformation about the problem made it extremely difficult to identify a solution, which led to it being described by the World Health Organisation (WHO 2020) as an infodemic.

There is also the indication that fake news and misinformation causes people to stop trusting in facts (Lewandowsky, Ecker, and Cook 2019). The sinister consequences from fake news and misinformation are predominantly noticeable when a half-truth is parcelled as a conspiracy theory. As defined by van Prooijen and van Vugt (2018), Swami et al. (2014), Simelela et al. (2015), and Sallam et al. (2020), conspiracy beliefs are unconfirmed and doubtful beliefs that involve the role of a malevolent force in scheming major events, when other explanations are more probable, with these authors stressing that such beliefs can have destructive health and social effects which are long-lasting, originating in the past and continuing to exist in the present. This striking case of the HIV epidemic in South Africa is a notable example of the effect of a conspiracy theory, which resulted in governmental policies with devastating effects on public health, as well as vaccination conspiracy theories (Benecke and De Young 2019), with sinister outcomes manifested by the re-occurrence of outbreaks of infectious diseases including measles, mumps and rubella (Jolley and Douglas 2014; Sallam et al. 2020). The mere exposure to conspiratorial discourse, even if the conspiratorial claims are dismissed, makes people less likely to accept official information (Lewandowsky, Ecker, and Cook 2019). Conspiracy theories often thrive during times of crisis, particularly in economic and political emergencies.

Irrespective of the swift rise in the number of publications regarding COVID-19, many characteristics of the disease have yet to be identified and/or explained, which has led to a torrent of fake news and misinformation about the virus and the disease. However, Sallam et al. (2020) detailed how these characteristics of the disease encompass the source of the virus and availability of a specific antiviral treatment and effective vaccine, including an enquiry into the correctness of the developed diagnostic modalities. Providing correct and appropriate information can result in the removal of ambiguity and apprehension.

Fake news is created around and by people themselves, with its dissemination dependent on people. In contemporary society, social media, which allows people to share information without proper verification, is one of the foremost sources of fake news. Mutsvairo and Bebawi (2019) specifically observed that in today's information-rich society, the main sources by which misinformation and fake news spread faster include but are not limited to social media such as Facebook and instant messaging such as WhatsApp. Regardless of whether fake news is shared intentionally or not, the information being disseminated is based on incorrect information; thus, opinions formed based on such information are not rooted in empirical fact, even though they are shared as such.

At the center of this is what motivates the continued sharing of fake news by individuals, irrespective of its destructive effects. Literature has not, to date, been able to explain the growing abundance of misinformation. What is the attraction to fake news and misinformation? Why do people believe in fake news and misinformation? If the information disorder is amended, do individuals update their views, and to what degree are individuals disturbed whether or not a piece of information is true? Talwar et al. (2019) suggest that prior literature lacks theoretical frameworks on understanding fake news sharing behaviour (knowingly or unknowingly) that takes into account the very nature of social media, though they acknowledged that there exists an individual inadequate understanding of the relationship between the use of social media and the spread of fake news, thus implying that there is a gap with regards to literature relating to the social characteristics of spreading false news by social media users. Equally, according to Di Domenico and Visentin (2020, n.p.), "the scientific literature still lacks in providing a convincing explanation of the determinants of creating and sharing problematic contents on social media."

Misinformation misleads, with a possible negative effect on personalities and humanity, suggesting that there are other treacherous and arguably more perilous rudiments of misinformation. Misinformation removes actualities (Lewandowsky, Ecker, and Cook 2019), and the widespread sharing of such has made it further challenging for many individuals to determine what a worthwhile and dependable piece of information is, and vice versa. This predisposition (of not being able to determine) makes it even more difficult for researchers to produce effective essays and research papers. Misinformation, therefore, is not just about being misinformed; it is also about the overall intellectual well-being of a society (Lewandowsky, Ecker, and Cook 2019) because misinformation is difficult to amend, with a bounce-back effect over time if it goes uncorrected. Lewandowsky, Ecker, and Cook (2019) went further, stating the deceptive effects of misinformation are mostly noticeable when misinformation is presented as a conspiracy theory, hence the sheer experience of conspirative addressing: even if the conspirative claims are dismissed, this leaves people less prone to accept official information.

A conspiracy theory can be seen as uncorroborated and improbable opinions that contain the role of a malicious power plotting major events when other clarifications are more credible (Sallam et al. 2020). As stated by van Prooijen and van Vugt (2018) and Swami et al. (2014), conspiracy beliefs can have adverse well-being and communal effects, which have been seen in the past and continue to exist to this day. The prominent illustration concerning the outcome of conspiracy theories is the HIV epidemic in South Africa, with the belief in conspiracy occasioned in governmental policies with disturbing effects on public health (Simelela et al. 2015). Elsewhere, Jolley and Douglas (2014) highlighted an instance regarding vaccination conspiracy theories, with evil outcomes manifested by the re-occurrence of outbreaks of infectious diseases including measles, mumps, and rubella (Benecke and DeYoung 2019). Conspiracy theories about vaccines, palm oil, and Coronavirus are the current examples of how they can cause harm and have a negative impact on consumers, companies, and democracy at large (Di Domenico and Visentin 2020). Conspiracy theories, as seen above, are mostly rampant during crises, such as those economic and political.

Fake news and misinformation epitomize only a facet of the current predicament of awkward information (Di Domenico and Visentin 2020), that is, "inaccurate, misleading, inappropriately attributed, or altogether fabricated information" (Jack 2017). Awkward information also includes hoaxes, conspiracy theories, propaganda, and true professional information offered in inaccurate ways in the sustenance of one's viewpoint (Di Domenico and Visentin 2020). Allcott and Gentzkow (2017) noted that two significant and unforeseen political consequences stimulated the attention on the likely effect of fake news and misinformation after 2016: the U.S. Presidential Elections and the Brexit referendum. The major propeller of such information sharing on social media is confirmation bias, namely individuals' inclination to only select information consistent with their idea of the world (Kim and Dennis 2019). Therefore, individuals are likely to continue sharing awkward information despite knowing that such information is false because such individuals "care more about the beliefs being promoted than the validity of the information" (Newhoff 2018). As fake news distribution is an undesirable characteristic of social media, it should be expected that there is a relationship with other undesirable features associated with social media. These unpleasant social media characteristics that negatively impact users (in areas such as performance, productivity, wellbeing) have frequently been seen as the murky aspects of social media (Talwar et al. 2019). Talwar et al. (2019) added to this, stating that the undesirable consequence characteristics of social media are compulsive social media use, social media fatigue, fear of missing out, and social comparison.

According to Zhang and Ghorbani (2019), information technology (IT) and computer science research have proposed new methods to automatically spot misinformation on social media. Spotting fake news and misinformation disseminated via social media early in its circulation before it spreads to a wider audience is extremely necessary and communally advantageous.

Significance/contribution of the study

This study is imperative as a contribution to the general body of knowledge in many areas but most significantly this research will make a contribution to the literature by evaluating the previous research in order to bring about new knowledge to resolving a trending problem and challenges in the area of information pollution, a threat to knowledge, and how to inject orderliness into the world of misinformation. This study has made a recommendation that will deepen the study of information science, information literacy, and how to institute information hygiene in the world that is bedeviled with the crisis of fake news. The research will make a noteworthy impact to knowledge by suggesting various areas of intervention leading to policy changes without the rigor of bureaucratic procedure.

Aim of the study

The study sought to broadly evaluate information pollution, a threat to knowledge, and the application of information hygiene as a panacea for combating information disorder with the following research questions:

Research questions

1. What is the correlation between the growing use of the internet and the world population?
2. What is the extent of the global use of social media in 2021?
3. What is the extent of reliability and accuracy of information of the internet?

Theoretical framework: Information laundering

Information laundering is the emergence of information and news, fabricated or otherwise, from unsubstantiated sources into a conventional platform. Information laundering is when information is disguised to make it appear legitimate, occurring when respectable media organizations report news on something published on a blog or platform with doubtful reliability. In presenting this news, its original source may be ignored, this is a way to lauder the information, whether or not its source was reliable or gains believability, particularly if it is used by news media reputable for high standards (Merelli 2016). Information laundering is an unethical method of manipulating information in an effort to disguise intention and make the information appear legitimate; the intention is to manipulate fact and distort reality by transiting the information in a way that is channeled toward deception for political gain, economic advantage, or social confusion.

Methodology

This study was a descriptive survey in which qualitative and quantitative research methods were used. The quantitative research method was used through a systematic empirical investigation in which existing data was obtained from reliable existing data and analyzed to reveal information relevant to this study. Quantitative research is the technique of gathering and scrutinizing numerical data, with this method often used in finding patterns and averages, making predictions and validating judgment, testing causative relationships, and generalizing or make an informed conclusion from results obtained from wider populations (Bhandari 2020). Qualitative research, includes observations, textual or visual analysis, or interviews (individual or focus group). In this case, the qualitative research method was used by evaluating the extant literature to reveal new knowledge.

Discussion

The discussion of this study focuses on the research questions emphasized at the beginning of this paper.

WORLD INTERNET USAGE AND POPULATION STATISTICS 2021 Year-Q1 Estimates						
World Regions	Population (2021 Est.)	Population % of World	Internet Users 31 Dec 2020	Penetration Rate (% Pop.)	Growth 2000-2021	Internet World %
Africa	1,373,486,514	17.4 %	590,296,163	43.0 %	12,975 %	11.7 %
Asia	4,327,333,821	54.9 %	2,707,088,121	62.6 %	2,268 %	53.6 %
Europe	835,817,917	10.6 %	728,321,919	87.1 %	593 %	14.4 %
Latin America / Caribbean	659,743,522	8.4 %	477,869,138	72.4 %	2,544 %	9.4 %
Middle East	265,587,661	3.4 %	188,132,198	70.8 %	5,627 %	3.7 %
North America	370,322,393	4.7 %	332,919,495	89.9 %	208 %	6.6 %
Oceania / Australia	43,473,756	0.6 %	29,284,688	67.4 %	284 %	0.6 %
WORLD TOTAL	7,875,765,584	100.0 %	5,053,911,722	64.2 %	1,300 %	100.0 %

Figure 1: Internet Usage and World Population Statistics estimates (March 31, 2021).

Figure 1 above shows the correlation between the growing use of the internet and the world population. The data shows the world population as of the first quarter of 2021 to have been approximately 7.9 billion with the corresponding estimate of Internet users above five billion representing an internet penetration rate of 64.2% and a significant growth rate of 1300% between 2000 and the first quarter of 2021. In relation to the population of each continent, the available data shows that North America had the highest internet penetration rate globally with a considerable increase of 89.9 (208%), followed by Europe with 87.1 (592%), Latin America and Caribbean with 72.4 (2,544%), Middle East with 70.8 (5,627%), Oceania/Australia with 67.4 (284%), Asia with 62.6 (2,268%) and Africa with 43% (12,975%). The use of the internet has improved the quality of education and access to knowledge remarkably. It has opened the floodgate to an abundance of information, knowledge, and educational resources, increasing prospects and opportunities for learning beyond the traditional classroom environment by using online resources to prepare lessons and help students to improve their spectrum of learning (Pursel 2020). However, irrespective of the indisputable benefits of the internet, this same technology has become a vehicle for misinformation. Narrations are twisted, quotes are misrepresented and perverted, and facts are skewed. Misinformation is unsettling politics, economy and health, which is particularly serious in the advent of Covid-19 where accurate information is a matter of life and death. Misinformation is also posing a severe challenge in tackling the likes of environmental problems and racial injustice. This difficulty is a threat to knowledge and, as a result, humanity cannot postpone critical intervention in addressing the issue of misinformation spreading worldwide. Imprecise information is intended to hijack human emotions, exploit prejudices, and manipulate sentiment. Knowingly or unknowingly, everybody can be a victim, especially if the information is from familiar sources. The solution to the proliferation of misinformation is for an individual to pause and ponder before sharing information, which will trigger a moment of critical thinking. Such is asserted by Durodolu, Ibenne, and Dube (2021), Durodolu and Ibenne (2020) and Durodolu, Chisita, and Dube (2021).

Figure 2: Global social media research summary 2021 (Chaffey 2021).

The illustration presents global social media usage statistics as of January 2021, highlighting key trends in adoption and growth. There are 4.2 billion active social media users, accounting for 53.6% of the global population. Social media saw a

13.2% annual growth, adding 490 million new users. Mobile access dominates, with 4.15 billion users accessing platforms via smartphones, representing 98.8% of total users. The data underscores the widespread use of social media, rapid user growth, and the importance of mobile-first strategies. The source of the infographic is We Are Social and Hootsuite.

Figure 2 above shows social media use all around the world in 2021, with the data revealing that there were 4.2 billion active social media users in the world at this time which represented 53.6% of global population and an approximately +490 Million (+13.2%) annual increase in population of users of social media globally. Based on the available data, 4.15 billion people have access to social media using a mobile phone which represents 98.9% of social media users in the world. The positive dimension to the use of social media can be identified in the dissemination of information across the globe. Platforms like Facebook, Twitter, WebChat, WhatsApp etc. have made access to information stress-free at a click of a button. Without this opportunity the world would have been remarkably slower, but notwithstanding the revolution in information access and interconnectivity promoted by social media, it has also come with considerable harm. Social media do assist in overcoming the barrier associated with knowledge management by facilitating the easing of knowledge problems through providing needed information, tracing knowledgeable and well-informed people, providing a virtual context, harnessing mutual wisdom, and building trust (Alzakhravi 2020). They unarguably have many positive implications for education and knowledge acquisition by providing timely information, communication, learning, improving skills, promoting career development. However, the same platforms have negative effects which include identity theft, cyber bullying, cyber propaganda, fake news, misinformation, and social isolation, some of the negative consequences can impair learning, discourage scholarship, and pose a major threat to knowledge. Other positive aspects of social media and internet technology is that they encourage online learning and enhance academic performance through collaboration. On the other side of the divide, they can also lead to distraction, reduce learning and research capability, and have a negative impact on health because of ergonometric situation based on machine-human interaction.

Figure 3 above shows the level of reliability and accuracy of information on the internet. The data reveals that many of the online information seekers considered established news websites more reliable compared to those who trust personal weblogs. Moyo, Mare, and Mabweazara (2020) asserted that fake news is usually seen from reportage in mainstream news outlets, while limited attention has been channeled to the role of mainstream media as a source of dissemination of disinformation, deception, half-truth and misinformation. Mavridis (2018) agrees that the subject of fake news and its negative impact has become conspicu-

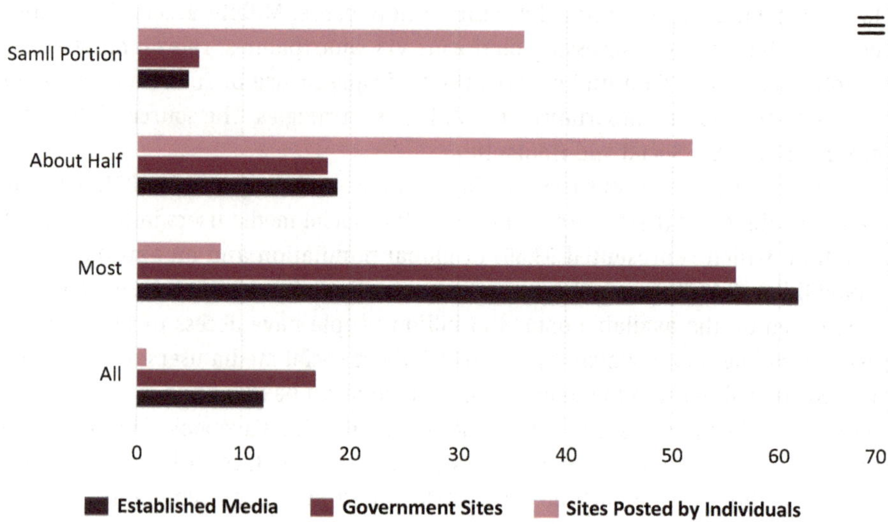

Figure 3: Reliable and accuracy of information on the internet? (Pew Research Center 2021).

ous in recent years. Notwithstanding the fact that the problem of fake news is not a new occurrence, advances in technological and popularity of social media have fashioned a luxuriant ground for the fake news to proliferate and flourish. Social media platforms, like Facebook, WhatsApp, WebChat, Twitter or YouTube, are fertile ground for the breeding and circulation of fake news.

Conclusion

Information pollution is a contemporary issue that affects a vast number of people in a variety of ways, ranging from politics to health, and has always existed, but due to the growing number of users of social media and instant messengers, the volume of information has tremendously increased. Individuals and businesses suffer direct damages as a result of information pollution, as it may include defamation of people, products, and businesses. As long as there has been information, there has also been information pollution. Although there is little that can be done to prevent information from being polluted, it is our view that efforts can be made to assist people in recognizing when information is polluted and being able to apply necessary measures to combat information disorder. For instance, librarians can employ creativity as one of their most effective weapons in this attempt. Identifying, developing, and disseminating memes and infographics that resonate with library patrons on the same level as fake news can help to turn the tide and sanitize the information space. Visual information can have a significant influence on library user, and it may stick with them longer than verbal or written instructions. Another hygienic measure that can be applied to address information pollution involves technology solutions that incorporate psychological concepts, an interdisciplinary approach called "technocognition" by Lewandowsky, Ecker and, Cook (2019). Critical ethical considerations that govern the pursuit of research topics should also be included since many people assume they are good at judging information, especially information received from the internet when the truth says otherwise.

Recommendations

As there is a link between worldview and susceptibility to information pollution as well as resistance to its correction, worldview should be separated from the proclivity to accept disinformation and reject correction. Therefore, it is hereby proposed that:

- Audiences with a wide range of personal viewpoints should be provided with the best possible avenues of communication,
- Information hygiene should not be implemented through mandates, censorship, or other undemocratic measures,
- Better information architectures must be created through completely transparent and democratic processes that are more conducive to the circulation of useful information rather than information chaos,
- Potential ways to break and filter bubbles associated with information pollution, as well as to educate individuals about information discernment, are also needed and must be politically neutral,
- To restore a more evidence-based information hygiene environment, cognitive science and applied research on memory and cognition, in combination with internet technology, will play a critical role,
- To attain the necessary information orderliness, it is critical to emphasize the relevance of specificity while evaluating information sources, whether offline or online, and lastly,
- Made-up stories may decrease as tools to sanitize the information arena develop and individuals become more conscious of the hazard of information pollution to knowledge acquisition and society at large, leading to the use of information hygiene as a panacea for addressing information disorder.

References

Allcott, H., & Gentzkow, M. (2016). Social media and fake news in the 2016 election. Retrieved from: https://web.stanford.edu/~gentzkow/research/fakenews.pdf

Alzakhravi, O (2020). What Is the Major Impact of Social Media? Retrieved 18/06/2021, from https://www.linkedin.com/pulse/what-major-impact-social-media-olena-alzakhravi?trk=related

Baccarella, C., Wagner, T., Kietzmann, J., McCarthy, I., (2018). SM? It's serious! Understanding the dark side of SM. *Eur. Manag. J.* 36, 431–438. https://doi.org/10.1016/j.emj.2018.07.002.

Bakir, V., & McStay, A. (2018). Fake news and the economy of emotions. Digital Journalism, 6(2), 154–175.

Batchelor, Oliver (2017) Getting Out the Truth: The Role of Libraries in the Fight against Fake News, *Reference Services Review*, (45)2. http://dx.doi.org/10.1108/RSR-03-2017-0006

Benecke, O. & DeYoung, S. E. (2019) Anti-Vaccine Decision-Making and Measles Resurgence in the United States. *Glob. Pediatric Health*, 6.

Berthon, P. R. & Pitt, L. F. (2018), "Brands, truthiness and post-fact: managing brands in a post-rational world", *Journal of Macromarketing*, Vol. 38(2), 218–227.

Bhandari, P. (2020). An introduction to quantitative research. Retrieved at 15/06/2021, from https://www.scribbr.com/methodology/quantitative-research/

Bhaskaran, H., Mishra, H., & Nair, P. (2017). Contextualising fake news in post-truth era: Journalism education in India. *Asia Pacific Media Educator*, 27(1), 41–50.

Bu, Z., Xia, Z., & Wang, J. (2013). A sock puppet detection algorithm on virtual spaces. Knowledge-Based Systems, 37, 366–377. doi:http://dx.doi.org/10.1016/j.knosys.2012.08.016

Chaffey, D. (2021). Global social media research summary 2021. Retrieved 17/06/2021, from https://www.smartinsights.com/social-media-marketing/social-media-strategy/new-global-social-media-research/

De Paor, S. & Heravi, B. (2020). Information literacy and fake news: How the field of librarianship can help combat the epidemic of fake news. *The Journal of Academic Librarianship* 46 (2020) 102218. https://doi.org/10.1016/j.acalib.2020.102218

Dewey, C., (2016). Facebook fake-news writer: 'I think Donald Trump is in the White House because of me.' the Washington Post. Available at: https://www.washingtonpost.com/news/the-intersect/wp/2016/11/17/facebook-fakenews-writer-i-think-donaldtrump-is-in-the-white-house-because-of-me/?tid=sm_tw.

Di Domenico, G., & Visentin, M. (2020). Fake news or true lies? Reflections about problematic contents in marketing. *International Journal of Market Research*, 62(4), 409–417.

Durodolu, O.O, Ibenne, S.K and Dube, TV (2021). New mental costumes: Leveraging ACRL's value of information frame in dealing with infodemics in a post-truth era. College & Research Libraries News (College & Research Libraries News). Vol. 82 (4). Pp. 175

Durodolu, O.O. and Ibenne, S.K. (2020). The fake news infodemic vs information literacy, Library Hi Tech News, Vol. 37 No. 7, pp. 13–14. https://doi.org/10.1108/LHTN-03-2020-0020

Durodolu, O.O, Chisita, C. T, and Dube, TV (2021). Flattening the curve of Fake News in the epoch of Infodemic: an epistemic challenge. Edited by Rebecca Blankenship. In Handbook of Research on Deep Fakes, Fake News, and Misinformation in Online Teaching and Learning Technologies. (pp. 1–300). IGI Global. ISBN13: 9781799864745.

Durodolu, O. O & Ocholla, D. O (2017). Search Strategy, Self-Concept and Metacognitive Skills of Secondary School Teachers in Selected Cities in Nigeria and South Africa. LIBRI: International

Journal of Library and Information Science. Vol. 67, Issue 2. from doi:10.1515/libri-2016-0104. https://www.igi-global.com/gateway/book/316130

Egelhofer, J. L. & Lecheler, S. (2019). Fake news as a two-dimensional phenomenon: a framework and research agenda, *Annals of the International Communication Association*, 43(2), 97–116, DOI: 10.1080/23808985.2019.1602782

Flood, A. (2016). 'post-truth' named word of the year by Oxford Dictionaries. Retrieved from https://www.theguardian.com/books/2016/nov/15/post-truth-named-word-of-the-year-by-oxford-dictionaries

Gallagher, B. (2016). What impact does fake news have on the real world? Retrieved from https://merryjane.com/culture/fake-news-internetimpact.

Gill, P., Stewart, K., Treasure, E. et al. Methods of data collection in qualitative research: interviews and focus groups. Br Dent J 204, pp. 291–295 (2008). https://doi.org/10.1038/bdj.2008.192

Giwa, A. L. & Desai, A. (2020). Novel Coronavirus COVID-19: An Overview for Emergency Clinicians. *Emergency Medicine Practice EXTRA*. http://www.ebmedicine.net

Hannah, D. R., McCarthy, I. P. and Kietzmann, J. (2015), "We're leaking, and everything's fine: how and why companies deliberately leak secrets", *Business Horizons*, 58(6), 659–667.

Horne, B. D., & Adalı, S. (2017). This just in: Fake news packs a lot in title, uses simpler, repetitive content in text body, more similar to satire than real news. Retrieved from https://arxiv.org/abs/1703.09398

Jack, C. (2017). *Lexicon of lies: Terms for problematic information*. Data and Society.

Jeffries, S., Kroondyk, J., Paolini, F., & Radisauskas, C. (2017). Says who? Librarians tackle fake news. *College & Research Libraries News*, 78(10), 538. DOI:10.5860/crln.78.10.538

Jin, Z., Cao, J., Zhang, Y., & Luo, J., (2016). News verification by exploiting conflicting social viewpoints in microblogs. In: AAAI. Proceeding AAAI'16 Proceedings of the Thirtieth AAAI Conference on Artificial Intelligence Pages 2972–2978 Phoenix, Arizona

Jolley, D. & Douglas, K. M. (2014**)** The effects of anti-vaccine conspiracy theories on vaccination intentions. *PLoS ONE* 9, e89177.

Kim, A., & Dennis, A. R. (2019). Says who? The effects of presentation format and source rating on fake news in social media. *MIS Quarterly*, *43*(3), 1025–1039.

Kuklinski, J. H., Quirk, P. J., Jerit, J., Schwieder, D., & Rich, R. F. (2000). Misinformation and the currency of democratic citizenship. *The Journal of Politics*, *62*(3), 790–816.

Kumar, S. and Shah, N. (2018), "False information on web and social media: a survey", presented at the Social Media Analytics: *Advances and Applications*.

Laurie, S (2019). Study Finds 'Fake News' Has Real Cost: $78 Billion. Retrieved 17/06/2021, from https://www.mediapost.com/publications/article/343603/study-finds-fake-news-has-real-cost-78-billion.html

Lewandowsky, S. (2011, March). Bitten by a sock puppet, but the climate is still changing. Retrieved from http://www.abc.net.au/unleashed/45638.html

Lewandowsky, S., Ecker, U. K. H. & Cook, J. (2019). Beyond Misinformation: Understanding and Coping with the "Post-Truth" Era. *Journal of Applied Research in Memory and Cognition* July. DOI: 10.1016/j.jarmac.2017.07.008

Mavridis, G (2018). Fake news and Social Media: How Greek users identify and curb misinformation online. Retrieved 21/06/2021, https://www.diva-portal.org/smash/get/diva2:1483157/FULLTEXT01.pdf

Merelli, A. (2016). The American media has become an "information-laundering scheme". Retrieved on 14/06/2021, from https://qz.com/850475/jon-stewart-the-american-media-has-become-an-information-laundering-scheme/

Moyo, D. Mare, A. Mabweazara, H.M. (2020) Editorial: Social Media, the Press, and the Crisis of Disinformation in Africa. Communicatio 46:4, pages 1–6.

Mustafaraj, E., & Metaxas, P. T. (2017). The fake news spreading plague: Was it preventable? Retrieved from http://arxiv.org/abs/1703.06988

Mutsvairo, B. & Bebawi, S. (2019). Journalism Educators, Regulatory Realities, and Pedagogical Predicaments of the 'Fake News' Era: A Comparative Perspective of the Middle East and Africa. *Journalism and Mass Communication Educator*, 74(2), 143–157.

Pursel, B (2020). Information, People, and Technology. Retrieved 17/06/2021, from https://psu.pb.unizin.org/ist110/chapter/1-4-internet-users-and-usage/

Rose-Wiles, Lisa M. (2018), "Reflections on Fake News, Librarians, and Undergraduate Research". *Praxis Publications*. 3. https://scholarship.shu.edu/praxis-publications/3

Sallam, M., Dababseh, D., Yaseen, A., Al-Haidar, A., Ababneh, N. A., Bakri, F. G. & Mahafzah, A. (2020) Conspiracy beliefs are associated with lower knowledge and higher anxiety levels regarding COVID-19 among students at the University of Jordan. *Int. J. Environ. Res. Public Health*, 17, 4915; doi:10.3390/ijerph17144915 www.mdpi.com/journal/ijerph

Newhoff, D. (2018, August 6). *Why do we share fake news?* https://illusionofmore.com/why-do-we-sharefake-news/

Shu, K., Sliva, A., Wang, S., Tang, J., & Liu, H., (2017). Fake news detection on SM. ACMSIGKDD *Explor. Newslett.* 19 (1), 22–36. https://doi.org/10.1145/3137597.3137600.

Simelela, N.; Venter, W.D.; Pillay, Y.; Barron, P. (2015) A Political and Social History of HIV in South Africa. *Curr. HIV/AIDS Rep.*, 12, 256–261.

Soll, J. (2017). The Long and Brutal History of Fake News. *POLITICO Magazine*. Retrieved from http://www.politico.com/magazine/story/2016/12/fake-news-history-long-violent-214535

Sullivan, M. C. (2019). Libraries and Fake News: What's the Problem? What's the Plan?. *Communications in Information Literacy*, 13 (1), 91–113. https://doi.org/10.15760/comminfolit.2019.13.1.7

Swami, V.; Voracek, M.; Stieger, S.; Tran, U.S.; Furnham, A. (2014**)** Analytic thinking reduces belief in conspiracy theories. *Cognition*, 133, 572–585.

Talwar, S., Dhir, A., Kaur, P., Zafar, N., & Alrasheedy, M. (2019). Why do people share fake news? Associations between the dark side of social media use and fake news sharing behavior. *Journal of Retailing and Consumer Services*, *51*, 72–82.

Thorson, E. (2016). Belief echoes: The persistent effects of corrected misinformation. *Political Communication*, 33(3), 460–480. https://doi.org/10.1080/10584609.2015.1102187

Van Duyn, E. & Collier, J. (2018): Priming and fake news: The effects of elite discourse on evaluations of news media, *Mass Communication and Society*, DOI:10.1080/15205436.2018.1511807

van Prooijen, J. W., & van Vugt, M. (2018**)** Conspiracy theories: evolved functions and psychological mechanisms. *Perspect. Psychol. Sci.*, 13, 770–788.

Waisbord, S. (2018). Truth is what happens to news: On journalism, fake news, and post-truth. *Journalism Studies*, 19(3), 1866–1878.

Ward, William; Cole-Miller, Katherine; Fruhling, Ann; and Cooper, Kathryn M., (2016). "Is Social Media a Threat or Can It Be a Trusted Agent?" Interdisciplinary Informatics Faculty Publications. 23. https://digitalcommons.unomaha.edu/interdiscipinformaticsfacpub/23

Watson, A. 2021. Factors that increase trust in online news worldwide 2020. Retrieved 17/06/2021, from https://www.statista.com/statistics/1228721/factors-trust-in-news-online-worldwide/

Weeks, B. E. (2015). Emotions, partisanship, and misperceptions: How anger and anxiety moderate the effect of partisan bias on susceptibility to political misinformation. *Journal of Communication*, 65(4), 699–719. https://doi.org/10.1111/jcom.12164

WEF. (2013). World economic forum: Outlook on the global agenda 2014. Retrieved from http://reports.weforum.org/outlook-14/
WHO (2020). UN tackles 'infodemic' of misinformation and cybercrime in COVID-19 crisis. Retrieved from https://www.un.org/en/un-coronavirus-communications-team/un-tackling-%E2%80%98infodemic%E2%80%99-misinformation-and-cybercrime-covid-19
Zhang, X., & Ghorbani, A. A. (2019). An overview of online fake news: Characterisation, detection, and discussion. *Information Processing and Management*, *57*, Article 102025.

Tella Adeyinka, Esther David Josiah Okai and
Oluyemisi Oluremi Oladapo

Dissemination of Information in the Era of Misinformation/ Disinformation: A Panoramic View of Literature Based on Empirical Sources

Abstract: Dissemination of information refers to the process of distributing information to a wide audience. With the rise of the internet and social media, the dissemination of information has become easier and more accessible than ever before, however, this also means that the spread of misinformation and disinformation has become a major problem. Against this background, this study examined the dissemination of information in the era of misinformation with the view to identify relevant citations that discuss the dissemination of information; present some case studies with relevant findings on the dissemination of misinformation; describe the dangers associated with the dissemination of misinformation in our society; and determine the strategies that can help promote accurate information, combat misinformation/disinformation, and determine ways to disseminate information effectively. A panoramic review of literature based on different empirical sources was conducted, which involved a search through numerous relevant databases such as the web of science, Scopus, and Google scholar. Manual searching for organizations such as World Health Organisation (WHO), World Economic Forum (WEF), and European Union's External Action Service was also embarked upon. The findings reveal numerous citations that discuss the dissemination of information and reveal different case studies on the dissemination of misinformation from across the globe, and also identified the dangers that are connected with the dissemination of misinformation and how these is harmful to society. In addition, the strategies to help promote accurate information and combat misinformation/disinformation and identification of ways to disseminate information effectively were also reported. The

Tella Adeyinka, Dept of Library and Information Science, University of Ilorin, Nigeria, Dept of Information Science, University of South Africa, Pretoria, e-mail: tellayinkaedu@yahoo.com, tella.a@unilorin.edu.ng
Esther David Josiah Okai, Abia State University, Uturu, Abia State, Nigeria, e-mail: e.okai@abiastateuniversity.edu.ng
Oluyemisi Oluremi Oladapo, The Library, Ladoke Akintola University of Technology, Ogbomoso, Nigeria

https://doi.org/10.1515/9783111176345-023

study concluded that the dissemination of misinformation can have serious consequences, including harm to individuals, harm to society, and damage to democratic institutions. It is therefore important for individuals to critically evaluate information and sources, and for policymakers and technology companies to take steps to combat the spread of misinformation. This can include promoting media literacy, investing in fact-checking and verification, and limiting the reach of false information on social media platforms.

Keywords: information, dissemination, information dissemination, misinformation, era of misinformation, fake news era, disinformation, panoramic review, empirical sources

Introduction

Information dissemination is the process of making information available to a large number of people. The internet and social media have made it easier and more accessible than ever before to disseminate knowledge, but in turn this has meant that the dissemination of false information and misinformation has grown to be a significant issue. There are multiple methods for the dissemination of information, including print media, television, and radio, while social media sites like Facebook, Twitter, and YouTube have grown in popularity as channels for sharing information in the digital age. With the use of these platforms, people can easily exchange knowledge with a worldwide audience.

The ability of information transmission to raise public awareness of significant issues is one of its main advantages. For instance, information on political rallies, social movements, and public health programs can be widely shared to foster support and increase awareness. However, there may be unfavourable effects from knowledge sharing as well, with the dissemination of false and misleading information among the most important problems. Disinformation is defined as incorrect information that is purposefully disseminated with the intention of influencing public opinion, as opposed to misinformation, which is defined as erroneous information that is disseminated accidentally.

Disinformation and false information can have detrimental effects on society, people, and democratic institutions, among other things. For instance, people may engage in risky behaviors or neglect to take the required safeguards as a result of receiving misleading information about a health problem. Elsewhere, misinformation regarding a political crisis has the potential to exacerbate division and erode confidence in democratic institutions. Therefore, stopping the transmission of false and misleading information is essential. This can involve curbing the spread of misleading information on social media platforms, investing in fact-checking and verification, and fostering media literacy.

Considering this context, information sharing might result in both advantages and disadvantages. While raising public awareness of significant issues can be beneficial, it is crucial to recognize the possibility of false information and take action to stop it from spreading. The growth of the internet and social media platforms has made knowledge more easily and widely available, but thisalso implies that the dissemination of false information and misinformation has grown to be a significant issue. The following pertinent citations address the spread of information.

According to a study that featuredin the journal *PLOS ONE*, many consumers get their news primarily from social media sites. According to the study, millennials most frequently obtain their news from social media sites like Facebook and Twitter, with over half of them citing social media as their main source (Shearer & Matsa, 2018). According to a different study that was published in the *Journal of Computer-Mediated Communication*, social media algorithms are important for the spread of knowledge, frequently favouring content that has already gained popularity, which might facilitate the spread of misleading information. The study also discovered that content is typically promoted by social media algorithms which can lead to the spread of sensational and misleading content (Guess et al. 2018).

Researchers alerted readers to the risks associated with the spread of false and misleading information in a World Economic Forum report (2024). The report emphasized how quickly misinformation can proliferate and lead to detrimental effects on people, society, and democratic institutions, among other things. In addition, the report recommended stepping up efforts to counter the dissemination of misinformation by investing in fact-checking and verification as well as media literacy promotion. Elsewhere, according to a study that was published in the *Journal of Medical Internet Research*, social media sites frequently contain misleading material on health-related issues. According to the study, misleading information regarding health issues has the potential to spread quickly on social media platforms and have detrimental effects, including injury to individuals. With this in mind, Ghezzi et al. (2020) advocated stepping up efforts to stop the propagation of misleading information regarding health issues on social media platforms. Overall, these studies show how critical it is to recognize the harm that misinformation and disinformation can do when they are disseminated and how urgent it is to step up efforts to stop their spread.

Literature review

Strategies to Promote Accurate Information and Combat Misinformation/ Disinformation

In the age of misinformation and disinformation, information dissemination can be difficult. The proliferation of social media and other digital communication channels has facilitated the rapid and widespread dissemination of inaccurate or misleading information. Several tactics that can aid in dispelling false information and promoting truthful knowledge include education and media literacy initiatives. People who possess critical thinking abilities and the capacity to assess sources are better able to distinguish reliable information from misinformation (Machete and Turpin 2020). Furthermore, promoting openness in the methods used to disseminate information, for example, by clearly attributing credit and sources, helps establish credibility and confidence.

Working together with other stakeholders, such as media outlets, tech businesses, and government agencies, is another tactic. When these groups collaborate, they may create and execute strategies and tools to recognize and combat false information, such as fact-checking campaigns and the use of AI systems to identify and report deceptive content. In addition, encouraging social media users to share and interact responsibly can help stop the spread of false information, such as double-checking material before sharing it and making reputable sources easily accessible (Aïmeur et al. 2023). Suggested strategies include but are not limited to:

1. Fact-checking: confirm the accuracy of any information from reputable sources before disseminating it, with numerous fact-checking groups available to assist in spotting misleading information. As mentioned by Liu et al. (2023), fact-checking, a technique for ensuring information is accurate, is especially useful in news reporting and public debate. It entails a thorough examination and validation of assertions, declarations, or news reports to ascertain their veracity. Usually, the procedure consists of multiple steps:

- Evaluation of the Source: fact-checkers evaluate the authority and dependability of the source from which the assertion is made. They take into account things like the source's credibility, accuracy history, and any possible biases (Guoand Schlichtkrull 2020).
- Verification of Claims: fact-checkers examine the particular assertion or statement being made. Examining the language, context, and any offered supporting documentation are necessary for this (Liu et al. 2023).

- Evidence Gathering: to support or contradict a claim, fact-checkers compile evidence from a number of trustworthy sources. Official reports, professional opinions, data analysis, and interviews with pertinent parties are a few examples of this (Micallef et al. 2022).
- Analysis: to ascertain the veracity and applicability of the gathered evidence, fact-checkers conduct a critical analysis of it. They evaluate the evidence's quality, the methods utilized to obtain it, and any potential biases or limits (Soprano et al. 2024).
- Fact-checkers draw a conclusion on the veracity of the claim based on their analysis, categorizing the allegation as truthful, erroneous, deceptive, or insufficiently supported. Typically, this conclusion is followed by a thorough justification of the decision (Nieminen and Sankari2021).
- Release: fact-checkers generally include links and references to the sources they consulted while conducting their research, and they publish their results in an understandable and transparent manner (Li and Chang2023; López-Marcos and Vicente-Fernández 2021). This enables users to independently confirm the data and comprehend the reasoning behind the fact-checking decision (Brodsky et al.2021).
- Promoting accurate information and thwarting misinformation/disinformation require fact-checking. It promotes reporting transparency, helps hold public officials and media outlets responsible for their words, and gives people the power to make decisions based on solid facts (Oladokun et al. 2024; Luengoand García-Marín 2020).

2. Educate the public: give individuals the knowledge and skills they need to identify false information and steer clear of it, and promote critical thinking and source evaluation. Promoting a culture of information literacy and critical thinking is a crucial part of public education, and entails giving people the abilities and information required to critically assess information, separate trustworthy sources from dubious ones, and make defensible conclusions (Supriyantiet al. 2020). People will learn how to fact-check information, comprehend media biases, and identify typical strategies used to disseminate false information as part of this instruction, as well as gain the ability to better traverse the complicated information landscape and make contributions to a more responsible and informed society through education.

3. Use a variety of communication channels: to reach as many people as you can, use a variety of channels including social media, email, text messaging, and traditional media. Using a range of communication channels to combat disinformation entails judiciously distributing factual information and refuting myths across

many platforms and media (Caled and Silva 2022). It involves using a variety of platforms, including social media, conventional media, community forums, educational initiatives, and public service announcements, to reach a wide range of consumers. Organizations may interact with a variety of demographics, expand the reach and impact of their messaging, and effectively combat misinformation by utilizing a multi-channel strategy.

4. Interact with the audience: invite questions and comments from them, pay attention to their worries, and address their inquiries, all of which will support credibility and trust-building. Creating meaningful and interactive interactions with the audience is essential to engaging them and promoting conversation, feedback, and engagement (Lansing et al.2023). It's about establishing relationships, encouraging a feeling of community, and providing avenues for two-way dialogue, which numerous venues, including social media, forums, live events, surveys, and interactive content, can be used for. Through active audience engagement, both individuals and organizations can acquire valuable insights, foster trust, and establish an atmosphere that is more inclusive and responsive.

5. Cooperate: cooperate with other groups, authorities, and people to advance truthful information and thwart false or misleading information. Working together and in a synchronized manner toward shared aims or objectives is the essence of collaboration (Alawida et al. 2022). Attain shared advantages and results comprise pooling resources, knowledge, and ideas. There are many different ways to collaborate, including alliances, teamwork, networking, and information exchange, with the utilization of diverse stakeholder viewpoints and strengths promoting innovation, creativity, and synergy. In the end, teamwork improves communication, encourages group problem-solving, and produces more lasting and successful results.

6. Encourage openness by being upfront about your sources and working procedures and distribute the information and resources that shaped your message. Encouraging accountability, openness, and honesty in decisions, activities, and communication is part of promoting transparency (Lee and Li 2021). This entails making information regarding procedures, guidelines, and results easily comprehensible so that interested parties can comprehend the thinking behind choices and actions. Transparency creates a culture of responsibility and integrity, increases credibility, and generates trust. Informed decision-making is also made easier, and constructive criticism and involvement from stakeholders are encouraged (Wang et al., 2023). All things considered, encouraging transparency fosters a more ethical and transparent atmosphere that is favourable to forging solid bonds and accomplishing common objectives.

7. Watch and react: keep an eye out for false or misleading content on social media and other platforms, and react quickly with correct information. Monitoring and responding entail keeping a close eye on pertinent circumstances or information, evaluating this, and then acting appropriately in light of the conclusions (Opdahl et al. 2023). This involves remaining watchful, obtaining information, examining patterns, and taking the initiative to seize new chances or problems. Such procedure entails putting up monitoring systems, gathering information from multiple sources, analyzing the information to draw conclusions, and then taking appropriate action. There are many different areas where monitoring and responding can be used, including public mood, market trends, social media interactions, and organizational performance. This helps people and organizations to stay informed, act quickly, and effectively adjust to changing conditions.

Finally, combating disinformation can also be accomplished through encouraging candid communication and interaction with opposing viewpoints, as promoting evidence-based conversations and courteous debate can result in a more knowledgeable populace that can tell reality from fantasy. In general, dispelling false information and misinformation calls for teamwork, even from those with different opinions. We can lessen the negative effects of misleading information on society by collaborating, educating the public, and disseminating accurate information.

Ways of Disseminating Information Effectively in the Misinformation Era

Information sharing is essential to the growth and advancement of society. Social media and the internet have made knowledge more accessible than ever in the current digital era, but it is often difficult to tell fact from fiction these days due to the proliferation of false and misleading information. The purpose of this chapter is to talk about how information is spread in the age of disinformation.

Use of Social Media:the use of social media platforms for knowledge dissemination has grown in popularity, withsixty-seven percent of Americans acquiring at least some of their news from social media, according to Pew Research Center (Gottfried and Shearer 2016). Social media sites like Instagram, Twitter, and Facebook give businesses the ability to produce interesting material that appeals to their target audience and can be shared widely, in the form of articles, infographics, and videos. Organizations can also interact with their followers on social media and address their questions and concerns. However, there are drawbacks to using social media as well. For example, it can be difficult to verify the veracity and accuracy of the information being disseminated (Bode and Vraga 2015).

Creation of Websites: websites can be made by organizations to disseminate information to a large audience, interact with them, and get feedback. To improve visibility, websites can be made more search engine friendly. The website's content ought to be interesting, simple to read, and give users useful information. In the age of disinformation, websites are another well-liked medium for information dissemination. Users frequently scan online sites, according to a Nielsen Norman Group study, therefore material needs to be interesting, simple to read, and give the viewer useful information (Nielsen 1997; Nielsen and Graves 2017). However, the problem with websites is that they require a significant investment of time and resources to create and maintain.

Use of Email Marketing: a cheap method of reaching a specific audience with information is through email marketing. Businesses may send interesting and personalized material right to the inbox of their target audience using email marketing, as well as engage with and get feedback from the audience. Campaign Monitor (2018) found that email marketing had a 4400% return on investment. However, to evade spam filters and make sure that the email content is interesting and pertinent to the recipient, email marketing can be difficult to get right.

Utilization of Search Engine Optimization (SEO): content can be made more search engine friendly by using SEO strategies, which may make it more visible on search engine result pages. Elsewhere, a BrightEdge analysis indicates that organic search, which accounts for 53% of website traffic, is the most important source of traffic (BrightEdge 2019). Nevertheless, SEO can be difficult to master since it necessitates a large time and resource commitment.

Leverage Traditional Media: information can still be effectively disseminated to a large audience through traditional media like radio, television, and newspapers. Establishments can work with such media to provide captivating content that informs and captivates viewers who might not have access to digital media. In contrast to 16% of Americans who obtain their news from newspapers, 65% of Americans get their news from television, according to Pew Research Center research (Mitchell et al. 2019). Traditional media does, however, come with several drawbacks, such as the difficulty in ensuring the veracity and accuracy of the material being distributed.

Use of Podcasts: podcasts can be used to distribute information to a specific audience and are growing in popularity. Podcasts that are entertaining and educational and reflect an organization's goals and values can be produced, and can facilitate audience engagement and allow companies to address questions and feedback from their listeners. Podcasts are growing rapidly in popularity; research by Edison Research indicates that 55% of Americans have listened to a podcast and that 75% of Americans are now aware of them(Edison Research 2019). Nevertheless, the difficulty with podcasts is that they take a lot of time and money to develop and run.

Engage with Influencers: social media influencers have the power to inform their following. In order to produce engaging content that connects with their audience, organizations can work with influencers that share their beliefs and goals (Geyser 2024). Influencers can be utilized to interact with the public and address questions and issues through informing their following.

Making sure the data being distributed is correct and trustworthy is crucial. Given the proliferation of false information, it is essential to double-check and confirm any material before releasing it to the general public. In the age of disinformation, organizations can effectively disseminate information by utilizing a variety of channels and interacting with their audience. Considering the aforementioned, it can be difficult to spread knowledge in the age of disinformation, but there are plenty of efficient methods to do it. To reach their audience, businesses should employ a variety of channels, such as social media, websites, email marketing, SEO, traditional media, podcasts, and influencer marketing. However, it is imperative to guarantee that the data being distributed is precise and trustworthy. In the age of disinformation, fact-checking and information verification are vital. In the current digital era, organizations can effectively communicate information and counteract misinformation by utilizing these techniques and interacting with the public.

Current State of Research on Disseminating Information in the Misinformation Era

It is now more difficult to provide the public with reliable information because of the misinformation era, and research in the past few years has thus concentrated on figuring out how to stop the spread of false information and developing practical countermeasures. The purpose of this survey of the literature is to give a broad overview of the state of the study on information dissemination in the age of disinformation.

The propagation of false information on social media platforms is a significant problem for the distribution of correct information. According to Vosoughi, Roy, and Aral's (2018) study, misleading news reports propagate more quickly and extensively on Twitter than accurate ones. Pennycook et al. (2018) also discovered in another study that those who are more likely to believe in conspiracies also tend to spread false information on social media. These results indicate that simply providing accurate information may not be enough to counteract the spread of misinformation.

Fact-checking can be a useful tactic in the dissemination of factual information. According to research by Guess, Nyhan, and Reifler (2020), people's confidence in false information can be decreased by exposing them to fact-checks. However, people's underlying prejudices and opinions may restrict the effectiveness of fact-checking; fact-checks were less successful in correcting disinformation among those with strong political convictions, according to a 2019 study by Wood and Porter).

Narrative storytelling is another method for delivering factual information. According to a study conducted in 2012 by Hart and Nisbet, people are more likely to recall and distribute information that is given in a narrative style. Moreover, narrative storytelling could be especially useful in capturing the interest of people who are political or scientific skeptics.

Along with these tactics, a growing corpus of study is examining the significance of source credibility in the dissemination of factual information. According to a study by Kim and Dennis (2019), people's decision to accept or reject information is largely influenced by the trustworthiness of the source. In particular, people are more inclined to believe information they learn from sources they consider reliable and trustworthy.

Lastly, studies have also been conducted on the influence of emotions on the spread of truthful information. According to a study by Bartos et al. (2022l Bartos et al., 2023) emotions have the potential to be an effective instrument for disseminating factual information. To be more precise, feelings of positivity can make people more motivated to look for reliable information, while feelings of negativity can make people more skeptical of false information.

In conclusion, distributing factual information in the age of disinformation is a difficult task with many facets. Fact-checking, narrative storytelling, source reliability, and emotional appeal are all examples of effective tactics. However, the dissemination of false information on social media sites and people's innate biases may restrict the effectiveness of these tactics.

Detailed Case Studies on Dissemination of Misinformation Citing Relevant Literature

The anti-vaccine movement, which has gained popularity recently, is a well-known case study of how false information may spread. The movement asserts, in spite of copious scientific evidence to the contrary, that vaccines cause autism and other health problems.

In a study released in 2021, Wawrzuta, Jaworski, Gotlib, and Panczyk looked at how social media contributes to the propagation of anti-vaccine sentiment. When the study examined tweets about vaccines, it discovered that about one-third of them contained false material, such as assertions that vaccines are dangerous or cause autism. The study also discovered that tweets opposing vaccinations were more likely to be retweeted than those in favour of them, suggesting that false information was disseminated more quickly than factual information.

Germani and Biller-Andorno (2021) conducted a second study that revealed the anti-vaccine movement used a number of strategies, such as selective data collection, presenting material in an untruthful manner, and appealing to emotions, to disseminate misleading information. Upon analyzing anti-vaccine posts on social media, the study discovered that these messages frequently referenced discredited or misinterpreted studies.

Another illustration of the perils of misleading information is the spread of erroneous information regarding the COVID-19 pandemic. According to a study by Ahmed & Rasul (2022), incorrect information regarding COVID-19 was disseminated more frequently than accurate information on social media, and spread quickly. Additionally, the study discovered that accounts with huge followings were more likely to share misinformation, suggesting that powerful people were aiding in the spread of false information.

According to a 2020 Oxford Internet Institute assessment, concerted efforts were made to promote misleading material across social media platforms as part of orchestrated campaigns to disseminate misinformation regarding COVID-19. The study also discovered that by endorsing treatments that have not been proven to work or downplaying the seriousness of the pandemic, certain political figures were aiding in the dissemination of false information. Similarly, Wonodi et al. (2022) discovered that false information about COVID-19 and the vaccine was widely disseminated, with the most common falsehood that COVID-19 is a fraud and that politicians utilize it to embezzle money. Monitoring public sentiment regarding COVID-19 will support initiatives to combat vaccine reluctance, and creative messaging and tactics will combat false information about the vaccine.

These case studies generally highlight the risks associated with the spread of false information, which can have detrimental effects on public health. False information spreading has the potential to cause vaccine reluctance and the resurgence of illnesses that were previously under control. Regarding the COVID-19 pandemic, disinformation campaigns may result in a decline in adherence to recommended practices, which could heighten the virus's spread and cause further fatalities. Governments, businesses, and individuals must all take action to stop the spread of misleading information and to advance the truth.

The Danger of Dissemination of Misinformation

In the current era of digital communication, misinformation is defined as misleading or inaccurate information that is disseminated accidentally or with the goal to mislead, which is an increasing concern. Misinformation spreading can have detrimental effects on society, individuals, and democratic institutions, among other things. The following is a comprehensive analysis, with citations to pertinent research, of the risks associated with the spread of false information.

Consequences for Public Health: spreading false information can have detrimental effects on public health, with the anti-vaccine movement, which has gained momentum recently, one example. The movement asserts, in spite of copious scientific evidence to the contrary, that vaccines cause autism and other health problems. Because of this, vaccination rates have dropped, which raises the possibility of illnesses that were previously under control reappearing. For instance, the World Health Organization (WHO) reported a 30% rise in measles infections worldwide in 2019, which they partially ascribed to vaccine hesitancy stoked by false information. Research has indicated that disinformation may have an impact on the adoption of novel vaccinations. Even among those who had previously supported vaccination, exposure to disinformation against the hypothetical new COVID-19 vaccine was linked to a lower willingness to get it, according to a 2020 study published in the Journal of Health Communication (Betsch et al. 2020). Similarly, exposure to anti-vaccine messaging on social media was linked to a lower intention to vaccinate against COVID-19, according to a 2020 study that was published in the journal *Vaccine* (Kata 2020).

Political Consequences: political repercussions from disinformation can also include harm to democratic institutions. Concerns over the dissemination of false information during political campaigns and elections have arisen in recent years, as election results can be influenced and public perception can be manipulated by spreading false information. In 2020, the European External Action Service published a report that revealed evidence of concerted efforts to disseminate false information throughout the EU in an effort to sway public opinion and affect the results of elections (European External Action Service 2020). In a similar vein, the U.S. Senate Select Committee on Intelligence reported in 2019 that the Russian government had waged a concerted effort to disseminate false information during the 2016 U.S. presidential election in an effort to sow discord and erode confidence in democratic institutions.

Misinformation can also have negative societal effects, such as injury to people and a breakdown in social cohesiveness. One instance of how bias and discrimination can be exacerbated is the dissemination of incorrect information about minority groups. According to a 2020 study that was published in the *Journal of Computer-Mediated Communication*, social media exposure to false information about refugees was linked to a rise in anti-refugee sentiment (Francheschi and Galli 2020). Similarly, prejudice and discrimination against people of Asian heritage have been connected to false information on COVID-19. According to a 2020 United Nations study, there has been a rise in instances of prejudice and hate speech directed towards people of Asian heritage as a result of the dissemination of false information on COVID-19.

Economic Repercussions: disinformation can also have negative effects on the economy, affecting both companies and the overall state of the economy. For instance, a company's reputation may suffer and revenues may decline if inaccurate information about the company's goods or services is spread. According to a 2020 *Journal of Business Research* study, consumers' belief in the reviewed product and willingness to purchase it were significantly impacted by exposure to unfavourable internet evaluations that were later proved to be incorrect (Jin and Phua 2020).

Similar to this, misleading information regarding the financial system or the economy can cause investor confidence to decline and market volatility to increase. According to a 2019 study that appeared in the *Journal of Economic Perspectives*, investors may react to misleading information by buying or selling assets, which would enhance market swings and potentially contribute to a rise in volatility in the financial markets (Gabaix et al. 2019).

Psychiatric Repercussions: psychological repercussions from false knowledge might also include heightened dread and anxiety. For instance, the public's concern and panic may escalate if misleading information about a health problem spreads. According to a 2020 study that appeared in the *Journal of Health Communication*, people's anxiety and worry levels rose when they were exposed to false information about COVID-19 (Petersen et al. 2020).

In a similar vein, misleading information on a political crisis may cause the general public to become more fearful and anxious. Even if information was subsequently updated, people's anxiety and worry levels increased when they were exposed to misleading information about a political crisis, according to a 2019 study published in the *Journal of Social and Political Psychology* (Jost et al. 2019). In conclusion, spreading false information can have detrimental effects on society, individuals, and democratic institutions, among other things. It is crucial that people assess information and sources cautiously, and that tech companies and legislators take action to stop the spread of false information. This can involve limiting the dissemination of misleading information, investing in fact-checking and verification, and encouraging media literacy.

Conclusion

According to the study, social media platforms, communication channels that are always changing, technical breakthroughs, and other variables all have an impact on the diverse and complicated nature of information distribution in the modern period. The study has brought attention to the important effects that misinformation and disinformation have on the public's perspective, decision-making procedures, and sources of confidence for information, which emphasizes how critical it is to successfully solve these issues. Understanding and combating misinformation/disinformation requires the use of empirical sources, such as data-driven research and evidence-based analysis, which offer a strong basis for spotting patterns, trends, and countermeasures for misleading information. The study has underlined how important it is for stakeholders, researchers, legislators, media outlets, and technological companiesto work together to devise tactics that effectively counteract misinformation and disseminate factual information. Programs for raising awareness and educating people were found to be essential in enabling people to assess information critically, spot misleading narratives, and make wise judgments. The issue's continuing nature was also emphasized in the conclusion, highlighting the necessity for continual study, strategy modification, and attention to detail in order to handle new obstacles in the distribution of information and counteract misinformation.

Recommendations

The following suggestions can be made in light of the study's findings about the spread of information in the age of misinformation and disinformation:

Improving Media Literacy Initiatives: create and execute extensive media literacy initiatives with the goal of teaching the general public, especially students and marginalized groups, how to critically analyze, verify information, and consume information responsibly.

Boost Cooperation: encourage cooperation between stakeholders to create and carry out coordinated plans for battling misinformation and disinformation. These stakeholders include government agencies, media outlets, tech corporations, educators, and civil society organizations. The sharing of resources, knowledge, and best practices should all be part of this partnership.

Encourage Transparency: provide explicit attribution, cite reliable sources, and declare any potential conflicts of interest to promote transparency in the methods used to disseminate information. Additionally, platforms and organizations must to be open and honest about the algorithms and standards they use for content moderation.

Encourage Fact-Checking Projects: invest in and lend support to impartial fact-checking projects that carefully assess and validate data. Encourage people to use fact-checking resources and tools so they can confirm information before spreading it.

Make Use of Technology: examine how to identify and flag false information using technology, such as machine learning algorithms and artificial intelligence (AI). Provide instruments and frameworks that can reduce the possibility of censorship by automating the detection and elimination of fraudulent information.

Encourage Responsible Social Media Use: encourage people to report questionable content, double-check information before sharing it, and have productive conversations on social media. Additionally, platforms must put policies in place to stop the dissemination of misleading material without sacrificing the right to free speech.

Spend Money on Research: provide funding for continuing investigations and studies on the prevalence, effects, and practical countermeasures of misinformation and disinformation. In order to combat false information, legislative choices, educational programs, and technology advancements should be informed by the results of this research.

Encourage Community Leaders: enable trusted speakers, community leaders, and influencers to effectively rebut false narratives and spread factual information. To assist them in navigating and resolving misinformation/disinformation issues, provide training and resources.

References

Ahmed, S., & Rasul, M. E. (2022). Social Media News Use and COVID-19 Misinformation Engagement: Survey Study. *Journal of Medical Internet Research*, *24*(9). https://doi.org/10.2196/38944

Aïmeur, E., Amri, S., & Brassard, G. (2023). Fake news, disinformation and misinformation in social media: A review. *Social Network Analysis and Mining*, *13*(1). https://doi.org/10.1007/s13278-023-01028-5

Alawida, M., Omolara, A. E., Abiodun, O. I., & Al-Rajab, M. (2022). A deeper look into cybersecurity issues in the wake of Covid-19: A survey. *Journal of King Saud University – Computer and Information Sciences*, *34*(10), 8176–8206. https://doi.org/10.1016/j.jksuci.2022.08.003

Bartoš, V., Bauer, M., Cahlíková, J., & Chytilová, J. (2022). Communicating doctors' consensus persistently increases COVID-19 vaccinations. *Nature*, *606*(7914), 542–549. https://doi.org/10.1038/s41586-022-04805-y

Bartos, A., Weinerova, J., & Diondet, S. (2023). Effects of human probiotics on memory and psychological and physical measures in community-dwelling older adults with normal and mildly impaired cognition: results of a bi-center, double-blind, randomized, and placebo-controlled clinical trial (CleverAge biota). *Frontiers in Aging Neuroscience*, *15*, 1163727. https://doi.org/10.3389/fnagi.2023.1163727.

Betsch, C., Wieler, L. H., & Habersaat, K. (2020). Monitoring behavioural insights related to COVID-19. The Lancet Public Health, 5(5), e249–e249.

Bode, L., & Vraga, E. K. (2015). In related news, that was wrong: The correction of misinformation through related stories functionality in social media. *Journal of Communication*, *65*(4), 619–638. https://doi.org/10.1111/jcom.12166.

BrightEdge (2019). Organic share of traffic increases by 53%. Retrieved from https://www.brightedge.com/blog/organic-share-of-traffic-increases-to-53 (accessed 6 April 2024).

Brodsky, J. E., Brooks, P. J., Scimeca, D., Todorova, R., Galati, P., Batson, M., Grosso, R., Matthews, M., Miller, V., & Caulfield, M. (2021). Improving college students' fact-checking strategies through lateral reading instruction in a general education civics course. *Cognitive Research: Principles and Implications*, *6*(1), 1–18. https://doi.org/10.1186/s41235-021-00291-4

Caled, D., & Silva, M. J. (2022). Digital media and misinformation: An outlook on multidisciplinary strategies against manipulation. *Journal of Computational Social Science*, *5*(1), 123–159. https://doi.org/10.1007/s42001-021-00118-8.

Campaign Monitor (2018). 70 email marketing stats every marketer should know. Retrieved from https://www.campaignmonitor.com/blog/email-marketing/70-email-marketing-stats-you-need-to-know/(accessed 6 April 2024).

Edison Research (2019). The Infinite Dial. Retrieved from https://www.edisonresearch.com/wp-content/uploads/2019/03/Infinite-Dial-2019-PDF-1.pdf(accessed 6 April 2024).

European External Action Service. (2020). Disinformation and COVID-19: Update. Retrieved from https://eeas.europa.eu/headquarters/headquarters-homepage/78489/disinformation-and-covid-19-update_en

Francheschi, L., & Galli, F. (2020). Effects of exposure to misinformation about refugees on attitudes towards refugees. Journal of Computer-Mediated Communication, 25(5), 357–375.

Gabaix, X., Laibson, D., & Madrian, B. C. (2019). Why has the financial sector grown so much? The role of misinformation and beliefs. *Journal of Economic Perspectives*, 33(3), 155–174.

Germani, F., & Biller-Andorno, N. (2021). The anti-vaccination infodemic on social media: A behavioral analysis. *PLoS ONE, 16*(3). https://doi.org/10.1371/journal.pone.0247642.

Geyser, W. (2024). What is an influencer? – social media influencers defined [Updated 2024]. https://influencermarketinghub.com/what-is-an-influencer/ (accessed 8 April 2024).

Gottfried, J. & Shearer, E. (2016). News Use Across Social Media Platforms 2016. Pew Internet Centre. Retrieved from https://www.pewresearch.org/journalism/2016/05/26/news-use-across-social-media-platforms-2016/

Guess, A., Nyhan, B., & Reifler, J. (2020). The effect of factual corrective information on misinformation. Political Behavior, 42(1), 155–176.

Guo, Z., & Schlichtkrull, M. (2020). Andreas Vlachos; A Survey on Automated Fact-Checking. *Transactions of the Association for Computational Linguistics*, 10 178–206. doi: https://doi.org/10.1162/tacl_a_00454

Hart, P. S., & Nisbet, E. C. (2012). Boomerang effects in science communication: How motivated reasoning and identity cues amplify opinion polarization about climate mitigation policies. *Communication Research*, 39(6), 701–723.

Jin, S. V., & Phua, J. (2020). False negative reviews on social media: The case of Yelp. *Journal of Business Research*, 116, 70–82.

Jost, J. T., Ledgerwood, A., & Hardin, C. D. (2019). Shared reality, system justification, and the relational basis of political ideology. *Journal of Social and Political Psychology*, 7(1), 276–301.

Kata, A. (2020). Anti-vaccine misinformation on social media: A critical political ecology of misinformation. *Social Science & Medicine*, 270, 113439.

Kim, S., & Dennis, A. R. (2019). Source credibility and the acceptance of new information. *MIS Quarterly*, 43(1), 1–23.

Lansing, A. E., Romero, N. J., Siantz, E., Silva, V., Center, K., Casteel, D., & Gilmer, T. (2023). Building trust: Leadership reflections on community empowerment and engagement in a large urban initiative. *BMC Public Health*, *23*. https://doi.org/10.1186/s12889-023-15860-z

Lee, Y., & Li, Y. Q. (2021). The role of communication transparency and organizational trust in publics' perceptions, attitudes and social distancing behaviour: A case study of the COVID-19 outbreak. *Journal of Contingencies and Crisis Management*, *29*(4), 368–384. https://doi.org/10.1111/1468-5973.12354.

Li, J., & Chang, X. (2023). Combating Misinformation by Sharing the Truth: a Study on the Spread of Fact-Checks on Social Media. *Inf Syst Front* **25**, 1479–1493. https://doi.org/10.1007/s10796-022-10296-z.

Liu, X., Qi, L., Wang, L., & Metzger, M. J. (2023). Checking the Fact-Checkers: The Role of Source Type, Perceived Credibility, and Individual Differences in Fact-Checking Effectiveness. *Communication Research*. https://doi.org/10.1177/00936502231206419

López-Marcos, C, & Vicente-Fernández, P. (2021). Fact Checkers Facing Fake News and Disinformation in the Digital Age: A Comparative Analysis between Spain and United Kingdom. *Publications*, *9*(3), 36. https://doi.org/10.3390/publications9030036

Luengo, M., & García-Marín, D. (2020). The performance of truth: Politicians, fact-checking journalism, and the struggle to tackle COVID-19 misinformation. *American Journal of Cultural Sociology, 8*(3), 405–427. https://doi.org/10.1057/s41290-020-00115-w

Machete, P., & Turpin, M. (2020). The Use of Critical Thinking to Identify Fake News: A Systematic Literature Review. *Responsible Design, Implementation and Use of Information and Communication Technology, 12067*, 235–246. https://doi.org/10.1007/978-3-030-45002-1_20.

Mitchell, A, Matsa, K.E., Weise, R., & Klein, H. (2019). For local news collection, Americans embrace digital but still want strong community connection. Per Internet Research. Retrieved from https://www.pewresearch.org/journalism/wp-content/uploads/sites/8/2019/03/PJ_2019.03.26_Local-News_FINAL.pdf(accessed 6 April 2024).

Nicholas Micallef, Vivienne Armacost, Nasir Memon, and Sameer Patil. 2022. True or False: Studying the Work Practices of Professional Fact-Checkers. Proc. ACM Hum.-Comput. Interact. 6, CSCW1, Article 127 (April 2022), 44 pages. https://doi.org/10.1145/3512974

Nieminen, S., & Sankari, V. (2021). Checking PolitiFact's Fact-Checks. Journalism Studies, 22(3), 358–378. https://doi.org/10.1080/1461670X.2021.1873818.

Nielsen, R.K & Graves, L. (2017). News you don't believe": Audience perspectives on fake news. Retrieved from: https://reutersinstitute.politics.ox.ac.uk/sites/default/files/2017-10/Nielsen%26Graves_factsheet_1710v3_FINAL_download.pdf (accessed 6 April 2024).

Nielsen, J. (1997). How users read on the web. Nielsen Normal Group, Retrieved from: https://www.nngroup.com/articles/how-users-read-on-the-web/ (accessed 6 April 2024).

Oladokun, Bolaji & Aruwa, Joel & Ottah, Gabriel & Ajani, Yusuf. (2024). Misinformation and Disinformation in the Era of Social Media: The Need for Fact-Checking Skills. *SRELS Journal of Information Management*. 61. 10.17821/srels/2024/v61i1/171266.

Opdahl, A. L., Tessem, B., Dang-Nguyen, D., Motta, E., Setty, V., Throndsen, E., Tverberg, A., & Trattner, C. (2023). Trustworthy journalism through AI. *Data & Knowledge Engineering, 146*, 102182. https://doi.org/10.1016/j.datak.2023.102182

Oxford Internet Institute (2020). Covid-related misinformation videos spread primarily through Facebook, as its fact-checkers fail to spot false information, finds new Oxford study. Retrieved from https://www.oii.ox.ac.uk/news-events/covid-related-misinformation-videos-spread-primarily-through-facebook-as-its-fact-checkers-fail-to-spot-false-information-finds-new-oxford-study/ (accessed 6 April 2024).

Pennycook, G., Cannon, T. D., & Rand, D. G. (2018). Prior exposure increases perceived accuracy of fake news. *Journal of Experimental Psychology: General*, 147(12), 1865–1880.

Petersen, E., Hui, D. S., Hamer, D. H., Blumberg, L., Madoff, L. C., Pollack, M., . . . & Nachega, J. B. (2020). Li wenliang, a face to the frontline healthcare worker? The first doctor to alert the outbreak of the COVID-19 virus. *International Journal of Infectious Diseases*, 93, 205–207.

Shearer, E., & Masta, E. K. (2018, September 10). News Use Across Social Media Platforms 2018. Pew Research Center. https://www.journalism.org/2018/09/10/news-use-across-social-media-platforms-2018/ (accessed 15 August 2024).

Soprano, M., Roitero, K., La Barbera, D., Ceolin, D., Spina, D., Demartini, G., & Mizzaro, S. (2024). Cognitive Biases in Fact-Checking and Their Countermeasures: A Review. *Information Processing & Management, 61*(3), 103672. https://doi.org/10.1016/j.ipm.2024.103672

Supriyanti, S., Permanasari, A & Khoerunnisa, F. (2020). Correlation Between Information Literacy and Critical Thinking Enhancement Through PjBL-Information Literacy Learning Model. *Journal of Educational Sciences*, 4 (4), 774. 10.31258/jes.4.4.p.774–784.

U.S. Senate Select Committee on Intelligence. (2019). Report of the Select Committee on Intelligence United States Senate on Russian active measures campaign and interference in the 2016 U.S. election. Retrieved from https://www.int

Vosoughi, S., Roy, D., Aral, S. (2018). The spread of true and false news online. *Science*. 359(6380):1146–1151, DOI. 10.1126/science.aap9559.

Wang, R., Bush-Evans, R., Arden-Close, E., Bolat, E., McAlaney, J., Hodge, S., Thomas, S., & Phalp, K. (2023). Transparency in persuasive technology, immersive technology, and online marketing: Facilitating users' informed decision making and practical implications. *Computers in Human Behavior, 139*, 107545. https://doi.org/10.1016/j.chb.2022.107545

Wawrzuta, D., Jaworski, M., Gotlib, J., & Panczyk, M. (2021). Characteristics of Antivaccine Messages on Social Media: Systematic Review. *Journal of Medical Internet Research, 23*(6). https://doi.org/10.2196/24564

Wonodi, C., Obi-Jeff, C., Adewumi, F., Keluo-Udeke, S. C., Gur-Arie, R., Krubiner, C., Jaffe, E. F., Bamiduro, T., Karron, R., & Faden, R. (2022). Conspiracy theories and misinformation about COVID-19 in Nigeria: Implications for vaccine demand generation communications. *Vaccine, 40*(13), 2114–2121. https://doi.org/10.1016/j.vaccine.2022.02.005

Wood, T., & Porter, E. (2019). The elusive backfire effect: Mass attitudes' steadfast factual adherence. *Political Behavior, 41*(1), 135–163. https://doi.org/10.1007/s11109-018-9443-y

World Economic Forum (2024). The global risks report 2024. Retrieved from: https://www3.weforum.org/docs/WEF_The_Global_Risks_Report_2024.pdf (accessed 15 August 2024).